AF805545

Nanotechnology in Wastewater Treatment

Remya Neelancherry · S. Mathava Kumar

Nanotechnology in Wastewater Treatment

Springer

Remya Neelancherry
School of Infrastructure
Indian Institute of Technology Bhubaneswar
Bhubaneswar, Odisha, India

S. Mathava Kumar
Department of Civil Engineering
Indian Institute of Technology Madras
Chennai, Tamil Nadu, India

ISBN 978-3-031-94245-7 ISBN 978-3-031-94246-4 (eBook)
https://doi.org/10.1007/978-3-031-94246-4

Jointly published with Capital Publishing Company.
The print edition is not for sale in Afghanistan, Bangladesh, Bhutan, India, Sri Lanka, the Maldives, Nepal and Pakistan. Customers from Afghanistan, Bangladesh, Bhutan, India, Sri Lanka, the Maldives, Nepal and Pakistan please order the print book from: Capital Publishing Company.

© Capital Publishing Company, New Dehli, India. 2025

This work is subject to copyright. All rights are solely and exclusively licensed by the Publisher, whether the whole or part of the material is concerned, specifically the rights of translation, reprinting, reuse of illustrations, recitation, broadcasting, reproduction on microfilms or in any other physical way, and transmission or information storage and retrieval, electronic adaptation, computer software, or by similar or dissimilar methodology now known or hereafter developed.
The use of general descriptive names, registered names, trademarks, service marks, etc. in this publication does not imply, even in the absence of a specific statement, that such names are exempt from the relevant protective laws and regulations and therefore free for general use.
The publishers, the authors, and the editors are safe to assume that the advice and information in this book are believed to be true and accurate at the date of publication. Neither the publishers nor the authors or the editors give a warranty, express or implied, with respect to the material contained herein or for any errors or omissions that may have been made. The publishers remain neutral with regard to jurisdictional claims in published maps and institutional affiliations.

This Springer imprint is published by the registered company Springer Nature Switzerland AG
The registered company address is: Gewerbestrasse 11, 6330 Cham, Switzerland

If disposing of this product, please recycle the paper.

Preface

Globally, population growth, industrialization and rising water demands have been the major challenge to human kind. To address these global challenges, innovative research has been continued to develop alternate source of water, safe water supply and wastewater treatment. Wastewater treatment stands at the forefront of protecting public health and the environment. Moreover, it has been given prime importance to meet several sustainable development goals such as clean water and sanitation, good health and well-being, life below water, etc.

The field of nanotechnology is gaining immense importance and attention in recent times, owing to the fact that certain materials possess properties at nanoscale quite different to those at macroscale. Employing these unique properties can help address various challenges of the modern world, including the potential applications in the field of water purification and wastewater treatment. Emerging nanotechnological developments provide additional advantages over the conventional water/wastewater treatment technologies adopted in meeting the water reuse and resource recovery goals. The overall goal of this book is to provide a comprehensive insight that are required to practically apply nanotechnology in various stages of water purification and wastewater treatment.

The role of nanotechnology in environmental engineering, and more exclusively its role in making various units of water and wastewater treatment plants more efficient is discussed in Chapter "Introduction to the Application of Nanotechnology in Water and Wastewater Treatment". Moreover, the challenges in the application of nanotechnology in water and wastewater treatment units is highlighted in this chapter. A comprehensive classification of nanomaterials based on composition, functionalization, and their specific application in wastewater treatment are provided in Chapter "Classification of Nanomaterials for Wastewater Treatment". Subsequently, the techniques used for the characterization of nanomaterials including morphological, structural, chemical, optical, and electrochemical are discussed in detail in Chapter "Characterisation of Nanomaterials". Chapters "Development of Polyaniline-Based Calcium Ferrite Nanocomposites for the Treatment of Eriochrome Black T Dye Laden Wastewater" and "Nanomaterials–Based Strategies for Addressing Arsenic Pollution in Water and Wastewater: Recent Developments" are devoted to provide the application of nanotechnology/nanomaterials for water and wastewater treatment. In Chapter "Development of Polyaniline-Based Calcium Ferrite Nanocomposites for the Treatment of Eriochrome Black T Dye Laden Wastewater", the application of calcium ferrite/polyaniline nanocomposite for dye laden wastewater treatment is explained whereas in Chapter "Nanomaterials–Based Strategies for Addressing Arsenic Pollution in Water and Wastewater: Recent Developments", the application of nanomaterials for addressing arsenic pollution from water and wastewaters through different techniques such as photocatalysis, adsorption, precipitation and membrane are discussed with examples. Recently, the green synthesis of nanomaterials has been receiving great attention. Therefore, the green synthesis of silver nanoparticles from plant extracts for wastewater remediation is discussed in Chapter "Silver Nanoparticles Production from Plant Extracts Using Green Synthesis Route for Wastewater Remediation". The advances in nano-filtration and the various materials used in the preparation of nano-filtration membranes are highlighted in Chapter "Advances in Nanofiltration". Chapter "Advanced Oxidation Processes Using Nanomaterials" is devoted to explain the utilization of nanomaterials in advanced oxidation processes (AOP)-based wastewater treatment, highlighting the innovations, challenges, and future perspectives. Exclusively, the application of nano-magnesium-oxide in AOP for the degradation of various organic pollutants are discussed in Chapter "Application of Nano Magnesium Oxide in Advanced Oxidation Processes for the Degradation of Organic Pollutants". Finally, the biochar, nano-biochar, and biochar-nanocomposites, examining their production methods, properties, and applications, particularly in environmental remediation and sustainable agriculture are reviewed in Chapter "Nanobiochar and Biochar Nanocomposites in Wastewater Treatment".

In summary, this book provides a comprehensive knowledge on synthesis, classification, characterization and application of nanomaterials in various stages/units of water and wastewater treatment plants.

Bhubaneswar, India — Remya Neelancherry
Chennai, India — S. Mathava Kumar

About This Book

This book aims at providing comprehensive insights focusing on the ongoing research in the application of nanotechnology in water purification and wastewater treatment. The challenges of pollutant removal from water and wastewater are addressed in this book, and constructively discusses the innovative and effective solutions that nanotechnology has to offer for the same. This book deals with synthesis, characterization and employment of nanomaterials for water purification and wastewater treatment. Moreover, the recently developed applications of nanotechnology in water purification and wastewater treatment are written in a comprehensive and structured manner. This book elaborately discusses the nanotechnology-based treatment methods for emerging contaminants and other refractory pollutants removal from water/wastewater. The emerging nanotechnological applications have been discussed in great depths in this book to address the challenges of various environmental engineering problems. This book would be of primary interest to academicians, researchers, scientists and engineers working in the field of nanomaterial applications in environmental engineering. Overall, the book serves as an additional reading material for students of Environmental Engineering, Chemical Engineering, Environmental Science and Civil Engineering.

Contents

Introduction to the Application of Nanotechnology in Water and Wastewater Treatment 1
1 Introduction 1
2 What is Nanotechnology? 2
3 Nanomaterial Synthesis and Characterisation 3
4 Role of Nanotechnology in Environmental Engineering 5
5 Water Treatment: Conventional Versus Advanced Treatment Techniques 8
5.1 Conventional Water Treatment Techniques 8
5.2 Advanced Water Treatment Techniques 8
5.3 Pros and Cons of Conventional and Advanced Water Treatment Techniques 9
6 Wastewater Treatment: Conventional and Advanced Treatment Techniques 10
7 How Nanotechnology Can Be Applied for Water and Wastewater Treatment 13
7.1 Nanomaterials in Photocatalysis-Based Tertiary Wastewater Treatment 14
7.2 Nanomaterials in Piezocatalysis and Piezo-Photocatalysis-Based Tertiary Wastewater Treatment 17
7.3 Nanomaterials in Microwave-Based Tertiary Wastewater Treatment 18
8 Challenges in the Application of Nanotechnology for Water and Wastewater Treatment 23
References 24

Classification of Nanomaterials for Wastewater Treatment 27
1 Introduction 27
2 Classification Based on Composition 28
2.1 Metal Nanoparticles 36
2.2 Metal Oxide Nanoparticles (MONPs) 38

2.3 Carbon-Based Nanomaterials 40
2.4 Polymeric Nanomaterials (PNPs) 43
3 Classification Based on Morphology 44
3.1 Nanoparticles 44
3.2 Nanostructured Materials 46
3.3 Nanocomposites 47
3.4 Nanotubes 48
3.5 Nanofibres 48
4 Classification Based on Functionalisation 49
4.1 Chemical Functionalisation 50
4.2 Biofunctionalisation 51
4.3 Responsive Nanomaterials 52
5 Classification Based on Specific Application 53
5.1 Photocatalytic Nanomaterials 53
5.2 Adsorbent Nanomaterials 54
5.3 Membrane-Based Nanomaterials 56
6 Challenges and Considerations 57
6.1 Toxicity 58
6.2 Long-Term Stability 58
6.3 Environmental Impact 58
7 Future Perspective 59
8 Conclusion 59
References 60

Characterisation of Nanomaterials 65
1 Introduction 65
2 Structural Characterisations of Nanomaterials 66
2.1 Electron Microscopy 66
2.2 X-Ray Diffraction (XRD) 67
3 Chemical Characterisations of Nanomaterials 69
3.1 Fourier Transform Infrared Spectroscopy (FTIR) 69
3.2 X-Ray Photoelectron Spectroscopy (XPS) 69
4 Optical and Optoelectronic Characterisations of Nanomaterials 71
4.1 Photoluminescence (PL) Spectroscopy 71
4.2 Ultraviolet Diffused Reflectance Spectroscopy (UV-DRS) 72
5 Electrical Characterisations of Nanomaterials 73
5.1 Cyclic Voltammetry (CV) 73
5.2 Electrochemical Impedance Spectroscopy (EIS) 74
6 Surface Characterisations of Nanomaterials 76
6.1 Atomic Force Microscopy (AFM) 76
6.2 Zeta Potential 77
7 Conclusion 78
References 78

Development of Polyaniline-Based Calcium Ferrite Nanocomposites for the Treatment of Eriochrome Black T Dye Laden Wastewater ... 81
1 Introduction ... 81
2 Experimental and Analytical Methodology ... 83
2.1 Reagents Used ... 83
2.2 Synthesis of $CaFe_2O_4$ Nanoparticles ... 83
2.3 Development of the $CaFe_2O_4$-PANI Nanocomposite ... 84
2.4 Batch Adsorption Experiment of the EBT Dye ... 85
2.5 Physiochemical Analysis of the $CaFe_2O_4$-PANI Nanocomposite ... 85
2.6 Experimental Design Based on the Central Composite Design ... 86
3 Results and Discussions ... 86
3.1 Characterisation Studies ... 86
3.2 Influence of Solution pH and the $CaFe_2O_4$-PANI Nanocomposite Dose ... 89
3.3 Effect of Initial EBT Dye Concentration and Contact Time ... 90
3.4 Adsorption Kinetic Analysis ... 91
3.5 Equilibrium Isotherm Analysis ... 92
3.6 Effect of Real Water Sample ... 95
3.7 Influence of Coexisting Ions ... 96
3.8 RSM-Based Optimisation and Analysis of Variance ... 97
3.9 Regeneration Study of the EBT Dye Uptake ... 99
3.10 Evaluation of Performance ... 100
4 Conclusions ... 101
References ... 102

Nanomaterials–Based Strategies for Addressing Arsenic Pollution in Water and Wastewater: Recent Developments ... 107
1 Introduction ... 107
2 Arsenic Chemistry in Aqueous Solutions ... 109
3 Nano–technology for Water and Wastewater Purification ... 110
4 Types of Nanomaterials for Water and Wastewater Purification Processes ... 111
4.1 Metal–Based NPs ... 111
4.2 Photocatalytic NPs ... 112
4.3 Carbon Nanomaterials ... 112
4.4 Composite Nanomaterials ... 113
4.5 Nano–Infused Sensors ... 113
5 Arsenic Removal Technologies ... 114
5.1 Oxidation ... 114
5.2 Photocatalysis ... 115
5.3 Adsorption ... 117

5.4 Chemical Precipitation 119
5.5 Membrane Filtration 120
6 Advantages and Limitations 121
7 Summary and Conclusions 121
References 122

Silver Nanoparticles Production from Plant Extracts Using Green Synthesis Route for Wastewater Remediation 127
1 Introduction 127
2 Synthesis of AgNPs 128
3 Extraction Methods of Plants for AgNPs Synthesis 129
3.1 Maceration Extraction 130
3.2 Solvent-Based Extraction 130
3.3 Microwave-Assisted Extraction 131
3.4 Ultrasound-Assisted Extraction 131
4 AgNPs Production from Plant Extracts Using the Green Synthesis Route 132
4.1 Polysaccharide Method 132
4.2 Tollens Method 133
4.3 Irradiation Method 133
4.4 Microwave-Assisted Synthesis 133
5 Characterisation of AgNPs 134
5.1 UV-Vis Spectroscopy Characterisation 134
5.2 FTIR 134
5.3 XRD 135
5.4 SEM 135
5.5 TEM 135
5.6 DLS 135
6 Factors Affecting AgNP Synthesis 136
6.1 Temperature 136
6.2 pH 136
6.3 Reaction Time 137
7 Applications of AgNPs in the Degradation of Pollutants 137
7.1 AgNPs for Disinfection 137
7.2 AgNPs for the Degradation of Emerging Contaminants (ECs) in Aqueous Solutions 138
8 Conclusion 143
References 143

Advances in Nanofiltration 147
1 Introduction 147
2 Fundamentals of NF 148
2.1 NF Separation Mechanisms 149
2.2 Conventional NF Membranes 152
2.3 Methodologies for the Preparation of the NF Membrane 153

2.4 Characterisation of NF Membranes 155
2.5 Membrane Modules 155
3 Novel Nanomaterial-Based NF Membranes 156
3.1 Metal/Metal Oxide-Based NF Membranes 157
3.2 Carbon Nanomaterial-Based NF Membranes 159
3.3 Metal Organic Framework (MOF)-Based NF Membranes 160
3.4 Catalytic NF Membranes 161
3.5 Block Copolymer NF Membranes 161
3.6 MXene-Based NF Membrane 162
3.7 Tannic Acid-Based NF Membranes 162
3.8 Aquaporin-Based NF Membranes 162
4 Conclusions and Future Prospects 163
References 163

Advanced Oxidation Processes Using Nanomaterials 171
1 Introduction 171
2 Advanced Oxidation Processes 172
2.1 Direct or Indirect Photolysis of Hydrogen Peroxide 174
2.2 Photocatalysis Using Semiconductors 174
2.3 Electrochemistry 174
3 Nanomaterials for AOPs 175
4 Application of Nanomaterials in AOPs 176
4.1 Photocatalysis 176
4.2 Sonocatalysis 178
4.3 Ozonation 179
4.4 Fenton 181
4.5 Photo-Fenton 182
4.6 Electro-Fenton 185
4.7 Hybrid Systems 186
5 Challenges and Future Perspectives 188
6 Final Remarks 189
References 190

Application of Nano Magnesium Oxide in Advanced Oxidation Processes for the Degradation of Organic Pollutants 199
1 Introduction 199
2 Properties of nMgO 200
2.1 Physicochemical Properties of nMgO 200
3 Application of nMgO in Photocatalysis 202
3.1 Photocatalytic Activity of nMgO 203
3.2 Photocatalytic Activity of Doped nMgO 207
3.3 Photocatalytic Activity of nMgO Photocomposites 208

4 nMgO in Fenton-Like Process 213
5 Catalytic Activity of nMgO in the Absence of Light 216
5.1 Activity of nMgO for the Generation of O_2^- in Dark 216
5.2 nMgO for Persulphate Activation in the Absence of Light 219
6 Ozone Activation by nMgO Catalyst 222
7 Summary and Knowledge Gaps 228
References 229

Nanobiochar and Biochar Nanocomposites in Wastewater Treatment 239
1 Introduction 239
2 Methods of Nanobiochar Production 241
3 Production of Biochar Nanocomposites 243
4 Application of Nanobiochar and Biochar Nanocomposites in the Treatment of Wastewater 245
5 Mechanism of Pollutant Removal by Nanobiochar and Biochar Nanocomposites 256
6 Challenges in the Application of Nanobiochar/Biochar Nanocomposites 256
References 257

Index 261

About the Authors

Dr. Remya Neelancherry is an Associate Professor at School of Infrastructure, Indian Institute of Technology Bhubaneswar, Odisha, India. She has an expertise in development of novel nano-catalysts and their application for bio-refractory pollutants removal, advanced oxidation processes for water and wastewater treatment and application of microwave for solid waste management. Dr. Remya has more than 15 years of teaching and research experience in various specialized areas of environmental engineering and science. She is a member of American Society of Civil Engineers, American Chemical Society, Indian Water Works Association, Institute of Engineers (India), a senior Member of Asia-Pacific Chemical, Biological and Environmental Engineering Society, etc.

Dr. S. Mathava Kumar is an Associate Professor in the Department of Civil Engineering, Indian Institute of Technology Madras, Chennai, Tamil Nadu, India. His research interest includes water and wastewater treatment using membrane reactors, new-materials for environmental applications, emerging pollutants removal by advanced oxidation processes, trace organics and nutrients removal for reuse and resource recovery from solid waste. Dr. S. Mathava Kumar has more than 20 years of teaching, research and consulting experience in the field of environmental engineering. He is a member of American Society of Civil Engineers, International Water Association, International Association for Hydro-Environment Engineering and Research, etc.

Introduction to the Application of Nanotechnology in Water and Wastewater Treatment

S. Mathava Kumar and Remya Neelancherry

1 Introduction

The intersection of population growth, the industrial revolution, escalating water demand and the need for effective water purification represents a pivotal and interconnected challenge in contemporary global water management. The continuous upsurge in global population, predominantly in urban areas, and the demand for clean water for essential needs, such as drinking, sanitation and industrial processes, is increasing dramatically. This demographic shift not only strains existing freshwater sources but also intensifies competition among critical sectors like agriculture, energy production and residential consumption, increasing water scarcity in many regions. Simultaneously, the industrial revolution, while fostering economic growth and technological advancement, has historically led to heightened water consumption and pollution. Industries require substantial volumes of water for manufacturing, cooling and waste disposal, often discharging pollutants that threaten water quality and aquatic ecosystems. This dual challenge of increasing demand and industrial impact underscores the urgent need for comprehensive water management strategies that balance developmental imperatives with environmental sustainability.

In response to these complex challenges, advanced water purification technologies emerge as indispensable tools. Innovations such as nanotechnology-enhanced filtration, membrane processes (including reverse osmosis (RO) and ultrafiltration (UF)) and advanced oxidation processes (AOPs) are at the forefront of efforts to treat

S. M. Kumar (✉)
Environmental Engineering Division, Department of Civil Engineering Indian Institute of Technology Madras, Chennai, Tamil Nadu, India
e-mail: mathav@iitm.ac.in

R. Neelancherry (✉)
School of Infrastructure, Indian Institute of Technology Bhubaneswar Argul, khurda, Odisha, India
e-mail: remya@iitbbs.ac.in

© Capital Publishing Company, New Dehli, India. 2025

R. Neelancherry and S. M. Kumar, *Nanotechnology in Wastewater Treatment*,
https://doi.org/10.1007/978-3-031-94246-4_1

contaminants effectively. These technologies enable the elimination of many pollutants from wastewater and potable water sources, including heavy metals, pathogens, pharmaceuticals and industrial chemicals. By enhancing treatment efficiency and water quality, these advancements play a critical role in safeguarding public health and environmental integrity amid increasing concerns over waterborne diseases and chemical hazards.

To achieve a sustainable water management, a holistic approach that integrates policy, technology and societal engagement is essential. Effective water governance frameworks are essential to regulate water allocation, manage competing demands and enforce environmental standards. Furthermore, investments in research and development are vital to driving continuous improvements in water treatment technologies, making them more efficient, cost-effective and adaptable to diverse water quality conditions worldwide. Public awareness and education also play pivotal roles in promoting water conservation practices and fostering community engagement in water stewardship. By increasing the awareness about the importance of managing water resources, encouraging responsible water use behaviours and advocating for equitable access to clean water, communities can contribute to collective efforts in achieving water security and sustainability.

In summary, addressing the intricate dynamics of population growth, industrialisation, rising water demands and water purification challenges requires collaborative and innovative approaches on a global scale. By leveraging technological advancements, strengthening governance frameworks, promoting public awareness and fostering international cooperation, societies can strive towards ensuring resilient water systems, preserving freshwater ecosystems and meeting the growing water needs of present and future generations.

2 What is Nanotechnology?

Nanotechnology, the manipulation of matter at the nanoscale, stands as a cornerstone of modern scientific and technological advancement. Operating between 1 and 100 nm, nanomaterials exhibit properties that defy conventional material behaviour, offering unique opportunities across diverse fields. At this scale, quantum effects become prominent, surface area-to-volume ratios increase significantly and materials gain enhanced strength, optical clarity and electrical conductivity. These attributes enable nanotechnology to revolutionise industries such as electronics, energy, environmental engineering and science, and medicine including drug manufacturing and delivery. Nanotechnology has revolutionised drug delivery systems by designing nanoparticles that can target specific cells or tissues, delivering medications precisely where needed while reducing systemic side effects. This precision is crucial for treating diseases like cancer, where traditional treatments often affect healthy tissues along with tumours. Nanoparticles also hold promise in diagnostics, enabling the detection of biomarkers at ultra-low concentrations, which is critical for early disease detection and monitoring treatment effectiveness. In electronics, nanotechnology has enabled the miniaturisation of devices and enhanced their performance. Nanoscale transistors and memory storage devices have pushed the limits of computing power, allowing for faster processing speeds and greater storage capacities in smaller devices. In addition, nanomaterials like graphene are paving the way for flexible electronics and wearable technologies that can revolutionise healthcare monitoring and consumer electronics.

The field of renewable energy has also benefited from nanotechnology, particularly in improving the efficiency of solar cells and advancing energy storage technologies. Nanomaterials such as quantum dots and perovskites are being integrated into photovoltaic cells to enhance light absorption and electron transport, leading to more efficient conversion of solar energy. In energy storage, nanotechnology facilitates the development of high-capacity batteries and supercapacitors with faster charging times and longer lifespans, essential for storing renewable energy and powering electric vehicles. In the realm of environmental science and water treatment, nanotechnology offers innovative solutions to address water scarcity and pollution challenges. Nanomaterial-enhanced membranes improve filtration efficiency by selectively removing contaminants while maintaining high flux rates and durability. AOPs using nanocatalysts like titanium dioxide (TiO_2) harness light energy to degrade pollutants in water, providing a sustainable approach to treating industrial effluents and municipal wastewater.

In short, nanotechnology represents a frontier of innovation with profound implications for the society, economy and environment. By means of capitalising the inimitable properties of nanomaterials, investigators and industries are poised to tackle some of the most pressing global challenges, from improving healthcare and advancing technology to enhancing renewable energy and ensuring clean water resources for our future generations.

3 Nanomaterial Synthesis and Characterisation

Depending on the material used for their synthesis, presently there are two types of synthesis procedures for obtaining nanomaterials, namely, top-down and bottom-up approaches (Fig. 1). The top-down approach for nanomaterial synthesis encompasses mechanical and chemical synthesis processes. In top-down approaches, nanoparticles are generated by deconstructing bulk materials through processes such as milling, etching, laser ablation, electro-explosion and sputtering (Baig et al., 2021). They are appropriate for the fabrication of thin films and nanomaterials with size exceeding 100 nm and hence are employed for making electrical circuits with optimal integration and connection (Xie et al., 2017). The primary limitation of the existing top-down methodologies is their resolution constraint (Ranjan et al., 2016). In contrast to top-down approaches, bottom-up methods facilitate the development of nanomaterials with fewer flaws. In these procedures, atoms, molecules and nanoparticles serve as the foundational materials for the fabrication of intricate nanostructures (Gherbi et al., 2022). Ionic and molecular self-assembly is a bottom-up methodology

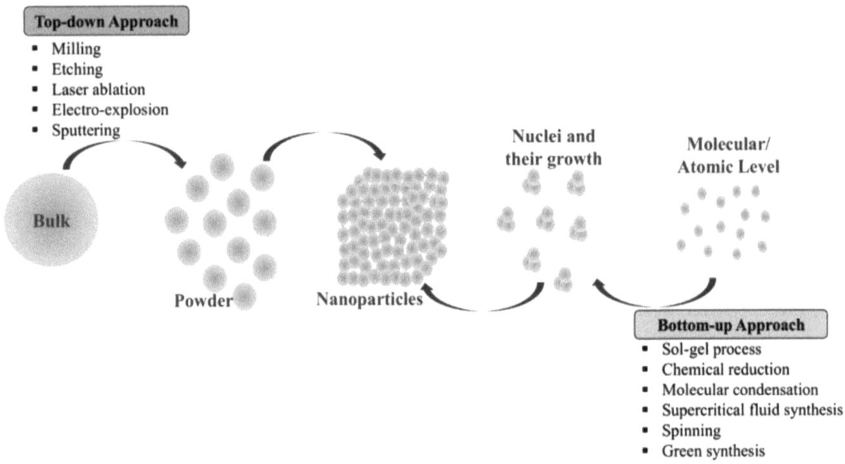

Fig. 1 Synthesis of nanomaterials from top-down and bottom-up approaches

wherein chemical or physical factors facilitate the aggregation of individual blocks or molecules into bigger structures via non-covalent interactions, including hydrogen bonds, ionic bonds, van der Waals forces and water-mediated hydrogen bonding. Bottom-up approaches provide physical regulation of the properties and composition of individual components. The dimensions of the building block are contingent upon the intended characteristics. Sol-gel process, chemical reduction, molecular condensation, supercritical fluid synthesis, spinning and green synthesis represent bottom-up strategies (Harish et al., 2022). Top-down and bottom-up fabrication methods are merged in hybrid methodologies to produce nanostructured platforms. Usually, these hierarchically structured systems are difficult or impossible to create with only top-down or bottom-up approaches (Gherbi et al., 2022). Photolithography depicts a hybrid approach, wherein etching and layer creation occurs through top-down and bottom-up approach respectively (Harish et al., 2022).

The predominant characterisations utilised to depict nanoparticles encompass UV–vis spectroscopy, X-ray powder diffraction (XRD), Fourier transform infrared spectroscopy (FT-IR), energy dispersive X-ray spectroscopy (EDX), scanning electron microscopy (SEM), transmission electron microscopy (TEM), dynamic light scattering (DLS), atomic force microscopy (AFM) and Brunauer–Emmett–Teller (BET) (Mourdikoudis et al., 2018). UV–vis spectroscopy is primarily employed to assess optical characteristics, size, concentration, aggregation state and indications of nanoparticle morphology (Singh et al., 2024). XRD is used to ascertain crystal structure, composition and crystalline grain size. FTIR spectroscopy is utilized to identify the specific functional group contained in nanoparticles. Furthermore, it can be employed for the analysis of diverse capping agents (Jahan Tamanna et al., 2024). EDX results help in determining the elemental composition of nanoparticles. SEM is majorly used for studying morphological characteristics. TEM is utilised to ascertain nanoparticle dimensions, assess size uniformity, characterise morphology, evaluate aggregation status, identify and localise/quantify nanoparticles within matrices and investigate growth kinetics (Vijayaram et al., 2024). DLS verifies the hydrodynamic size and further detects agglomerate (Chaturvedi et al., 2024). Nanoparticle size and form in the three-dimensional mode, nanoparticle morphology covering degree of surface, nanoparticle dispersion in cells and other matrices, nanoparticle precision in lateral dimensions and elemental composition were all assessed using AFM (Jagadeesh et al., 2024). BET analysis is employed to assess the surface area and porosity of materials through the quantification of gas adsorption on their surfaces (Al-Harbi & Abd-Elrahman, 2024).

4 Role of Nanotechnology in Environmental Engineering

Nanotechnology significantly impacts environmental engineering by providing advanced solutions to numerous environmental issues. The distinctive features of nanomaterials, such as large surface area, high reactivity and customisable molecular structure, make them particularly effective in areas like pollution mitigation, resource management and environmental monitoring. An overview of the contribution of nanotechnology in environmental engineering is given in Table 1.

Despite its advantages, nanotechnology in environmental engineering also presents the following challenges: (i) **Health and Safety Concerns**: Understanding the potential health and environmental impacts of nanomaterials is essential, as their small size and high reactivity could pose risks. (ii) **Regulation and Standardisation**: Establishing comprehensive regulations and standards for the production, use and disposal of nanomaterials is necessary to ensure environmental safety. (iii) **Cost and Scalability**: While nanotechnology offers advanced solutions, the cost and scalability of these technologies can be limiting factors for widespread adoption.

The future of nanotechnology in environmental engineering is promising, with ongoing investigations intended at overcoming present challenges and expanding applications. Innovations in nanomaterials and nanodevices are expected to provide more efficient, cost-effective and sustainable solutions for environmental protection and resource management. In conclusion, nanotechnology enhances environmental engineering by offering advanced methods for pollution control, resource management, environmental monitoring and waste management. As research progresses, it holds the potential to address some of the most critical environmental issues, contributing to a cleaner and more sustainable world.

Table 1 Overview of the contribution of nanotechnology in environmental engineering

Broad area	Field	Application type	Description
Pollution control and clean-up	Water treatment	Nanofiltration membranes	These membranes are capable of eliminating different pollutants, including bacteria, viruses, heavy metals and organic pollutants, from water. They are more efficient and selective than traditional filtration methods
		Nanosorbents	Materials such as graphene oxide (GO) and carbon nanotubes (CNTs) can absorb pollutants like heavy metals, dyes and organic chemicals from wastewater, providing a large surface area and high reactivity for better removal efficiency
		Photocatalysts	Nanoparticles like titanium dioxide (TiO2) can break down organic pollutants in water through photocatalytic reactions when exposed to sunlight or ultraviolet light, converting harmful substances into less toxic forms
	Air pollution reduction	Nanocatalysts	These catalysts used in catalytic converters and air purifiers can effectively break down pollutants such as carbon monoxide, oxides of nitrogen and many different volatile organic compounds (VOCs) into harmless gases
		Nanocoatings	These coatings can be applied to surfaces in urban areas to degrade airborne pollutants through photocatalytic reactions, thereby improving air quality
	Soil decontamination	Nanoscale zero-valent iron (nZVI)	This material is used for insitu remediation of contaminated soils and groundwater. It can reduce and immobilise contaminants like chlorinated organic compounds and heavy metals, preventing their spread and reducing environmental hazards

(continued)

Table 1 (continued)

Broad area	Field	Application type	Description
Resource management	Water desalination	Nanomaterials in desalination	Nanotechnology improves desalination efficiency and reduces costs. For example, graphene-based membranes filter salt from seawater more effectively and consume less energy compared to traditional methods
	Energy efficiency and renewable energy	Nanomaterials in solar cells	Quantum dots and perovskite nanoparticles enhance the efficiency of solar cells, making renewable energy more practical and efficient
		Nanotechnology in batteries	Nanomaterials improve battery performance and energy storage, facilitating better use of renewable energy sources
Environmental monitoring	Nanosensors	Chemical and biological detection	Nanosensors can detect tiny amounts of pollutants in water matrices, air and soil matrices with higher accuracy and sensitivity, enabling real-time monitoring of environmental conditions
		Wireless sensor networks	These networks use nanosensors to collect extensive environmental data over large areas, aiding in pollution tracking and environmental management
	Contaminant detection	Nanostructured materials	These are used in analytical techniques to detect and measure environmental contaminants like heavy metals, pesticides and pathogens with high accuracy and precision
Waste management	Recycling enhancements	Catalytic processes	Nanocatalysts can improve recycling processes by breaking down complex waste materials into simpler, reusable forms, thereby increasing the efficiency of material recovery
		Degradable nanomaterials	Research aims to develop nanomaterials that degrade safely, minimising their long-term environmental impact

5 Water Treatment: Conventional Versus Advanced Treatment Techniques

5.1 Conventional Water Treatment Techniques

Water treatment is indispensable for ensuring clean and safe water for various applications, including drinking, industrial processes and irrigation. Conventional methods of water treatment include plain sedimentation, sedimentation after the addition of chemicals (i.e., coagulation-flocculation), filtration (slow-sand, rapid-sand and pressure-sand filtrations) and disinfection with chlorine. In coagulation and flocculation, chemicals like aluminium sulphate (i.e., alum) and ferric chloride are added to water, causing suspended particles to aggregate into larger clusters (flocs), which can then be removed more easily by gravitational settling. The process of sedimentation is intended for the removal of colloidal particles contributing to turbidity. Water softening is carried out simultaneously in the sedimentation tanks/basins by adding lime and soda ash or caustic soda. Filtration processes, utilising materials such as sand, gravel and activated carbon, further remove suspended particles, unsettled flocs coming from sedimentation tanks and some microorganisms. However, these filters require regular maintenance and replacement, and their effectiveness can be limited by the nature of the contaminants present. Disinfection, employing chemicals like chlorine or physical methods like UV light, is crucial for killing or inactivating pathogenic microorganisms, although it can produce harmful disinfection by-products (DBPs) and may not be effective against all pathogens.

5.2 Advanced Water Treatment Techniques

Advanced water treatment technologies leverage modern scientific advancements to enhance efficiency and effectiveness. Membrane filtration techniques, including microfiltration (MF), UF, nanofiltration (NF) and RO, force water through semi-permeable membranes, effectively removing particles based on size. MF and UF remove larger particles and pathogens, while NF and RO are capable of filtering out smaller dissolved substances, such as salts and organic compounds. These methods, though highly effective, are associated with high energy costs, membrane fouling and significant maintenance requirements.

In addition, AOPs utilise powerful oxidants like ozone, hydrogen peroxide and ultraviolet (UV) light to generate highly reactive radicals, such as hydroxyl radicals, which can oxidise and break down a wide range of organic pollutants. A wide range of AOPs have been developed for advanced water treatment like photocatalysis, piezo-photocatalysis, electrokinetic oxidation, microwave and microwave-assisted processes, sonolysis, ozone-assisted technologies, gamma ray based, combination of chemical processes and combination of chemical and irradiation processes (UV, visible-light, microwave, plasma-based processes, etc.).These processes are particularly effective for degrading persistent organic pollutants (POPs) and emerging contaminants (ECs), such as pharmaceuticals and endocrine disruptors. Despite their effectiveness, AOPs are energy-intensive and can potentially form secondary pollutants.

Nanotechnology introduces innovative materials such as CNTs, GO of different forms and metal-nanoparticles into water treatment. These materials, characterised by their large surface area and reactivity, can adsorb or catalytically degrade a broad spectrum of contaminants, including heavy metals, pathogens and organic pollutants. For instance, carbon nanotubes (CNTs) and graphene oxide (GO) have large surface areas and high reactivity and are excellent adsorbents for removing heavy metals and organic pollutants. However, the potential environmental and health risks associated with nanomaterials and their high production costs are major challenges.

5.3 Pros and Cons of Conventional and Advanced Water Treatment Techniques

Conventional water treatment techniques are effective for basic purification needs and are generally more cost-effective. However, they produce huge amounts of sludge and may not remove all contaminants efficiently. Advanced methods, though more costly and complex, remove a wider range of contaminants and adhere to stringent water quality standards. Moreover, advanced water treatment techniques remove and treat higher volumes of contaminants compared to conventional water treatment techniques. For instance, advanced membrane filtration techniques ensure higher levels of purification by removing smaller particles and dissolved substances compared with conventional methods. Similarly, AOPs can break down complex organic pollutants that are resistant to conventional treatments.

Integrating conventional and advanced water treatment technologies can optimise water treatment processes. For example, combining coagulation-flocculation with membrane filtration can enhance the removal of both particulate and dissolved contaminants. Likewise, coupling biological treatment with AOPs can improve the overall efficiency of pollutant degradation. Such integrated approaches ensure the provision of clean and safe water for various needs, addressing both traditional and emerging water quality challenges. The choice of treatment method depends on specific water quality requirements, cost considerations and sustainability goals, with ongoing research and development continuously improving the effectiveness and affordability of advanced water treatment technologies.

6 Wastewater Treatment: Conventional and Advanced Treatment Techniques

Wastewater treatment is of paramount importance in protecting public health and the environment. It is essential for purifying sewage and industrial effluents before safe discharge or reuse. Conventional methods of wastewater treatment form the backbone of initial treatment stages, where manual processes like screening and grit removal eliminate large debris, ensuring that downstream processes are not hindered. Sedimentation then allows for the settling of heavier solids and the floating of lighter materials, thereby reducing the load of suspended solids and organic matter for further treatment. Secondary treatment employs biological processes, such as the activated sludge method, where microorganisms break down organic pollutants in aerated tanks, and trickling filters, where wastewater percolates through media coated with microbial biofilms that degrade organic substances. Oxidation ponds harness natural microbial and algal processes in shallow basins to further purify wastewater. This is particularly effective in warmer climates where sunlight aids in biological activity.

Advanced wastewater treatment technologies aid these conventional methods to deal with more complex contaminants and meet stringent regulatory standards. The scheme for advanced wastewater treatment plant is shown in Fig. 2, and processes that can be employed in various stages of conventional and advanced wastewater treatment plants are listed in Table 2.

In advanced wastewater treatment plants, the tertiary treatment/polishing of the secondary effluent is carried out by the combination of AOPs, adsorption, ion-exchange, membrane technologies, etc. As mentioned earlier, AOPs utilise highly reactive radicals, which could effectively degrade POPs, pharmaceutical residues,

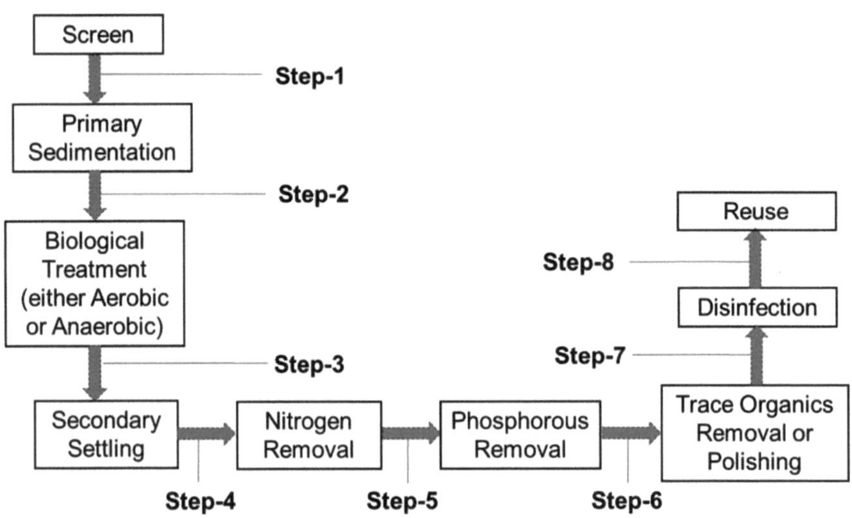

Fig. 2 Scheme for advanced wastewater treatment

Table 2 List of processes that can be used at various stages of conventional and advanced wastewater treatment plants

Steps (Type of pollutant removed)	List of processes used at various stages of wastewater treatment plants	
	Conventional treatment plant	Advanced treatment plant
Step-1 (Removal of floating substances, excluding oil and grease)	• Screening	• Screening
Step-2 (Removal of suspended solids)	• Primary sedimentation	• Primary sedimentation
Step-3 (Removal of dissolved organics)	• Activated sludge process • Trickling filters Rotating biological contactors	• MBRs (aerobic or anaerobic) • Sequential batch reactor (SBR) • Fluidised-bed bioreactor (FBBR) • Moving bed biofilm reactors (MBBR) • Upflow anaerobic sludge blanket (UASB) reactors • Attached (or) suspended grow
Step-4 (Removal of suspended solids)	• Secondary settling tank	• Membrane filtration • Nitritation-denitritation • Centrifugation
Step-5 (Removal of dissolved nutrients)	• Nitrification- denitrification	• Nitritation-denitritation • Anammox, Sharon • OLAND • Electrodialysis and ionexchange
Step-6 (Removal of dissolved nutrients)	• No attempt	• Precipitation • Electrodialysis • Distillation
Step-7 (Removal of dissolved trace organics and inorganics)	• No attempt	• Adsorption • Advanced oxidation process (AOP) • Chemical/electrochemical oxidation • Reverse osmosis • Electrodialysis
Step-8 (Disinfection)	• Chlorination	• Ultraviolet • Plasma technologies
Handling of sludge	• Dried and disposed in the landfill	• Dewatering • Sludge anaerobic digestion Composting • Biogas production

(continued)

Table 2 (continued)

Steps (Type of pollutant removed)	List of processes used at various stages of wastewater treatment plants	
	Conventional treatment plant	Advanced treatment plant
Assessment of risk	• Not attempted	• Need-based risk assessment conducted
Assessment of risk	• Not attempted	• Need-based risk assessment
Reuse of water	• Disposed into land or water	• Supplied back for drinking • Supplied to industries • Supplied for agriculture • Used for GW recharge • Used for the development of green belt

ECs and other stable organic compounds susceptible to conventional biodegradation. Despite their efficacy, AOPs require significant energy inputs and careful operational control to manage by-products and ensure optimal performance. Membrane filtration technologies, including MF, UF, NF and RO, leverage semi-permeable membranes to remove suspended solids, bacteria, viruses and dissolved substances at various scales, from municipal wastewater treatment to industrial applications. While these methods offer high removal efficiency and smaller footprints, they also face a lot of challenges including: (i) membrane fouling, (ii) high operational costs, (iii) need for periodic maintenance, (iv) skilled labour for supervision and (v) high initial investment/cost.

Moreover, biological nutrient removal (BNR) processes are also used for tertiary treatment in advanced wastewater treatment plants, which focus on reducing the levels of nitrogen and phosphorus through specialised microbial processes like nitrification–denitrification and the removal of biological phosphorus. These processes play a critical role in mitigating nutrient pollution that contributes to eutrophication in receiving waters. Nanotechnology enhances wastewater treatment through enhanced adsorption, catalytic degradation of contaminants and antimicrobial properties. Despite promising capabilities of nanomaterials, concerns about their environmental impact and cost-effectiveness necessitate further research and development for their safe and sustainable deployment in wastewater treatment facilities. Constructed wetlands and biofilters utilise natural ecosystems to treat wastewater, providing sustainable alternatives that promote biodiversity and ecological balance while effectively removing contaminants through microbial degradation processes. However, these systems require adequate land area and careful management to optimise treatment efficiency under varying environmental conditions.

In a nutshell, the integration of conventional and advanced wastewater treatment methods represents a comprehensive approach to achieving water quality goals and ensuring environmental sustainability. The selection of treatment technologies depends on specific treatment objectives, regulatory requirements, economic considerations and environmental impacts. Continued advancements in technology and research are crucial for enhancing the efficiency, sustainability and affordability of wastewater treatment processes, thereby safeguarding water resources for future generations.

7 How Nanotechnology Can Be Applied for Water and Wastewater Treatment

Nanotechnology continues to advance the field of water and wastewater treatment by offering groundbreaking solutions that improve efficiency, effectiveness and sustainability across various stages of the treatment process. In membrane filtration, nanotechnology has revolutionised the capabilities of NF and RO systems. These membranes, engineered with nanoscale pores and advanced materials like graphene and ceramic nanoparticles, exhibit enhanced selectivity and durability. They efficiently remove contaminants, such as heavy metals, pathogens, pharmaceutical residues and organic pollutants, providing a critical barrier against pollutants that traditional filtration methods struggle to handle adequately. This technological advancement is pivotal for ensuring safe drinking water supply in urban areas and enabling efficient reuse of water in industrial processes, agriculture and municipal applications.

Nanoparticle-based adsorbents represent another frontier in water treatment innovation. CNTs, GO and various metal nanoparticles offer high surface area and tailored surface chemistry, which enable them to adsorb a wide range of contaminants from water. These nanomaterials remove heavy metals like arsenic, cadmium and chromium, as well as organic compounds such as pesticides, dyes and industrial solvents. Their versatility extends to addressing ECs and micro pollutants, which enables them to comprehensively purify wastewater effluents from industrial facilities and urban wastewater treatment plants. The application of these nanomaterials in decentralised water treatment systems and point-of-use devices further enhances their potential to provide clean water solutions in remote or resource-constrained regions.

AOPs that are driven by nanotechnology offer robust solutions for degrading POPs and refractory compounds in wastewater. Nanoparticles like TiO_2, zirconia, iron oxides, other catalytic nanomaterials and several types of nanocomposites activate under specific conditions to produce reactive oxygen species (ROS). These ROS exhibit high oxidative potential, effectively breaking down complex organic molecules into harmless by-products like carbon dioxide and water. AOPs are particularly valuable in treating industrial wastewater streams containing pharmaceutical residues, pesticides and chemical by-products, where conventional treatment methods may be inadequate.

Nano sensors represent a significant advancement in real-time water quality monitoring and management. These miniature devices, integrated with nanomaterials for enhanced sensitivity and selectivity, enable continuous monitoring of water parameters such as pH, turbidity, dissolved oxygen levels and specific contaminants. Pollutants and microbial pathogens can be timely detected using nano sensors, which facilitates early intervention and proactive maintenance of water treatment facilities. This capability supports rapid response strategies to mitigate contamination events, ensuring compliance with regulatory standards and safeguarding of public health.

Moreover, nano-enabled coatings and surface modifications play a crucial role in mitigating operational challenges in water treatment systems. These coatings, applied at the nanoscale on membranes and equipment surfaces, enhance anti-fouling properties, reduce biofilm formation and improve resistance to scaling and corrosion. By minimising fouling and enhancing operational efficiency, nanocoatings contribute to prolonged membrane lifespan, reduced energy consumption and lower maintenance costs in water treatment plants. While nanotechnology offers promising avenues for advancing water treatment capabilities, several challenges need to be addressed. These include scaling up nanomaterial production to meet industrial demands, assessing the long-term environmental impacts of nanomaterials released during water treatment processes, ensuring the safety and regulatory compliance of nanotechnological applications and addressing cost-effectiveness considerations for widespread implementation. Continued research and development efforts, interdisciplinary collaborations and strategic investments are essential for harnessing the full potential of nanotechnology in addressing global water challenges and achieving sustainable water management practices worldwide.

7.1 Nanomaterials in Photocatalysis-Based Tertiary Wastewater Treatment

Photocatalysis is one of the most promising AOPs utilising a semiconductor catalyst for the tertiary treatment of secondary effluent of water or wastewater. The inherent advantages of photocatalysis compared with other AOPs are as follows: (a) pertinence to a wide pH range, (b) no generation of sludge, (c) generation of highly redox potential species (OH is 2.8 V vs NHE), (d) utilisation of natural solar light and (e) easy recovery and reusability of catalyst. Photocatalysis has been extensively researched as the tertiary treatment for contaminants present in the secondary effluent (Fig. 3). In photocatalysis, light of wavelength (hv) is superior than the band gap of the photocatalyst (i.e., difference in energy level among valence band (VB) and conduction band (CB)) incident on its surface. The outermost shell electrons (e^-) absorb that energy and drive to a higher energy level to VB, leaving behind a hole (h^+) in the CB. ROS are generated by water and dissolved oxygen from these e^- and h^+ depending on the redox potential of the generated e^- and h^+. These ROS ($O_2^{-\cdot}$, OH, H_2O_2) have one unpaired electron. Therefore, they have a short life span and they readily combine with pollutants and degrade them into simpler by-products. Generally, semiconductors are used as photocatalysts because metals have VB and

CB overlapping, and so after e^--h^+ production, they would readily recombine. On the other hand, the band gap is very high in insulators. This necessitates high energy light, which makes the process uneconomical. Photocatalysis can be either homo- or hetero-photocatalysis based on whether the catalyst gets dissolved or remains suspended in the reaction medium, respectively.

Metal oxides, metal sulphides, graphene-based compounds, bismuth compounds, iron-based compounds, silver compounds, etc., are the majorly studied semiconductors for photocatalysis. The different photocatalysts are unique in their own way, as they have unique bandgaps and band positions. They can also be combined to form binary junctions, tertiary junctions, ternary junctions and heterojunctions, which improve their performance. Further, the performance of a photocatalyst can be improved by utilising membrane sheets or beads, or adding magnetic properties into it. To further improve the photocatalytic activity, the process can be combined with other AOPs to make hybrid technology.

However, in spite of the advantages of photocatalysis, the following challenges have to be addressed: (i) wide bandgap requiring UV light and therefore inefficient utilisation of visible light (sunlight wavelength ranges from 250 to 2500 nm, and the UV region is only around 4–5%), (ii) rapid electron–hole recombination causing a decrease in charge carriers' concentration, (iii) aggregation of photocatalyst nanoparticles causing light shielding to the active surface sites of photocatalysts, (iv) separability issues because of small-sized nanoparticles and (v) low adsorption capacity because of low specific surface area. These issues are addressed by different strategies like doping, surface deposition, heterojunction formation between two or more semiconductors, surface/chemical modification, substrate coating, etc. For example, iron-doped zirconium oxide nanoparticles, carbon fibre cloth supported by bismuth oxybromide, TiO_2–g–C_3N_4 heterojunction photocatalyst immobilised on walnut shell biochar, ZnO and WO_3 heterojuction photocatalysts supported on hemp herd biochar and TiO_2 coatings on ordinary stainless steel and pyrex glass have shown enhanced photocatalytic activity and post-treatment separability (Ali et al., 2024; Bannier et al., 2011; Parvees et al., 2024; Patil et al., 2023; Verma et al., 2024).The treated water after photocatalysis can be checked for toxicity if any degradation by-product is remaining. If not, then the treated water can be safely disposed to existing water bodies like surface water or groundwater recharge. If the treated water is still contaminated, then some polishing unit can be given finally. For instance, either the retention time in the photocatalysis unit can be increased or some other AOP technology like pulse plasma or ozonation can be employed to make the water fully clean.

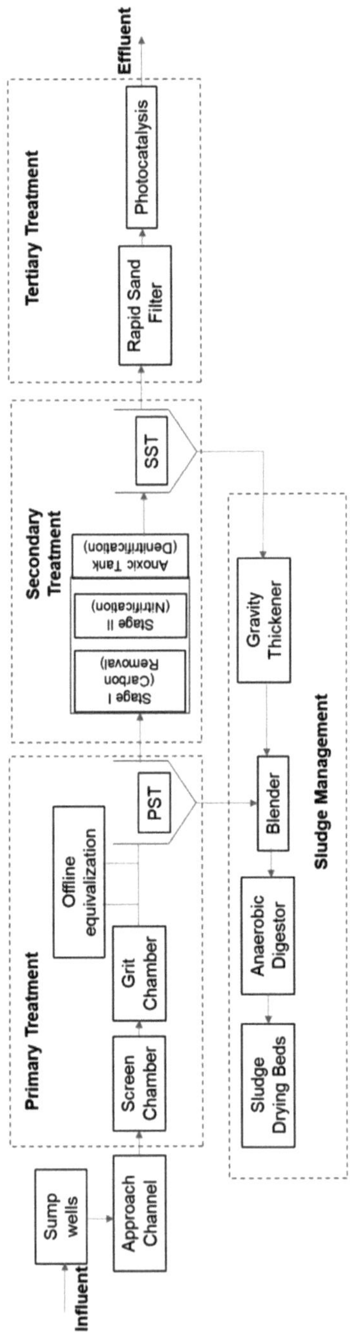

Fig. 3 Scheme for photocatalysis-based tertiary wastewater treatment

7.2 *Nanomaterials in Piezocatalysis and Piezo-Photocatalysis-Based Tertiary Wastewater Treatment*

Piezocatalytic materials have a non-centrosymmetric crystal structure that allows the electrons/holes to participate in redox reaction when mechanically excited due to the induced polarisation (Zhang et al., 2023). In the beginning, Pb-based materials were used as piezocatalysts because of their higher piezoelectric coefficient. However, due to the toxicity effects of Pb-based materials, recently the piezocatalytic efficiency of other materials like MoS_2, $KNbO_3$, $NaNbO_3$, $BaTiO_3$, ZnO and BiOX (Cl, I, Br) are being explored (Meng et al., 2023; Zhang et al., 2023). Piezocatalysis has proven to be an effective AOP technology for sustainable and environmentally friendly water remediation (Zhang et al., 2023). The application of piezocatalysis in water remediation can be broadly classified into two: removal of emerging contaminants and disinfection. The fundamental mechanism of piezocatalysis for water remediation is the production of ROS like ·OH for the mineralisation of pollutants or removal of bacteria (Masekela et al., 2023; Singh et al., 2021). Recently, different synergetic systems have been adopted along with piezocatalysis for the enhanced production of ROS. Figure 4 schematically represents the possible augmentations that can be adopted to piezocatalysis.

The redox reactions taking place in a piezocatalytic process is very similar to the redox reactions in conventional photocatalytic technology. Equations (1 & 2) represent the formation of ·OH from single oxidation of H_2O and OH^-. Equation (3) represents the production of $\cdot O_2^-$ from O_2. Equations (4–6) represent the production of ·OH from double reduction reaction of O_2.

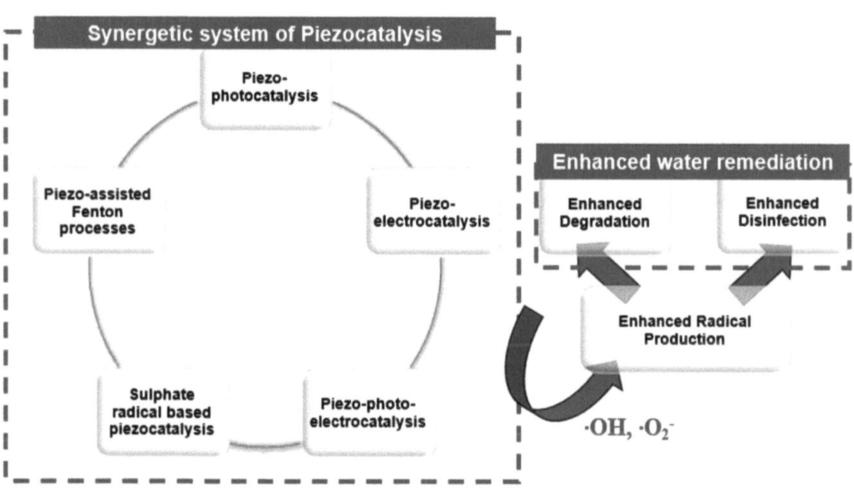

Fig. 4 Possible augmentations available for increasing the radical yield of apiezocatalytic system for enhanced water treatment

$$\text{holes} + H_2O \rightarrow \cdot OH + H^+ \tag{1}$$

$$\text{holes} + OH^- \rightarrow \cdot OH \tag{2}$$

$$\text{electrons} + O_2 \rightarrow \cdot O_2^- \tag{3}$$

$$\text{electrons} + \cdot O_2^- + 2H^+ \rightarrow H_2O_2 \tag{4}$$

$$H_2O_2 + \text{electrons} \rightarrow \cdot OH + OH^- \tag{5}$$

$$H_2O_2 + \text{holes} \rightarrow 2 \cdot OH \tag{6}$$

Apart from the above redox reaction's direct oxidation of pollutants by holes, direct reduction of pollutants by electrons can also take place. Most of the current scientific studies use ultrasonication-based piezoexcitation for the mechanical activation of piezocatalysts to induce redox reactions. However, the mechanical excitation required for the redox reaction can be provided by flowing water, waves, wind energy, etc. This ensures the sustainability of the adopted technology as the energy consumed for the treatment process can be minimised unlike highly energy-driven membrane processes. Recently, many catalysts have been prepared to enhance the mechanical sensitivity of piezocatalysts like doping, heterojunction formation and defect development (Meng et al., 2023).

The schematic representation of treatment chains containing piezocatalytic and piezo-photocatalytic technologies for the secondary treated wastewater is shown in Fig. 5. Though piezocatalysis is an effective method of sustainable water remediation, the main challenge in the practical implementation of the piezocatalytic process is the usage of the powdered nanocatalyst that can impose a secondary level of contamination to the remediated water. Immobilisation of piezocatalysts is an effective way to prevent secondary contamination. Polyvinylidene fluoride (PVDF) is the most commonly used material for immobilising a piezocatalytic material due to its inherent piezocatalytic nature (Zhang et al., 2023, Singh et al., 2021). PVDF has been used in various forms like pipes, membranes and films for immobilising piezocatalysts.

7.3 Nanomaterials in Microwave-Based Tertiary Wastewater Treatment

Microwave-induced catalytic degradation (MICD) has tremendous potential for becoming a rapid, eco-friendly and cost-effective response by enhancing the reaction rate and catalyst reusability, and successfully reducing the footprint area. Microwave

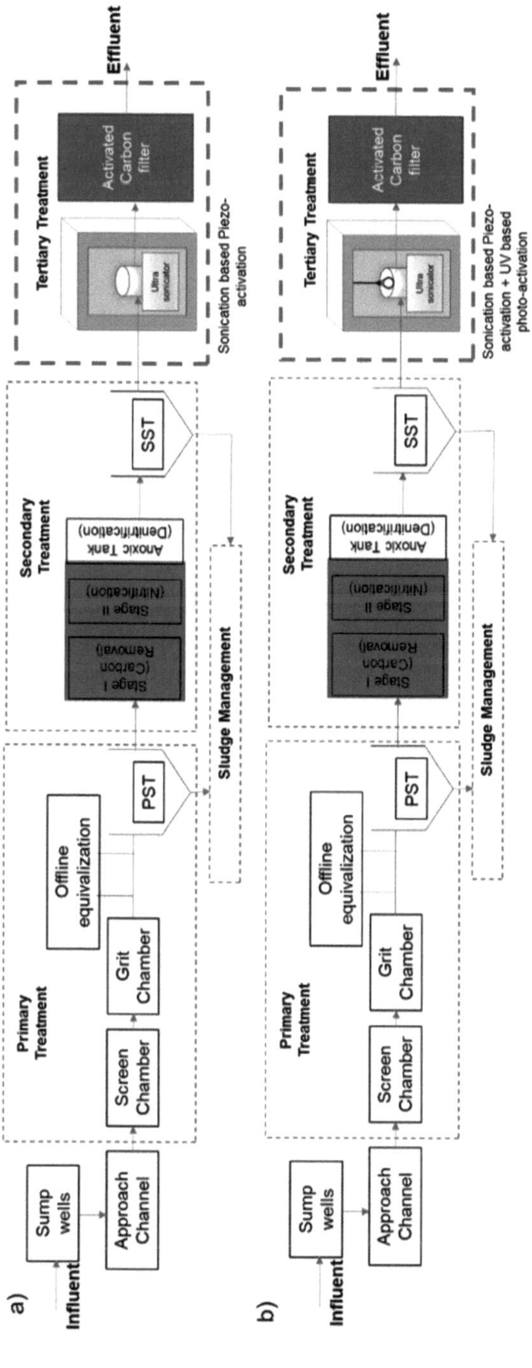

Fig. 5 Schematic representation of **a** sonication-assisted piezocatalysis for tertiary treatment followed by carbon filter to remove the powdered catalyst and **b** sonication-assisted piezocatalysis with UV-assisted photocatalysis for tertiary treatment followed by carbon filter to remove the powdered catalyst

(MW) is a non-ionising alternating electromagnetic wave with a wavelength of 1–0.001 m (frequency from 300 MHz to 300 GHz) that can be absorbed by a microwave-responsive catalyst. Dielectric materials such as water, activated-carbon, CNTs and zinc-oxide can absorb the incident energy and dissipate it when the rotation of the dipole moment fails to match the fast-altering electromagnetic field created by MW irradiation (Kumar et al., 2020). The scheme for the application of nanomaterial-assisted MW-based technologies for tertiary wastewater treatment is shown in Fig. 6.

On the other hand, in the case of magnetic materials such as metal oxides, spinel ferrites, etc., energy dissipation may occur due to hysteresis loss, eddy current loss and electron spin loss (Wei et al., 2020). However, individual dielectric or magnetic lossy materials typically face a few shortcomings in terms of narrow MW bandwidth absorption capability and increment of the reflective loss (RL), which causes energy wastage. A catalyst having both types of materials creates a well-matched relation between dielectric ($\tan\delta_{\mu}$) and magnetic loss tangent ($\tan\delta_{\varepsilon}$) to enhance the impedance matching (Bose & Kumar, 2022). Therefore, combining both materials retains their properties and introduces a synergistic effect in MW absorption and pollutant degradation. Recently, combined heterojunction materials such as spinel ferrite/carbon material, metal oxide/activated carbon, metal oxide/oxides of titanium or zinc or bismuth, etc., are grabbing the attention of researchers. Excellent MW absorber metal–organic framework (MOF)-derived catalysts (cobalt, iron, nickel-based), which retain MOF crystal structure even after pyrolysis along with magnetic and dielectric counterparts, have proven to be one of the best groups of MW-responsive catalysts (Feng et al., 2022). The dissipated energy may form localised 'hotspots' owing to extremely high temperatures (>1000 °C) on the surface, which may degrade the pollutant by thermal decomposition and break water molecules into hydroxyl radicals ($^{\cdot}OH$). The intrinsic semiconductor nature of the heterojunction catalysts promotes the separation of electron–hole pairs and, eventually, the formation of superoxide ion radicals ($O_2^{\cdot-}$) and $^{\cdot}OH$ (Bose & Kumar, 2022). To cut down power consumption, additional oxidants, such as persulphate and peroxide, can be activated rapidly in the presence of a transitional metal-type catalyst under MW irradiation by a redox coupling electron transfer pathway. Even in the absence of acatalyst, MW can activate PS in a multi-oxidant system to produce highly redox potential (E_0 = 2.5-3.1) sulphate radicals ($SO_4^{\cdot-}$) for the degradation of organic pollutants (Tripathy et al., 2020).

Even though this traditional MICD process has various advantages over other AOPs, this system also has a few drawbacks. Usually, powdered catalysts containing nanoparticle crystals were implemented in the lab-scale batch studies in the past literature, which may cause loss of costly catalysts in the continuous mode of operation. Therefore, for implementing MICD in the continuous field scale process, the powdered catalyst may be transformed into catalytic beads with the help of thermally stable polymers (i.e., polyaniline, polypyrrole) or by using carbon or glass beads as templates. Another approach was to paint the inner surface of the quartz MW reactor using nanoparticles dispersed in a polymeric solution. This probably was an efficient method for immobilising catalysts. MW responsive beads might be

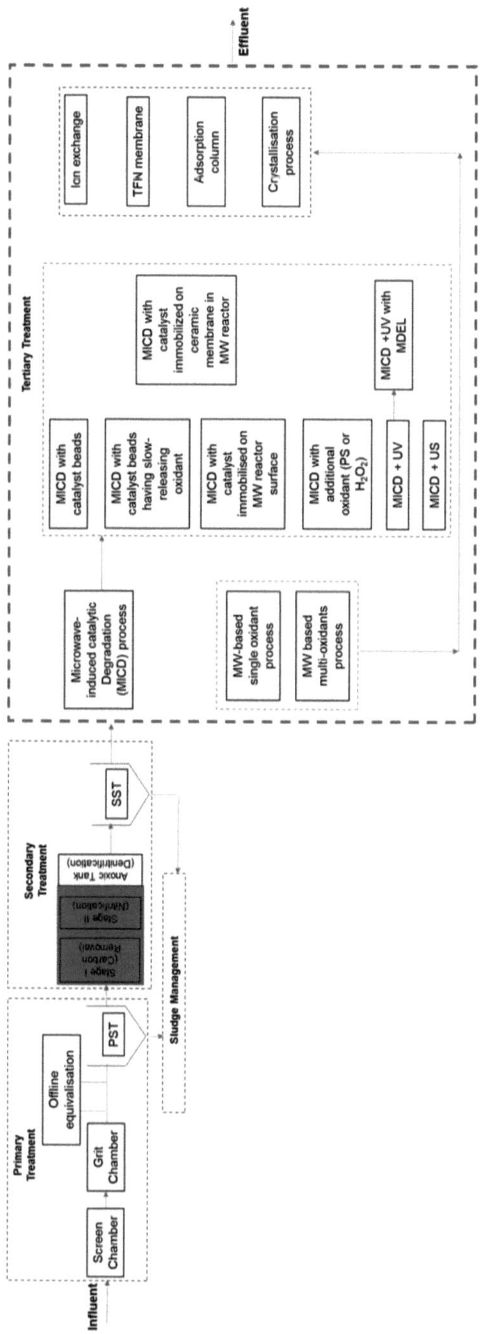

Fig. 6 Schematic diagram of MW-based wastewater treatment processes and future possibilities of post-treatment modifications

modified to develop the catalytic slow-releasing oxidant beads, which may release oxidant dose continuously for a certain reaction time, promoting a steady reaction rate and inhibiting the self-scavenging effect of free radicals.

Moreover, catalyst immobilisation on the ceramic membrane may reduce the loss of catalyst and increase the activation of any additional oxidants that come in contact with the membrane surface. Overall, many heterojunction MICD catalysts, except a few (i.e., only carbon-based), contain intrinsic semiconductor properties that can be activated under UV and ultrasound (US) ambient for the separation of electron-holes pair to generate free radicals. Thus, apart from traditional MICD processes, hybrid MW-based advanced processes may accommodate the UV and US in the same system to further reduce operation time and to increase mineralisation efficiency. In addition, microwave discharge electrodeless lamps (MDEL) may trap the fraction of reflected MW energy to convert it into UV light without any requirement of extra energy input for continuing the UV process. MW can disinfect the water with dielectric dipole movement and generation of 'hotspots' even at very minimum power input and exposure time, which may eliminate the supplementary cost of disinfection after the biological treatment process to remove living microbial cells. However, like other AOPs, MICD generates intermediates after the oxidation of parent pollutant molecules, resulting in secondary pollution. Therefore, NF polymeric membranes, especially thin-film nanocomposite membranes with nanoparticle fillers such as metal oxide, GO, CNTs, metal–organic frameworks (MOFs), covalent-organic frameworks, etc., may separate the intermediates from treated effluent water sustaining a higher flux rate. If the intermediates are non-toxic, the power consumption may be lowered by reducing operating time in the actual MW reactor, followed by an adsorbent column for reducing TOC load comprising simpler carbonaceous organic fractions. Although the MW-based catalytic process may remove heavy metals such as hexavalent chromium (Cr (VI)) by reduction pathway, MICD is not effective in the case of other heavy metals (mercury, lead, cadmium, etc.) and toxic metals (nickel, cobalt, etc.) usually present in wastewater. In that case, the above mentioned post-MW treatment adsorption column can remove heavy metals from the treated effluent. Furthermore, industrial wastewater, mostly industrial wastewater containing dyes, also contains different inorganic salts and ions (i.e., chloride, nitrate, phosphate, carbonate and sulphate), which may be precisely eliminated using an ion exchange column as a post-treatment process after MICD. Valuable salts and metals might be recovered by combining crystallisation techniques such as evaporation crystallisation or cooling crystallisation. In a nutshell, the MICD process and other hybrid MW-based treatment processes, including several post-treatment modifications followed by MW-based treatment, constitute a potential tertiary treatment process for the mineralisation of emerging contaminates and removal of inorganic counterparts while reducing operational time, cost and footprint area.

8 Challenges in the Application of Nanotechnology for Water and Wastewater Treatment

The challenges in the application of nanotechnology in water and wastewater treatment, which necessitate comprehensive solutions to maximise its potential benefits, are listed below.

One of the primary concerns is the economic feasibility of nanomaterial production and deployment in treatment processes. The high costs associated with scaling up production and ensuring consistent quality pose barriers, particularly for smaller treatment facilities and regions with limited financial resources. Finding ways to optimise production efficiencies and reduce costs will be crucial for broadening access to nanotechnology-enhanced water treatment solutions.

Secondly, environmental considerations also play a role in the adoption of nanotechnology. Nanomaterials used in water treatment, such as nanoparticles and nanocomposites, have the potential to accumulate in natural environments and pose risks to ecosystems and human health. Understanding their behaviour in aquatic systems, including their transport, transformation and long-term impacts, is essential for developing effective risk assessment frameworks and environmental management strategies.

Strategies for the safe disposal or recycling of nanomaterials after their useful life in treatment processes will also need careful attention to prevent unintended environmental consequences.

Regulatory frameworks currently face challenges in keeping pace with the rapid development of nanotechnology applications in water treatment. There is a need for robust regulatory oversight that addresses potential health and environmental risks associated with nanomaterials while promoting innovation and ensuring compliance with safety standards. Developing clear guidelines for the manufacturing, handling and disposal of nanomaterials in water treatment contexts will provide clarity and confidence to stakeholders involved in their deployment.

Operational challenges, such as fouling of nanotechnology-enhanced membranes and maintaining their long-term integrity, require innovative solutions. Strategies to mitigate fouling, such as surface modifications and anti-fouling coatings at the nanoscale, are critical for sustaining high filtration efficiency and extending membrane lifespan.

Optimising energy consumption in nanotechnology-driven processes, such as advanced oxidation and high-pressure membrane filtration, is essential for reducing operational costs and minimising environmental footprint.

Moreover, public perception and acceptance of nanotechnology in water treatment also play a crucial role for its effective employment. Informing the community about the profits, safety procedures and likely risks associated with nanotechnology applications can help build trust and support for these innovative solutions. Transparent communication about research findings, regulatory safeguards and real-world applications will be essential for fostering public confidence and acceptance. Overcoming these multifaceted challenges requires collaborative efforts across academia, industry, government agencies and the public. By addressing economic, environmental, regulatory, operational and societal considerations, stakeholders can collectively advance nanotechnology's role in improving water quality, ensuring sustainable water management practices and meeting the growing global demand for safe and clean water resources. Continued research, innovation and responsible deployment will be pivotal in answering the complete potential of nanotechnology in water and wastewater treatment, and also realising its benefits on a broader scale.

References

Al-Harbi, N., & Abd-Elrahman, N. K. (2024). Physical methods for preparation of nanomaterials, their characterization and applications: A review. *Journal of Umm Al-Qura University for Applied Sciences,* 0123456789. https://doi.org/10.1007/s43994-024-00165-7

Ali, A. S., Hassan, A. M., Issa, S. A. M., Alsufyani, S. J., Tekin, H. O., & Zakaly, H. M. H. (2024). Structural, optical and attenuation properties analysis of iron-doped zirconium oxide nanoparticles for photocatalytic and radiation shielding applications. *Inorganic Chemistry Communications, 170*(July), Article 113133. https://doi.org/10.1016/j.inoche.2024.113133

Baig, N., Kammakakam, I., Falath, W., & Kammakakam, I. (2021). Nanomaterials: A review of synthesis methods, properties, recent progress, and challenges. *Materials Advances, 2*(6), 1821–1871. https://doi.org/10.1039/d0ma00807a

Bannier, E., Darut, G., Sánchez, E., Denoirjean, A., Bordes, M. C., Salvador, M. D., Rayón, E., & Ageorges, H. (2011). Microstructure and photocatalytic activity of suspension plasma sprayed TiO_2 coatings on steel and glass substrates. *Surface and Coatings Technology, 206*(2–3), 378–386. https://doi.org/10.1016/j.surfcoat.2011.07.039

Bose, S., & Kumar, M. (2022). Microwave-assisted persulfate/peroxymonosulfate process for environmental remediation. *Current Opinion in Chemical Engineering, 36*, Article 100826. https://doi.org/10.1016/J.COCHE.2022.100826

Chaturvedi, S., Maheshwari, D., Chawathe, A., & Sharma, N. (2024). Current analytical approaches for characterizing nanoparticle sizes in pharmaceutical research. *Journal of Nanoparticle Research, 26*(1), 19. https://doi.org/10.1007/s11051-023-05924-x

Feng, Y., Tao, Y., Meng, Q., Qu, J., Ma, S., Han, S., & Zhang, Y. (2022). Microwave-combined advanced oxidation for organic pollutants in the environmental remediation: An overview of influence, mechanism, and prospective. *Chemical Engineering Journal, 441*, Article 135924. https://doi.org/10.1016/j.cej.2022.135924

Gherbi, B., Laouini, S. E., Meneceur, S., Bouafia, A., Hemmami, H., Tedjani, M. L., Thiripuranathar, G., Barhoum, A., & Menaa, F. (2022). Effect of pH value on the bandgap energy and particles size for biosynthesis of ZnO nanoparticles: Efficiency for photocatalytic adsorption of methyl orange. *Sustainability (Switzerland), 14*(18), 11300. https://doi.org/10.3390/su141811300

Harish, V., Ansari, M. M., Tewari, D., Gaur, M., Yadav, A. B., García-Betancourt, M. L., Abdel-Haleem, F. M., Bechelany, M., & Barhoum, A. (2022). Nanoparticle and nanostructure synthesis and controlled growth methods. *Nanomaterials, 12*(18), 1–30. https://doi.org/10.3390/nano12183226

Jagadeesh, P., Rangappa, S. M., & Siengchin, S. (2024). Advanced characterization techniques for nanostructured materials in biomedical applications. *Advanced Industrial and Engineering Polymer Research, 7*(1), 122–143. https://doi.org/10.1016/j.aiepr.2023.03.002

Jahan Tamanna, N., Sahadat Hossain, M., Mohammed Bahadur, N., & Ahmed, S. (2024). Green synthesis of Ag_2O& facile synthesis of ZnO and characterization using FTIR, bandgap energy & XRD (Scherrer equation, Williamson-Hall, size-train plot, Monshi-Scherrer model). *Results in Chemistry*, **7**(December 2023), 101313. https://doi.org/10.1016/j.rechem.2024.101313

Kumar, B., Kumar, S., Kumar, M., & Debnath, A. (2020). Environmental nanotechnology, monitoring & management microwave induced catalytic treatment of brilliant green dye with carbon doped zinc oxide nanoparticles: Central composite design, toxicity assessment and cost analysis. *Environmental Nanotechnology, Monitoring & Management, 14*(July), Article 100361. https://doi.org/10.1016/j.enmm.2020.100361

Masekela, D., Hintsho-Mbita, N. C., Sam, S., Yusuf, T. L., & Mabuba, N. (2023). Application of $BaTiO_3$-based catalysts for piezocatalytic, photocatalytic and piezo-photocatalytic degradation of organic pollutants and bacterial disinfection in wastewater: A comprehensive review. *Arabian Journal of Chemistry, 16*(2), Article 104473. https://doi.org/10.1016/j.arabjc.2022.104473

Meng, N., Liu, W., Jiang, R., Zhang, Y., Dunn, S., Wu, J., & Yan, H. (2023). Fundamentals, advances and perspectives of piezocatalysis: A marriage of solid-state physics and catalytic chemistry. *Progress in Materials Science, 138*(June), Article 101161. https://doi.org/10.1016/j.pmatsci.2023.101161

Mourdikoudis, S., Pallares, R. M., & Thanh, N. T. K. (2018). Characterization techniques for nanoparticles: Comparison and complementarity upon studying nanoparticle properties. *Nanoscale, 10*(27), 12871–12934. https://doi.org/10.1039/c8nr02278j

Parvees, A., Priyadarshini, U., & Remya, N. (2024). Visible light induced photocatalytic degradation of methylene blue using carbon fibre cloth–Bismuth oxybromide–water hyacinth derived silver nanocomposite. *Environmental Research, 262*(August), Article 119787. https://doi.org/10.1016/j.envres.2024.119787

Patil, A., Remya, N., Singhal, N., & Dubey, B. K. (2023). Walnut shell biochar supported TiO2–g-C3N4 heterojunction photocatalyst for solar photocatalytic degradation of congo red. *Biomass Conversion and Biorefinery, 13*(11), 9537–9547. https://doi.org/10.1007/s13399-023-04375-7

Ranjan, S., Dasgupta, N., Rajendran, B., Avadhani, G. S., Ramalingam, C., & Kumar, A. (2016). Microwave-irradiation-assisted hybrid chemical approach for titanium dioxide nanoparticle synthesis: Microbial and cytotoxicological evaluation. *Environmental Science and Pollution Research, 23*(12), 12287–12302. https://doi.org/10.1007/s11356-016-6440-8

Singh, G., Sharma, M., & Vaish, R. (2021). Flexible Ag@LiNbO3/PVDF composite film for piezocatalytic dye/pharmaceutical degradation and bacterial disinfection. *ACS Applied Materials and Interfaces, 13*(19), 22914–22925. https://doi.org/10.1021/acsami.1c01314

Singh, S., Gade, J. V., Verma, D. K., Elyor, B., & Jain, B. (2024). Exploring ZnO nanoparticles: UV–visible analysis and different size estimation methods. *Optical Materials, 152*, Article 115422. https://doi.org/10.1016/j.optmat.2024.115422

Tripathy, B. K., Johnson, I., & Kumar, M. (2020). Melanoidin removal in multi-oxidant supplemented microwave system: Optimization of operating conditions using response surface methodology and cost estimation. *Journal of Water Process Engineering, 33*, Article 101008. https://doi.org/10.1016/j.jwpe.2019.101008

Verma, A., Priyadarshini, U., & Remya, N. (2024). Solar photocatalytic degradation of ciprofloxacin using biochar supported zinc oxide-tungsten oxide photocatalyst. *Environmental Science and Pollution Research,* 0123456789. https://doi.org/10.1007/s11356-024-33764-2

Vijayaram, S., Razafindralambo, H., Sun, Y. Z., Vasantharaj, S., Ghafarifarsani, H., Hoseinifar, S. H., & Raeeszadeh, M. (2024). Applications of green synthesized metal nanoparticles—A review. *Biological Trace Element Research, 202*(1), 360–386. https://doi.org/10.1007/s12011-023-03645-9

Wei, R., Wang, P., Zhang, G., Wang, N., & Zheng, T. (2020). Microwave-responsive catalysts for wastewater treatment: A review. *Chemical Engineering Journal, 382*(73), Article 122781. https://doi.org/10.1016/j.cej.2019.122781

Xie, P., Cao, X., Lin, Z., & Javanmard, M. (2017). Top-down fabrication meets bottom-up synthesis for nanoelectronic barcoding of microparticles. *Lab on a Chip, 17*(11), 1939–1947. https://doi.org/10.1039/c7lc00035a

Zhang, Q., Jia, Y., Wu, W., Pei, C., Zhu, G., Wu, Z., Zhang, L., Fan, W., & Wu, Z. (2023). Review on strategies toward efficient piezocatalysis of $BaTiO_3$ nanomaterials for wastewater treatment through harvesting vibration energy. *Nano Energy, 113*(May), Article 108507. https://doi.org/10.1016/j.nanoen.2023.108507

Classification of Nanomaterials for Wastewater Treatment

K. S. Shameem, M. Bency, and P. C. Sabumon

1 Introduction

Nanomaterials are imperative for the advancement of wastewater treatment techniques. The intrinsic qualities of nanomaterials, such as high surface area, catalytic activity and adsorption capacity, offer a pedestal to solve problems related to water quality and formulate effective and long-lasting wastewater treatment procedures. Nanomaterials have revolutionised wastewater treatment in response to the growing concerns of water contamination. This chapter explores the intricate realm of nanomaterials, emphasising how they are categorised according to composition, morphology, functionalisation and specific applications. By dissecting these tiny but potent entities, this chapter aims to contribute to the discourse on sustainable water management and highlight their potential applications in addressing the problems faced by the current treatment technologies. The need to investigate nanomaterials stems from the intrinsic constraints of conventional wastewater treatment techniques, which are unable to handle complex industrial effluents, emerging contaminants and escalating waste volumes. Nanomaterials have dimensions between 1 and 100 nm (Lu et al., 2016). They are remarkable because of their enormous surface area, distinct chemical and electrical characteristics and their ability to suit particular applications. This is of paramount importance because of nanomaterials' ability to overcome the pitfalls of conventional treatment techniques by offering a flexible and effective way to handle various water treatment issues globally.

This chapter explores the diversity of nanomaterials by using a comprehensive classification framework based on composition, morphology, functionalisation and

K. S. Shameem · M. Bency
School of Civil Engineering, Vellore Institute of Technology, Chennai, India

P. C. Sabumon (✉)
Centre for Climate Change and Environment, Vellore Institute of Technology, Chennai, India
e-mail: pcsabumon@vit.ac.in

© Capital Publishing Company, New Dehli, India. 2025

R. Neelancherry and S. M. Kumar, *Nanotechnology in Wastewater Treatment*,
https://doi.org/10.1007/978-3-031-94246-4_2

specific application. The investigation starts with a thorough analysis of composition, dissecting the fundamental components that define the initial layer of a nanomaterial's identity. Metal-based nanoparticles are excellent scavengers that remove organic contaminants and heavy metals, while metal oxides are excellent catalysts (Aragaw et al., 2021; Ghosh et al., 2022; Palani et al., 2023; Weidner et al., 2022). Carbon-based nanomaterials, including graphene sheets and nanotubes, have superior adsorption capacities and unique electrical properties (Sheoran et al., 2022). With their complex chains and functional groups, versatile polymeric nanoparticles offer platforms that can be tailored to target specific contaminants. Beyond composition, the study explores the role of morphology in nanomaterials. Nanostructured materials provide increased surface area, ubiquitous nanoparticles are excellent at adsorption and nanotubes are effective filtration channels (Kurwadkar et al., 2019). Nanopolymers offer flexible platforms for a wide range of applications, while nanocomposite nanomaterials combine distinct characteristics in a synergistic way. Surface modifications define the tactical advantage of nanomaterials. Targeted interactions with pollutants can be achieved through chemical functionalisation, biofunctionalisation provides biomimetic coatings for selective pollutant recognition and removal, and responsive nanomaterials dynamically optimise performance under different treatment circumstances (Kumar et al., 2021). Their various abilities make nanoparticles ideal for different tasks in the wastewater treatment industry. Adsorbent nanoparticles function as sponges, collecting and concentrating pollutants. Photocatalytic nanomaterials use sunlight to break down organic pollutants, while membrane-based nanomaterials precisely filter out pollutants so that only clean water may pass through (Amdeha, 2023).

This comprehensive classification framework guides researchers across the vast field of nanomaterials (Fig. 1 and Table 1). By understanding the interplay of composition, morphology, functionalisation and application, this chapter aims to highlight the full potential of these nanomaterials, thereby paving the way for efficient wastewater treatment in the future.

2 Classification Based on Composition

Based on their composition, for wastewater treatment, nanomaterials can be categorised into a number of groups, each with particular features for the efficient removal of pollutants. These classifications are mostly based on the fundamental composition and structural characteristics of nanomaterials. The most prevalent categories include carbon-based nanomaterials, polymer-based nanomaterials, metal oxides and metal nanoparticles.

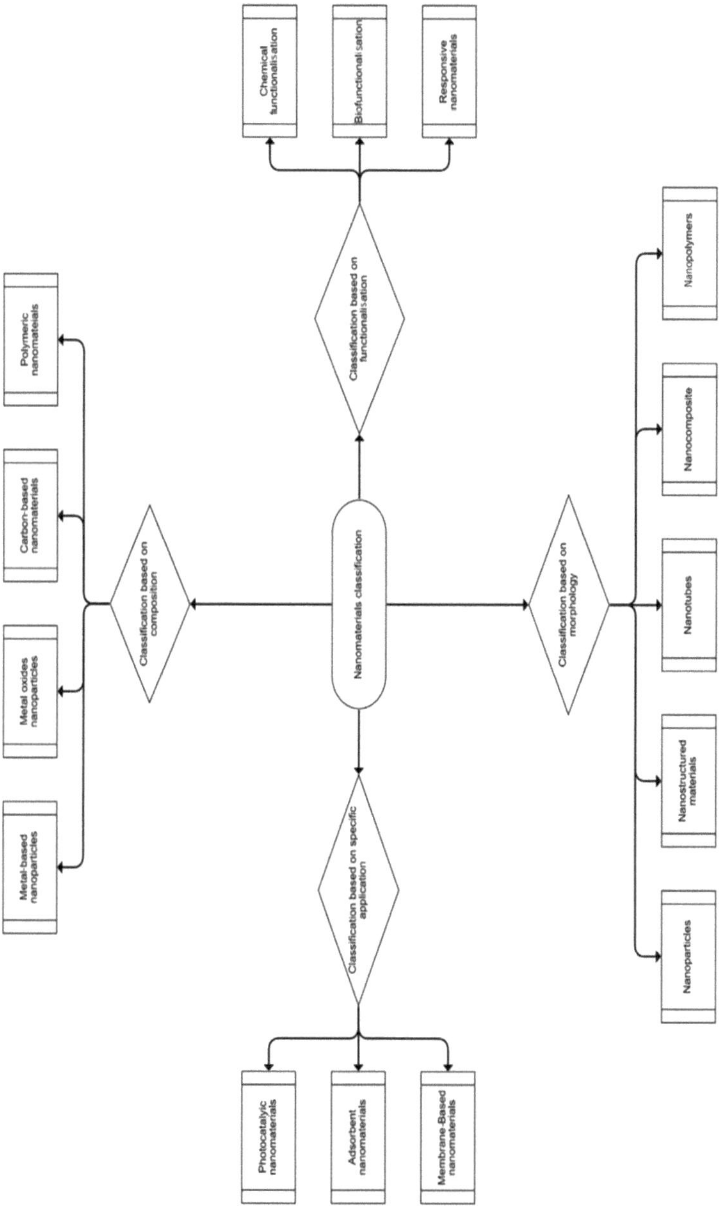

Fig. 1 Classification of nanomaterials

Table. 1. Classifications, applications, advantages, challenges and current research trends of nanomaterials

Classification	Subcategory	Examples	Applications	Advantages	Challenges	Current research trends	References
Composition	Metal-based nanoparticles	Silver (Ag), copper, iron (Fe) and hybrid metallic nanoparticles	Disinfection, removal of heavy metals and organic pollutants	High antimicrobial activity, selective targeting	Potential toxicity, cost, aggregation	Safer materials, controlled release strategies, bimetallic nanoparticles	(Ighalo et al., 2021; Pryshchepa et al., 2020; Wang et al., 2021)
	Metal oxides	Titanium dioxide, zinc oxide, iron oxide, manganese oxide nanoparticles	Photocatalytic degradation of organic pollutants, disinfection, removal of heavy metals	Efficient pollutant degradation, renewable energy integration	Low photostability, limited light utilisation	Doping with other elements, he tero structures for enhanced activity, visible light-activated catalysts	(Burhan et al., 2023; Ghosh et al., 2022; Gudkov et al., 2021; Islam et al., 2018; Martin et al., 2022)
	Carbon-based nanomaterials	Carbon nano tubes (CNTs), graphene-based nanomaterials, Carbonaceous nanocomposites, carbon quantum dots, biochar nanoparticles, fullerene	Adsorption of organic pollutants, heavy metals and dyes	High surface area, tunable properties, sustainability	Potential environmental impact, complex synthesis, waste disposal	Novel carbon nanostructures, composite materials with improved adsorption, functionalised carbon for specific pollutants	(Amdeha, 2023; Joshi & Gururani, 2022; Kurwadkar et al., 2019; Mishra & Sundaram, 2023; Raghav et al., 2021; Rajendran & Kumar, 2021; Rani et al., 2020)

(continued)

Table. 1. (continued)

Classification	Subcategory	Examples	Applications	Advantages	Challenges	Current research trends	References
	Polymeric nanomaterials	Polymeric nanofibres	Membrane filtration, adsorption of heavy metals and organic pollutants with specific targeting	Tailorable structure, antifouling properties, cost-effective	Biofouling susceptibility, mechanical strength limitations	Super hydrophobic membranes, functionalised polymers for specific pollutants, self-healing membranes	(Asif et al., 2021; Sahu et al., 2023)
Morphology	Nanoparticles	Silver nanoparticles (AgNPs), titanium dioxide nanoparticles (TiO_2NPs)	High surface area for adsorption and reactivity	High surface area to volume ratio, tunable properties based on size and shape	Potential toxicity (some nanoparticles)	Targeting specific particle sizes and shapes for optimised performance	(Aragaw et al., 2021; Kiruba Daniel et al., 2014; Martin et al., 2022; Pryshchepa et al., 2020)
	Nanostructured materials	Nanoscale membranes, hierarchical nanostructures	Tailored pore structure for selective adsorption, catalysis and targeted filtration of contaminants	High porosity, enhanced mass transfer, potential for selective separations	Difficulty in controlling pore size and uniformity	Development of advanced synthesis methods for precise pore structures	(Bhol et al., 2021; Kausar et al., 2022; Sahu et al., 2023; Yu et al., 2015)

(continued)

Table. 1. (continued)

Classification	Subcategory	Examples	Applications	Advantages	Challenges	Current research trends	References
	Nanotubes	CNTs, metal oxide nanotubes	Synergistic effects for enhanced adsorption, photocatalysis and antibacterial activity	Excellent mechanical strength, efficient electron transport	Complex fabrication, potential environmental impact	Hybrid nanotubes with improved functionality, sustainable synthesis methods	(Benamor et al., 2019; Kurwadkar et al., 2019; Mishra & Sundaram, 2023)
	Nanocomposite nanomaterials	Polymer nanocomposite, hybrid nanocomposite	Membrane filtration, adsorption and catalytic applications	Combined advantages of different materials	Difficulty in achieving uniform dispersion and interfacial compatibility	Development of new matrix materials and compatibilisers, multifunctional nanocomposites	(Kausar et al., 2022; Rajendran & Kumar, 2021; Sahu et al., 2023; Shih et al., 2022)
	Nanofibres	Polymeric nanofibres, inorganic nanofibres	Membranes for filtration and desalination, adsorbents for the removal of pollutants, controlled release of disinfectants/enzymes	Potential introduction of harmful byproducts	Development of greener functionalisation methods, targeted functionalisation for specific pollutants	Optimisation of nanofibre properties for the targeted removal of pollutants and sustainable wastewater treatment solutions	(Ghosh et al., 2022; Liao et al., 2018; Lu et al., 2016)

(continued)

Table. 1. (continued)

Classification	Subcategory	Examples	Applications	Advantages	Challenges	Current research trends	References
Functionalisation	Chemical functionalisation	AgNPs, graphene oxide (GO)	Enhanced adsorption of specific pollutants, improved dispersion of nanoparticles, catalytic activity for pollutant degradation	High specificity and efficiency, environmentally friendly	Potential unintended consequences of introducing biological agents	Design of stable and controllable biofunctionalised materials, integration with microbial fuel cells	(Elghamry et al., 2023; Lakshmi et al., 2024; Rahman et al., 2022)
	Biofunctionalisation	TiO_2NPs, polymeric nanofibres	Biodegradation of pollutants with enzymes/bacteria, biosensing for real-time pollutant detection, biofouling control on membranes and filters	Tailored response to environmental changes	Limited control over release mechanisms, potential unintended consequences	Development of new stimuli-responsive materials, fine-tuning response mechanisms	(Bhatt et al., 2023; Rafeeq et al., 2022; Wang et al., 2011)

(continued)

Table. 1. (continued)

Classification	Subcategory	Examples	Applications	Advantages	Challenges	Current research trends	References
	Responsive nanomaterials	Responsive nanofiltration membranes, ion-responsive, light-responsive, temperature-responsive, pH-responsive nanoparticles	Controlled release self-cleaning membranes for regeneration, targeted removal of pollutants with switchable binding, light-activated nanomaterials for specific capture	Renewable energy integration, efficient removal of pollutants	Low efficiency under visible light, need for co-catalysts	Visible light-responsive photocatalysts, catalyst regeneration strategies	(Birmod et al., 2021; Mohan et al., 2020; Zahmatkesh et al., 2023)
Application	Photocatalytic nanomaterials	TiO_2NPs	Degradation of organic pollutants and disinfection using light	High capacity, tunable selectivity	Regeneration of potential, potential environmental impact	Design of adsorbents with high regeneration capacity and low environmental footprint, development of adsorbent membrane hybrids	(Burhan et al., 2023; Iyyappan et al., 2023)

(continued)

Table. 1. (continued)

Classification	Subcategory	Examples	Applications	Advantages	Challenges	Current research trends	References
	Adsorbent nanomaterials	AgNPs, GO	Removal of heavy metals, organic pollutants and dyes through physical or chemical adsorption	High efficiency, tunable pore size	Biofouling, membrane fouling	Biofouling-resistant membranes, self-cleaning membranes, energy-efficient desalination methods	(Aragaw et al., 2021; Rani et al., 2020; Sheoran et al., 2022; Thakur et al., 2024)
	Membrane-based nanomaterials	Polymeric nanofibres	Filtration of suspended solids and microbes and desalination	Enhanced filtration efficiency and removal of pollutants	Maintaining stability and preventing fouling	Integration of novel nanomaterials for improved performance and durability	(Sahu et al., 2023; Sawunyama et al., 2023)

2.1 *Metal Nanoparticles*

Metal nanoparticles are capable of dealing with heavy metals and destroying microbes. Examples of these nanoparticles are silver (Ag), copper (Cu) and zerovalent iron (Fe^0). Toxic metal ions are removed from water systems more easily, and antimicrobial characteristics are demonstrated by hybrid metallic nanoparticles made of iron (Fe), Ag and Cu (Baby et al., 2022).

2.1.1 Silver Nanoparticles (AgNPs)

Argent or AgNPs are of great importance in wastewater treatment because of their unique features and numerous applications. AgNPs are recognised for their antibacterial activity, applications in disinfection and controlling growth of microorganisms in wastewater systems. The improved surface features and chemical reactivity of nanoparticles make them very effective in removing or dissolving dyes during wastewater treatment. Several studies have demonstrated the efficacy of AgNPs in the treatment of dye effluents, and their well-known antibacterial qualities against a variety of pathogens are extensively recognised in the fields of agriculture and health (Palani et al., 2023). Their antibacterial, catalytic and adsorption qualities make them very useful for wastewater treatment applications. AgNPs are known for their cost-effectiveness, ease of manufacture and adaptability. Reproducibility and cost-effectiveness can be prioritised in the production of AgNPs. Throughout the process, variations in reactants and reaction circumstances lead to variations in synthesis techniques. The sustainable synthesis of AgNPs with plants or microbes includes both intra- and extracellular techniques. Extracellular methods are generally chosen among these because they negate the difficulties in removing AgNPs from cells during subsequent processing. This biological synthesis is a more sustainable option than physical or chemical synthesis because of its cost- and environment-friendly nature (Baby et al., 2022; Pryshchepa et al., 2020). The fundamental benefit of AgNPs in eradicating microbes stems from their antimicrobial capabilities. They also help in removing organic compounds, pollutants and dyes by adsorption, as well as exhibit catalytic and photocatalytic activity for wastewater treatment.

2.1.2 Copper Nanoparticles (CuNPs)

CuNPs can serve as an effective adsorbent in wastewater treatment. They are relatively small in size, possess a large surface area and are relatively affordable to synthesise. CuNPs are also non-toxic, and their raw components are naturally abundant (Ighalo et al., 2021). Particularly, CuNPs have strong antibacterial, catalytic and adsorption properties that enable them to remove various kinds of contaminants from water. The catalytic qualities of CuNPs have been accepted, and these capabilities can be extremely important in the breakdown of organic contaminants found in wastewater. The catalytic activity aids in the breakdown of pollutants, improving the general effectiveness of wastewater treatment procedures. Their innate antibacterial qualities are beneficial for preventing the growth of microorganisms in water, thereby disinfecting wastewater (Madeła, 2020). Their distinctive form and large surface area provide them with substantial adsorption capacity. They can also remove dye from wastewater without producing any toxic compounds, and they have the potential to effectively adsorb heavy metals found in wastewater (Kumar et al., 2017). Overall, they show significant potential in wastewater treatment, providing diverse solutions for the removal of pollutants, catalytic degradation of organic compounds and water disinfection. Researchers are still exploring and optimising the applications of CuNPs in wastewater treatment processes.

2.1.3 Iron Nanoparticles (FeNPs)

FeNPs have received a lot of attention due to their potential applications in wastewater treatment. These nanoparticles made of Fe^0 provide numerous benefits for tackling water pollutants. They effectively remove heavy metals, organic contaminants and inorganic substances by mechanisms like adsorption, co-precipitation and redox reactions (Wang et al., 2021). Their efficacy in remediating water containing heavy metals, where Fe^0 reduces metal ions to immobilise or precipitate them, is of special significance. Furthermore, FeNPs have the potential to significantly enhance water quality since they aid in the degradation of organic contaminants including nitroaromatics and chlorinated solvents. The reduction of nitrate content in wastewater has demonstrated potential with the use of zerovalent FeNPs. They can be synthesised by using techniques like chemical reduction or green synthesis, which increases their reactivity for use in wastewater treatment. Even though they have a lot of potential for treating wastewater, research is still on to optimise their use in wastewater treatment processes. This includes taking stability, aggregation behaviour and potential long-term environmental impacts into account (Aragaw et al., 2021).

2.1.4 Hybrid Metallic Nanoparticles (Cu, Ag, Fe)

Hybrid metallic nanoparticles, a combination of Cu, Ag and Fe, have demonstrated substantial potential for wastewater treatment applications. Recent studies have focussed on developing entities that have several functionalities, including multitasking, through the use of nanotechnology. Hybrid metal nanomaterials have gained attention as prospective options for effective treatment of wastewater by combining the unique benefits of each metal to increase efficiency. The combined effect of multiple constituents bestows multifunctionality, photocatalytic attributes, rendering them efficacious in the degradation of an extensive array of contaminants. FeNPs improve the adsorption of contaminants, especially heavy metals, whereas CuNPs prevent the growth of microorganisms. AgNPs break down organic pollutants and inhibit microorganisms. By hybridising these metallic nanoparticles, multipurpose materials are produced, like a blend of Fe for better adsorption in wastewater treatment and Cu and Ag for antibacterial and catalytic properties, respectively (Kiruba Daniel et al., 2014). These clusters were intended to handle a variety of industrial effluents, including sewage, and waste from tanneries, textile plants and pharmaceutical plants. These hybrid clusters have higher amounts of Cu and Fe, and lower content of Ag to reduce treatment expenses. In addition, these hybrid clusters remove excess salts, chlorides, fluorides, heavy metals and colours from the effluent as well as disinfect microorganisms. Hybrid nanomaterials must be carefully engineered and designed, and their safe and efficient use in wastewater treatment must be ensured by extensive testing.

2.2 Metal Oxide Nanoparticles (MONPs)

MONPs are of great value in wastewater treatment because of their distinctive physicochemical properties. They exhibit effectiveness in eliminating harmful substances such as azo-dyes, heavy metals, pesticides and colourants. These MONPs have adsorption, photocatalytic and antifungal and antibacterial qualities. They also show chemical stability without having any negative impacts. MONPs are widely used to remove heavy metal ions and dyes from wastewater in an efficient manner (Ghosh et al., 2022). The following sections provide an explanation of the MNOPs commonly utilised in wastewater treatment.

2.2.1 Titanium Oxide Nanoparticles

Titanium oxide nanoparticles, particularly titanium dioxide nanoparticles (TiO_2NPs), have received a lot of attention in recent years for their different applications in the treatment of wastewater. These nanoparticles have extraordinary photocatalytic characteristics, making them extremely effective at degrading organic contaminants and facilitating water treatment operations. When subjected to ultraviolet (UV) radiation, TiO_2NPs undergo photocatalysis, which converts organic pollutants into harmless byproducts. This method is very effective in removing organic substances such as dyes, medications and pesticides, making titanium oxide nanoparticles an important tool in advanced wastewater treatment technology (Burhanet al., 2023). Moreover, TiO_2NPs can be immobilised on various supports or employed in hybrid nanostructures to improve their stability and performance, thus broadening their potential in solving water-related concerns (Rigoletto et al., 2022). TiO_2NPs can be synthesised using physical, chemical or biological processes. Chemical and physical methods frequently use hazardous chemicals, and these techniques often demand high-end and costly equipment. But, biological methods that utilise plant extracts like *Jatrophacurcas* are environment-friendly and reduce toxic chemicals (Hudlikar et al., 2012).

2.2.2 Zinc Oxide (ZnO) Nanoparticles

Zinc oxide nanoparticles (ZnONPs) are known for their eco-friendly nature, robust oxidising capacity and a broad spectrum of wavelengths. They also demonstrate superior photocatalytic properties, capturing more light compared to other metal oxides. ZnONPs, which have natural antibacterial properties, are of great importance in regulating the growth of microbes in water and provide an effective disinfection solution (Gudkov et al., 2021). In addition, they play a key role in minimising organic contaminants through their photocatalytic activity, which is especially active under UV light and allows for the effective degradation of organic pollutants in wastewater. This dual activity makes ZnONPs an attractive option for complete water treatment solutions. ZnONPs were used as photocatalysts to decompose methylene blue (MB) under UV light. The duration of UV light, as well as the size and shape of the ZnONPs, has an impact on MB degradation efficacy (Nguyen & Nguyen, 2020). ZnONPs can be synthesised using chemical precipitation, sol–gel and hydrothermal, microwave-assisted and template-assisted methods. These technologies provide control over particle size and shape, allowing for more personalised wastewater treatment applications.

2.2.3 Iron Oxide Nanoparticles (FeONPs)

Iron oxide nanoparticles (FeONPs) are emerging as a powerful entity in the fight against wastewater contamination because of their versatile magnetic characteristics, biocompatibility and numerous uses in environmental research and technology. These nanoparticles that have an extremely large surface area for their size behave like minuscule sponges to collect and remove pollutants. FeONPs show potential in addressing emerging pollutants (EPs) such as microplastics and personal care products, which conventional treatment approaches fail to remove. Their magnetic characteristics enable the easy separation of microplastics and nanoplastics from treated wastewater, which helps in reusing microplastics and nanoplastics and reducing waste (Martin et al., 2022). They can handle a range of pollutants, including heavy metals (lead, arsenic and mercury), organic pollutants (pesticides, pharmaceuticals and dyes) and microbes. FeONPs, being naturally abundant and economically feasible, represent a sustainable and scalable solution for wastewater treatment from a scientific perspective. In addition, they aid in the removal of nitrogen compounds from wastewater by enhancing both nitrification and denitrification pathways (Desireddy & Pothanamkandathil Chacko, 2021). FeONPs can be hybridised with other materials or functionalised with specific groups to improve their efficacy and customise them to specific wastewater treatment applications (Xu et al., 2012). They can be synthesised using a variety of techniques, including co-precipitation of $Fe2^{+}$ and $Fe3^{+}$ ions accompanied by calcination or green techniques such as plant extract reduction (Aragaw et al., 2021). With further research and development, FeONPs can revolutionise wastewater treatment and contribute to a cleaner, healthier environment.

2.2.4 *Manganese Oxide Nanoparticles (MnO_2NPs)*

MnO_2NPs are gaining considerable interest in wastewater treatment because of their remarkable properties and functionalities. MnO_2NPs have a very large surface area for their size, which allows them to effectively adsorb heavy metals, organic compounds and emerging contaminants (Shih et al., 2022). They serve as adaptable catalysts for a wide range of degradation processes and are capable of overseeing oxidation reactions, breaking down organic pollutants like pesticides and dyes, and effectively removing resilient microplastics. Due to their exceptional metal-binding affinity, these catalysts emerge as promising candidates for the sequestration of toxic heavy metals from water contaminated with such pollutants. MnO_2NPs catalyse the degradation of diverse organic contaminants, mitigating their harmful effects and enhancing the suitability of water for various applications, including irrigation and industrial processes (Islam et al., 2018). They can be synthesised using co-precipitation, hydrothermal synthesis and sol–gel processing. Superparamagnetism is exhibited in certain MnO_2 phases, such as maghemite, which makes it easy to separate from treated water with the use of an external magnet. This promotes sustainability by streamlining the treatment procedure and enabling the recycling of used nanoparticles.

2.3 *Carbon-Based Nanomaterials*

Carbon-based nanomaterials are being extensively researched for potential applications in the treatment of wastewater due to their distinct properties, large surface area and higher adsorption capacities. Moreover, they are generally non-toxic. Following is the categorisation of carbon-based nanomaterials tailored to their applications in the context of wastewater treatment.

2.3.1 Carbon Nanotubes (CNTs)

Carbon nanotubes (CNTs) are cylindrical structures composed of carbon atoms arranged in honeycomb lattices at a nanometre scale. Their extraordinary characteristics, such as high surface area, exceptional strength and adjustable electrical conductivity, make them effective materials for scientific innovations in the fields of energy storage, electronics and environmental remediation. CNTs exist in two primary forms: single-walled carbon nanotubes (SWCNTs), composed of a single rolled-up layer of carbon atoms, and multiwalled carbon nanotubes (MWCNTs), with multiple layers nested like Russian dolls. This dual structure provides strength and versatility, leading to diverse applications. CNTs, characterised by substantial surface area (up to 2600 m^2/g), offer numerous binding sites for the adsorption of pollutants. They demonstrate effectiveness in removing diverse contaminants, including heavy metals (Pb, Cd and Cr), organic pollutants (dyes, pharmaceuticals and pesticides) and emerging contaminants (microplastics, PFAS) (Kurwadkar et al., 2019). Chemical modification, such as amine-functionalised CNTs, enhances their adsorption capacity for specific pollutants, particularly heavy metals. In membrane filtration, CNT-based membranes exhibit advantages over conventional membranes, featuring high water permeability, strong antifouling properties and tailorable pore sizes for the removal of specific contaminants. Applications being investigated comprise bacteria and virus removal, as well as desalination of brackish water. CNTs serve as catalysts for catalytic degradation in wastewater, employing mechanisms like photocatalysis, Fenton-like reactions and electrochemical oxidation (Mishra & Sundaram, 2023). These CNT-based catalysts offer efficiency, reusability and the potential for targeted degradation of specific pollutants. In disinfection, CNTs with antimicrobial properties can directly inactivate bacteria and viruses or immobilise antimicrobial agents like AgNPs. CNT-based disinfection methods show promise for efficient and environment-friendly wastewater treatment but challenges include environmental impact, potential nanoparticle release, production costs and scalability. Ongoing research aims to optimise their use in practical applications.

2.3.2 Graphene-Based Nanomaterials

Graphene-based nanomaterials are swiftly gaining prominence in wastewater treatment due to their exceptional properties. Their remarkable surface area, chemical adaptability and outstanding electrical conductivity equip them to address a variety of water purification challenges (Bhol et al., 2021). For example, graphene oxide (GO) nanosheets exhibit a strong affinity for heavy metals and organic pollutants, facilitating efficient adsorption and removal of pollutants from contaminated water. Graphene-based nanomaterials for wastewater treatment include GO with effective adsorption capabilities, reduced graphene oxide (rGO) exhibiting improved electrical conductivity and enhanced adsorption, graphene-based nanocomposites for synergistic effects, graphene quantum dots (GQDs) with applications in sensing and photocatalysis, 3D graphene-based structures for efficient adsorption and graphene nanoribbons with unique electronic properties (Joshi & Gururani, 2022; Yu et al., 2015). Studies have highlighted the impressive capabilities of these materials, showcasing their effectiveness in removing over 95% of lead ions from wastewater (Javanmirpourshirzadi et al., 2023; Yang et al., 2019). Beyond adsorption, graphene can be tailored into membranes for effective filtration, utilised in electrochemical degradation of organic pollutants and employed as catalysts for targeted breakdown of pollutants. Ongoing research suggests that the potential of graphene-based nanomaterials for wastewater treatment is continually expanding, offering opportunities for more sustainable water management.

2.3.3 Carbonaceous Nanocomposites (CNCs)

CNCs are hybrid materials formed by integrating carbon nanomaterials (CNMs) like graphene or CNTs with other components such as polymers, metals or oxides. In this composite structure, CNMs serve as the foundation, while additional materials act as building blocks, resulting in a synergistic combination with enhanced properties (Rajendran & Kumar, 2021). The resultant CNCs exhibit remarkable characteristics that surpass those of their individual components. CNCs emerge as potent tools for wastewater remediation, leveraging unique properties to address diverse wastewater treatment challenges. By combining the exceptional surface area and customisable functionality of CNMs like graphene and CNTs with various support materials, CNCs offer a tailored solution for capturing pollutants such as heavy metals, microplastics and organic molecules. Their extensive internal porosity facilitates effective adsorption, while strategic material selection allows them to selectively bind to specific environmental contaminants (Sheoran et al., 2022). Beyond adsorption, CNCs exhibit versatility in membrane filtration, catalytic degradation and antimicrobial disinfection (Raghav et al., 2021). Studies demonstrate their effectiveness in removing heavy metals, degrading dyes and pharmaceuticals, and filtering microplastics (Raghav et al., 2021). However, challenges like cost optimisation, environmental impact and large-scale implementation persist. Ongoing research efforts addressing these issues and harnessing the potential of CNCs offer promise for clean and sustainable water resources in the future.

2.3.4 Carbon Quantum Dots

Carbon quantum dots (CQDs) represent a novel category of carbon nanoparticles, each measuring less than 10 nm in size. Characterised by strong fluorescence, these particles exhibit unique optical characteristics, including tunable photoluminescence and excitation-dependent multicolour emission (Wang et al., 2019). CQDs exhibit tremendous potential for revolutionising wastewater treatment. Created from controlled carbon precursor pyrolysis, these luminescent nanoparticles offer an exceptional surface area, acting as powerful adsorbents for heavy metals, organic pollutants and microplastics (Rani et al., 2020). Their tunable fluorescence allows for real-time sensing for targeted interventions for the treatment of pollutants. Beyond adsorption, CQDs demonstrate versatility, engaging in catalytic degradation, photocatalysis and disinfection. While studies highlight their efficacy in removing heavy metals and degrading harmful pollutants, challenges like cost-effective synthesis, environmental fate assessment and scalable integration persist.

2.3.5 Biochar Nanoparticles

Biochar nanoparticles (BCNPs) are ultrafine particles synthesised from the pyrolysis of organic materials. Their nanoscale dimensions increase their surface area, making them useful in agriculture, water treatment and environmental sustainability (Amdeha, 2023). This carbon-rich material, derived from the pyrolysis of organic waste, will undergo a metamorphosis into a nanoscale for wastewater treatment, typically under 100 nm in size. BCNPs possess a spongy, high-surface area with intricate pores, acting as a powerful pollutant adsorbent, effectively removing heavy metals, organic contaminants and microplastics from wastewater (Noreen and Abd-Elsalam, 2021). Their customisable nature allows researchers to create BCNPs with tailored functionalities, targeting diverse pollutants with precision. BCNPs exhibit a multifaceted arsenal, including adsorption, catalytic degradation and intrinsic antimicrobial properties for disinfection (Weidner et al., 2022). Research on BCNPs for wastewater treatment demonstrates promising results, such as their efficacy in removing toxic metals and degrading harmful dyes and pharmaceuticals. Despite challenges like cost-effective production and environmental fate, ongoing research efforts aim to unlock the full potential of BCNPs.

2.3.6 Fullerenes

Fullerenes are carbon allotropes that are made up of 60 carbon atoms arranged in hexagons and pentagons. Unlike CNTs, which exhibit tube-like structures, fullerenes present unique cage-shaped configurations. These unique carbon structures have garnered attention in wastewater treatment and purification when incorporated into polymer nanocomposite membranes (Kausar et al., 2022). Fullerene-based membranes exhibit the capability to remove toxic metals, microorganisms, chemical compounds and salts. In addition, fullerenes can be employed for photocatalytic oxidation of organic contaminants or for inhibiting and inactivating microbes. Despite possessing advantageous properties, fullerenes are hampered by their high cost. However, even considering the challenges of cost, their potential role in wastewater treatment remains promising.

2.4 Polymeric Nanomaterials (PNPs)

PNPs are minute, solid and colloidal particles, spanning from a few nanometres to 1000 nm in size, and are composed of repeating units of polymer molecules. These diminutive marvels exhibit distinctive properties that render them appealing for diverse applications, including drug delivery, imaging, sensors and notably, wastewater treatment (Zielinska et al., 2020). PNPs are composed of polymer chains chemically or physically linked together, utilising both natural (e.g., chitosan, cellulose) and synthetic (e.g., polyethylene glycol, polylactic acid) polymers. PNPs exhibit high surface area, enabling effective interaction and capture of pollutants in wastewater (Sahu et al., 2023). Their tunable functionality allows for modification with diverse chemical groups, targeting specific pollutants and functionalities. Biodegradability is designed into some PNPs, minimising environmental impact post-use, and they can be engineered for controlled substance release. In wastewater treatment, PNPs function in adsorption, catalytic degradation, coagulation, disinfection and sensing, addressing pollutants like heavy metals, organic contaminants and microplastics (Asif et al., 2021). Challenges include cost-effective production, investigating long-term environmental impacts and scaling laboratory successes to large treatment plants. They can be synthesised through methods like precipitation and emulsion polymerisation, dispersion polymerisation, microfluidic techniques and green synthesis, each with unique advantages and challenges, catering to diverse applications and emphasising on-going research for enhanced efficiency and sustainability.

3 Classification Based on Morphology

Emerging as a viable solution for sustainable water management, nanomaterials have gained considerable attention because of their potential to transform the treatment of wastewater. The nuances of their structure and form, intricately intertwined in their morphology, ensure their effectiveness and determine their applicability across a range of treatment methods. In an era marked by escalating concerns about water pollution, scientists and engineers are increasingly adopting nanotechnology to develop materials with heightened capabilities to combat this pressing issue. The diverse array of nanomaterial morphologies opens up numerous possibilities for customised applications, providing the means to selectively eliminate specific pollutants from wastewater. This examination delves into the profound importance of nanomaterial morphology in wastewater treatment, highlighting its pivotal role in optimising crucial processes like adsorption and catalysis essential for water purification. The subsequent sections present a comprehensive taxonomy of nanomaterials, organised according to their morphological characteristics.

3.1 Nanoparticles

Nanoparticles, characterised by their nanoscale dimensions, typically ranging from 1 to 100 nm, serve as foundational units within the nanoscale domain (Pryshchepa et al., 2020). Demonstrating a diverse array of shapes and compositions, including spheres, rods and more, these particles exhibit unique properties that extend beyond their diminutive size. This inherent versatility positions them as crucial entities in various scientific and technological disciplines, encompassing fields such as catalysis, drug delivery and, as examined here, wastewater treatment (Palani et al., 2023). The optimisation of their performance in these multifaceted applications requires a nuanced understanding and deliberate manipulation of their characteristics, with a specific emphasis on their morphological attributes. They are subdivided as follows.

3.1.1 Nanospheres

Spherical nanoparticles, characterised by their compact structure and expansive surface area, represent a significant class of nanomaterials employed in wastewater treatment. This morphology optimises their performance in both adsorption and catalytic processes. The high surface-to-volume ratio facilitates efficient capture of and interaction with waterborne contaminants, significantly contributing to the treatment's overall effectiveness. In wastewater treatment, they are excellent at removing heavy metals, degrading organic pollutants and promoting microbial treatment (Aragaw et al., 2021; Kiruba Daniel et al., 2014; Kumar et al., 2017). Despite being in their early phases, their ability to address a wide range of contaminants, from pharmaceuticals to microplastics, holds hope for cleaner water in the future. In the realm of wastewater treatment, diverse families of spherical nanoparticles possess impressive capabilities for addressing contamination challenges (Asif et al., 2021). Metal-based spherical nanoparticles, including Ag and Fe varieties, play a crucial role, demonstrating exceptional adsorption capacities, particularly in the removal of heavy metals. Polymeric spheres, composed of various polymers, offer distinct advantages by effectively capturing organic pollutants through adsorption and facilitating efficient filtration processes. Meanwhile, silica-based spherical nanoparticles act as catalysts, overseeing the breakdown of contaminants into less harmful components (Nayl et al., 2022a, b). This varied collection of spherical nanoparticle types allows for the customisation of applications to meet specific needs in wastewater treatment, underscoring the versatility and potential of nanotechnology in mitigating water pollution issues.

3.1.2 Nanorods

Nanoparticles categorised as nanorods feature an elongated form, exemplified by nanorods and nanowires. This unique morphology, characterised by extended length, offers distinct advantages in the treatment of wastewater. The elongated shape facilitates deeper penetration and interaction within the water matrix, enhancing contact with contaminants. This characteristic renders rod-shaped nanoparticles particularly valuable in scenarios demanding efficient dispersal and thorough coverage of the treatment medium, ultimately optimising the removal of pollutants. Their extended surface area serves as potent catalysts, accelerating chemical reactions, and is likened to miniature factories with elongated production lines for breaking down complex contaminants (Anjum et al., 2019). The elongated shape facilitates efficient movement and dispersion, preventing clumping and ensuring extensive encounters with pollutants, reminiscent of purposeful eels gliding effortlessly through water. Nanorods offer targeted customisation based on composition and surface modifications, tailoring solutions for specific wastewater challenges, from heavy metals to microplastics (Baby et al., 2022).

3.2 *Nanostructured Materials*

Nanostructured materials, characterised by intricately arranged nanoscale features (1–100 nm), prove formidable in wastewater treatment, offering unique advantages for the removal of contaminants. Their small size grants them expansive surface areas, facilitating efficient adsorption and capturing a range of pollutants, from organic molecules to heavy metals (Raghav et al., 2021). With confined geometry and distinct electronic properties, they exhibit catalytic prowess, accelerating the breakdown of stubborn contaminants through targeted chemical reactions. Customisable through precise engineering, these materials can be tailored for specific functionalities, enabling selective adsorption, photocatalysis or disinfection processes (Sahu et al., 2023; Yu et al., 2015). Two prominent categories showcase their versatility and are described below.

3.2.1 Nanoscale Membranes

These ultrathin films, just microns thick, act as highly selective gatekeepers with nanoscale-sized pores, designed for targeted filtration of contaminants. Engineered for precise filtration, they feature pores tailored to specific contaminants, offering more efficiency than conventional filters (Bhol et al., 2021; Kausar et al., 2022; Raghav et al., 2021; Sahu et al., 2023). As water pressure requirements are reduced, consumption and operational costs are minimised, ensuring rapid and thorough removal of contaminants. Different materials and pore structures allow for customised treatment solutions, as seen in reverse osmosis membranes for removing salts and microfiltration membranes for pre-treating water before further purification.

3.2.2 Hierarchical Nanostructures

These intricately organised assemblies leverage diverse nanoscale features like nanoparticles, nanowires and nanosheets, optimising pollutant capture, enhancing mass transport and promoting catalytic activity. This complex architecture enables the degradation of challenging contaminants such as organic dyes and pharmaceutical residues (Xu et al., 2012; Yu et al., 2015). Harnessing the unique physicochemical properties and multifaceted functionalities of nanostructured materials heralds a new era in wastewater treatment, promising enhanced precision, efficiency and sustainability. Examples include photocatalytic nanostructures utilising sunlight for organic pollutant degradation, adsorbent nanocomposites creating effective contaminant sponges and magnetic nanocarriers ensuring easy separation and recovery of adsorbed pollutants (Amdeha, 2023; Nguyen & Nguyen, 2020).

3.3 *Nanocomposites*

Nanocomposites in wastewater treatment resemble tiny particles made of various materials with their collective strength. These microscopic marvels create a remarkable impact, utilising their extensive surface area as a playground for pollutants, effortlessly capturing undesirable molecules (Rajendran & Kumar, 2021). Some nanocomposites even house miniature chemical factories, effectively breaking down stubborn contaminants into harmless components. The true strength of nanocomposites lies in their customisation, allowing researchers to blend materials selectively to combat specific pollution threats (Noreen and Abd-Elsalam, 2021). The two categories of nanocomposite are explained below.

3.3.1 Polymer–Nanomaterial Composites

Polymer–nanomaterial composites represent a strategic integration of nanoparticles within polymer matrices, combining the advantageous properties of both constituents. The polymer component provides structural stability and flexibility, serving as a robust matrix, while nanomaterials contribute to a substantial increase in surface area, augmenting adsorption capabilities for the capture of pollutants (Sahu et al., 2023). The customisation of these composites through the selection of specific nanomaterials and polymers allows for the design of tailored functionalities, including targeted adsorption, catalytic degradation or antimicrobial properties (Kausar et al., 2022). Examples include the incorporation of CNTs in polymers for heightened adsorption and the utilisation of MONPs in polymer membranes to enhance the efficiency of filtration.

3.3.2 Hybrid Nanocomposites

Hybrid nanocomposites entail the integration of diverse nanomaterials, presenting a multifaceted approach to address waterborne contaminants. These composites leverage the synergistic effects derived from the collective capabilities of different nanomaterials, resulting in enhanced multifunctionality (Rajendran & Kumar, 2021; Shih et al., 2022; Wang et al., 2021). This approach achieves superior treatment efficiency compared to single-material technologies by concurrently addressing multiple aspects of the removal of contaminants. Examples include magnetic–photocatalytic nanocomposites, uniting magnetic nanoparticles for facile separation and photocatalytic materials for solar-driven pollutant degradation, and CNT–metal oxide hybrids, which combine the high surface area of CNTs with the catalytic power of metal oxides for effective adsorption and breakdown of complex contaminants. In addition, layered double hydroxide-based hybrids provide tunable functionalities through layered structures, enabling targeted adsorption and controlled release for specific molecules in water treatment applications.

3.4 Nanotubes

Nanotubes are cylindrical structures with diameters ranging from 1 to 100 nm. These structures can be made of different materials, including CNTs, which are among the most well-known (Kurwadkar et al., 2019; Mishra & Sundaram, 2023). Their unique properties, such as high aspect ratio, exceptional strength and conductivity, and their adsorption and catalytic capabilities, make them versatile and valuable in a variety of scientific and technological applications, including wastewater treatment. The two different types of nanotubes are listed below.

3.4.1 Carbon Nanotubes

CNTs, tubes of pure carbon, incredibly thinner than a strand of hair, possess an unparalleled surface area. Their vast surface functions as a magnet, efficiently trapping various pollutants. The comprehensive depiction of CNTs is presented in Sect. 2.3.1.

3.4.2 Metal Oxide Nanotubes

Metal oxide nanotubes go beyond mere adsorption by introducing catalysis into their functionality. These nanotubes act as catalysts, accelerating chemical reactions and effectively dismantling persistent contaminants. Specifically, their catalytic efficacy extends to the degradation of intricate pollutants, such as pesticides and pharmaceuticals, converting them into benign substances and thereby elevating the purity of water (Benamor et al., 2019). Furthermore, select metal oxide nanotubes exhibit inherent antibacterial and antiviral properties, augmenting the safeguarding measures for treated water. The synergy of adsorption and catalytic capabilities establishes nanotubes as trailblazers in the forefront of advancing water treatment technologies.

3.5 Nanofibres

Nanofibres are incredibly thin with sizes measured in nanometres, often ranging from 1 to 100 nm. These fibres can be made from polymers, carbon and inorganic compounds. Nanofibres have unique capabilities because of their small dimension and high surface area-to-volume ratio. The word 'nanofibres' is frequently used to describe materials with at least one nanoscale dimension, making them useful for a wide range of technical and scientific applications (Nayl et al., 2022a, b). The two types of nanofibres are polymeric nanofibres and inorganic nanofibres.

3.5.1 Polymeric Nanofibres

Polymeric nanofibres exhibit a remarkable capacity for water purification by functioning as microscale sorbents. Their substantial surface area, akin to a mesoscale playground, facilitates the efficient adsorption of diverse contaminants, spanning heavy metals to organic molecules. The intricate structure of these nanofibres creates a labyrinthine filtration network, imparting a high degree of efficacy in purifying water (Liao et al., 2018). Notably, their tailorability stands out as a distinctive feature; scientists can engineer these nanofibres with precision by selecting distinct materials and structures, thus creating bespoke filtration mechanisms. This adaptability renders polymeric nanofibres highly effective in addressing varied water contamination challenges with a targeted and customisable approach.

3.5.2 Inorganic Nanofibres

Inorganic nanofibres develop as dynamic entities in wastewater treatment, transcending their conventional functions. These nanofibres, made from metal oxides or clay, contain nanoparticles that act as catalysts, speeding processes and effectively breaking down challenging pollutants (Ghosh et al., 2022). This catalytic ability, similar to microscopic chefs at work, improves water purification by turning hazardous compounds into benign substances. Inorganic nanofibres demonstrate amazing adaptability by targeting a wide range of water-based risks, from persistent chemicals to harmful microbes, serving as a specialised force with tailored tools for specific pollutants (Lu et al., 2016). Their ability to provide customised solutions makes them key actors in expanding wastewater treatment technologies, offering a sophisticated and efficient approach for removing contaminants.

4 Classification Based on Functionalisation

In the realm of wastewater remediation, tailoring nanomaterial surfaces via functionalisation emerges as a cornerstone for optimising removal of pollutants. This translates to modifying nanomaterial interfaces through chemical, biological or stimuli-responsive approaches, thereby amplifying their affinity towards targeted contaminants. By categorising nanomaterials based on their functionalisation schemes, researchers unravel a systematic approach to explore their water treatment applications. Delving into the distinct functionalisation strategies helps to make informed choices when selecting nanomaterials for diverse treatment modalities, including adsorption, catalysis and separation (Sadegh et al., 2017). The classification based on functionalisation is summarised in Table 2 and Fig. 2.

A detailed classification of functionalisation is given below.

Table 2 Classification based on functionalisation

Category	Description	Materials	Applications	References
Chemical functionalisation	Forming strong chemical bonds with functional groups	CNTs, graphene, metal oxides and quantum dots	Enhanced adsorption, stability and catalytic activity	(Elghamry et al., 2023; Lakshmi et al., 2024)
Biofunctionalisation	Utilising biomolecules for surface modification	Magnetic nanoparticles, metal oxides and carbon nanofibres	Selective removal, biodegradation and enhanced biocompatibility	(Bhatt et al., 2023; Wang et al., 2011)
Responsive nanomaterials	Exhibiting property changes in response to stimuli	Polymers, hydrogels and metal nanoparticles	Controlled release, reusability and targeted removal	(Birmod et al., 2021; Zahmatkesh et al., 2023)

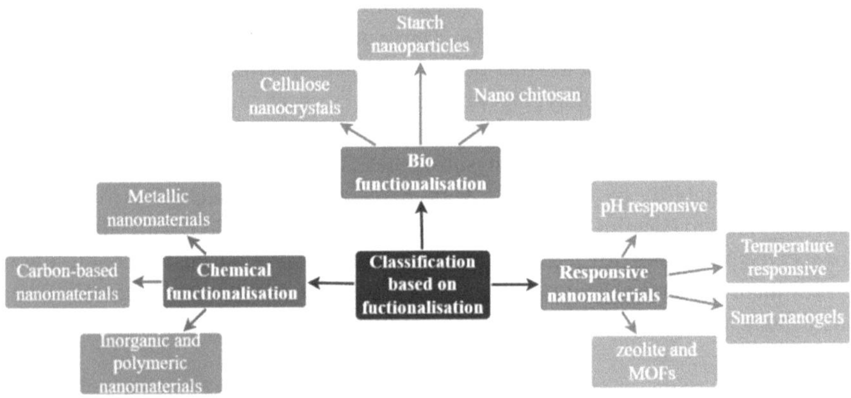

Fig. 2 Classification of nanomaterials based on functionalisation

4.1 *Chemical Functionalisation*

Chemical functionalisation is an important method for improving the efficacy of nanomaterials in the remediation of wastewater. This focussed method uses the distinct features of metallic and carbon-based nanomaterials to reddress the complex difficulties posed by water contamination. Chemical modifications using organic ligands or polymers improve the stability and reactivity of metal nanoparticles such as Ag and iron oxide (Lakshmi et al., 2024). These protective coatings enhance their longevity by preventing aggregation and oxidation, allowing for long-term performance in water treatment operations. Similarly, chemical functionalisation enhances carbon-based nanomaterials such as nanotubes and graphene derivatives by introducing certain surface groups such as hydroxyl, carboxyl or amine. This rigorous tuning enhances their natural features (high surface area and conductivity) and enables customisation to precisely target certain pollutants in water.

Chemical functionalisation extends beyond metallic and carbon-based nanomaterials, demonstrating its effectiveness across a wide range of material classes. This increases the versatility of complex water treatment systems. Chemical modifications, for example, can significantly improve the performance of polymeric nanoparticles. The addition of tailored functional groups improves adsorption capacity and the overall performance of pollution removal (Elghamry et al., 2023). By methodically adjusting the surface chemistry of polymeric nanoparticles, their selectivity and affinity for certain contaminants may be finely controlled, providing a personalised solution to the diverse challenges offered by various wastewater pollutants.

Chemical functionalisation is also effective in changing inorganic materials such as zeolites and metal–organic frameworks (MOFs) for better water treatment applications. Zeolites, with their porous structures, benefit from specific alterations to improve their ion exchange capabilities (Rahman et al., 2022). Similarly, MOFs, which are known for their large surface area, can be chemically adjusted to selectively adsorb certain contaminants. This extensive application across nanomaterial classes demonstrates the adaptability of chemical functionalisation as a universal tool. It enables to customise nanomaterials to the specific requirements of different contaminants and water treatment scenarios.

4.2 *Biofunctionalisation*

Biofunctionalisation emerges as a transformative approach in wastewater treatment, providing nanomaterials with a biocompatible dimension consistent with sustainability and environmental principles. This novel strategy involves integrating biomolecules or biocompatible substances onto nanomaterial surfaces, which improves environmental compatibility and performance. This method is especially important in the field of bio-based nanomaterials like cellulose nanocrystals and starch nanoparticles. These bio-based materials are precisely designed with biofunctional groups, resulting in surfaces with a high affinity for pollutants (Wang et al., 2011). The inclusion of such groups allows for selective binding and removal of contaminants, underlining these materials' potential as efficient absorbents in the treatment of wastewater.

Chitosan, a naturally occurring biopolymer made from chitin with distinct features, is a key player in the field of biofunctionalisation. Its incorporation into the process of biofunctionalisation greatly aids in the synthesis of nanomaterials with improved adsorption and flocculation properties (Bhatt et al., 2023). The presence of amino groups inside the chitosan structure promotes strong interactions with a wide range of contaminants, resulting in effective removal from aqueous solutions. Besides its practical applications, the use of chitosanin biofunctionalisation demonstrates a dedication to the principles of green chemistry, advocating for sustainable and economically acceptable materials in environmental remediation. This comprehensive approach not only underlines the relevance of merging nanotechnology with biological components but also broadens the breadth of sustainable and eco-friendly solutions within the complex domain of wastewater treatment.

The significance of biofunctionalisation extends beyond its immediate potency in improving nanomaterial performance. Integrating biocompatible components is consistent with the larger aim of building sustainable techniques in water treatment (Rafeeq et al., 2022). This strategic alignment with green chemistry principles represents a paradigm change towards holistic approaches that prioritise mitigation of the environmental effect while also tackling water pollution issues. As the complex interplay between nanotechnology and biofunctionalisation plays out, it paves way for advances in environmental remediation, promising both improved pollution removal and a smaller ecological imprint.

4.3 Responsive Nanomaterials

Responsive nanomaterials are a revolutionary invention in the dynamic field of wastewater treatment, adding a transformational edge to traditional approaches. These materials have intrinsic dynamic and adaptive properties that transform the approach to tackling contaminants in water treatment operations. A classic example is pH-responsive nanomaterials, which have been precisely developed with pH-sensitive coatings or functional groups. This innovative design enables these nanomaterials to produce personalised responses to changes in solution acidity (Mohan et al., 2020). As a result, a controlled release mechanism is formed, allowing these nanomaterials to selectively release adsorbed contaminants in response to certain pH conditions, ushering in unparalleled levels of precision in the removal of pollutants.

Another type of responsive nanomaterials is temperature-responsive variants. These professionally engineered materials include thermosensitive polymers into their structures, allowing them to respond to stimuli based on temperature (Birmod et al., 2021). This temperature-sensitive trait reveals its true potential in situations where temperature fluctuations have a substantial impact on pollutant solubility and transportation. This helps in photocatalytic processes like degradation of dye. Temperature-responsive nanoparticles provide a dynamic framework for maximising the removal of pollutants based on the current environmental circumstances, demonstrating their extraordinary adaptability to a wide range of wastewater treatment situations.

The introduction of smart nanogels marks a watershed moment in the field of responsive nanomaterials (Zahmatkesh et al., 2023). These ingeniously engineered nanogels, which are sensitive to environmental cues, have extraordinary swelling or shrinking capabilities in response to external stimuli. This sensitive behaviour enables the on-demand release of encapsulated chemicals, resulting in a highly regulated and adaptive method to the removal of pollution. Furthermore, smart nanogels, with their intrinsic ability to fine-tune treatment processes, provide unparalleled prospects for improving the efficiency of wastewater treatment. In essence, responsive nanomaterials improve the precision of the removal of pollutants while also paving the way for novel and adaptable solutions in the ever-changing environment of sustainable wastewater treatment methods.

5 Classification Based on Specific Application

Wastewater treatment is being revolutionised by the large spectrum of novel materials with special features that nanotechnology offers. Based on their unique use in the treatment process, these materials can be categorised broadly.

5.1 Photocatalytic Nanomaterials

Materials with the capacity to absorb visible or UV light, such as ZnS, TiO_2, AgCl and ZnO, are frequently used in photocatalysis processes. This assimilation results in the production of electrons and holes, accelerating the rate of degradation. The effectiveness of a variety of photo-catalysts, such as semiconductors, metal oxides, dendrimers and MOFs, in eliminating organic pollutants from water has been extensively investigated (Iyyappan et al., 2023).

5.1.1 Nanoscale Semiconductors

The physicochemical characteristics of materials undergo a fundamental alteration at the nanoscale in comparison to their bulk counterparts. This phenomenon is associated with the concept of quantum entrapment. It happens when the size of the particles falls below the exciton's Bohr radius, making the electron in the nanocrystal behave like it were inside a three-dimensional potential box. The excited band's energy drops with an increase in the exciton's Bohr radius. In relation to the shift in the excited band's energy, the quantum confinement effect is crucial for the characteristics of nanocrystals (Feliczak-Guzik, 2023).

5.1.2 Titanium Dioxide (TiO_2)

TiO_2 has been widely employed for water splitting, water treatment, air purification and self-cleaning surfaces for its special photocatalytic capabilities (including photodegradation and photo-induced superhydrophilicity), stability, ease of manufacture and commercial availability (Yin & Deng, 2015). There are three known natural polymorphs of TiO_2: rutile, anatase and brookite. Each of these polymorphs has unique characteristics. Brookite and anatase are stable at low temperatures, while rutile becomes the thermodynamically preferred phase when the temperature approaches 500 °C. Due to their strong photocatalytic capabilities, anatase and rutile TiO_2 can destroy organic contaminants and microorganisms, preventing or postponing the formation of a gel layer on the membrane surface (El-Kady et al., 2023).

5.1.3 Zinc Oxide

Zinc oxide is one of the most widely used photocatalysts, along with TiO_2. Its capacity to create a variety of nanostructures, including nanowires, nanobelts, nanoscratches, nanospheres, nanofibres and nanotetrapods, has made it a material of growing interest. On the other hand, zinc oxide nanowires are currently of major interest, particularly when stacked in layers perpendicular to the conducting substrate. The overall surface area of the deposited ZnO can be up to 100 times larger than the geometric surface area on which this deposition happens because of the high diameter-to-height ratio of the nanowires produced in this manner. As a result, the ZnO surface can be covered with a significant amount of photosensitive material, which raises the light absorption efficiency value (Feliczak-Guzik, 2023).

5.2 Adsorbent Nanomaterials

Adsorption is brought about by van der Waals forces and electrostatic interactions. The length of time an adsorbent is exposed to a pollutant determines the adsorption rate, so a material's adsorption capacity changes depending on the system in which it is utilised. To be effective, adsorption materials need to be neutral, biocompatible, have a high adsorption capacity and be mechanically force-resistant. Three types of nanoadsorbents are available such as carbon based, oxide based, nano zero-valent iron (nZVI) based, and they are described below.

5.2.1 Carbon-Based Nanoadsorbent

Activated carbon (AC) has been studied extensively as an adsorbent for the removal of various pollutants from wastewater, such as heavy metal ions, phenols and synthetic colours. It is recognised for its high surface area, mechanical stability and abundance of adsorption sites. The unique qualities of carbon-based nanostructured materials, like hydrophilicity, low density, ease of regeneration and reusability, make them highly promising for use in wastewater treatment applications. Examples of these materials are CNTs, GO and carbon dots. The capacity of carbon-based adsorbents to functionalise with other functional groups, such as amidoxime, thiourea, pyridine and oxime, which have a strong attraction for various organic and inorganic contaminants, is one of their most intriguing features (Mittal et al., 2024).

Carbon Nanotubes

The cylindrical carbon structures known as CNTs have a high aspect ratio and superior mechanical strength. CNTs have a high adsorption capability because of their enormous surface area and special structural characteristics. They have high adsorption efficiency and chemical stability, which makes them very useful for removing different EPs such as heavy metals, organic pollutants and medicinal chemicals (Thakur et al., 2024). The hollow interiors of the nanotubes known as inner spots, intermediate channels (ICs) between different nanotubes, grooves on the outermost nanotube's upper shell and the perimeter of the nanotube aggregation are among the possible adsorption sites within CNT aggregates for different pollutants. Likewise, additional adsorption opportunities are provided by the individual nanotube's curved surfaces on the aggregate's outside (Sheoran et al., 2022). A detailed account of CNTs is given in Sect. 2.3.1.

Graphene

Two-dimensional carbon compounds made of a single layer of sp2 hybridised carbon atoms are called graphene and GO. Due to its high conductivity and vast surface area, graphene is a desirable adsorbent for the elimination of EPs. Its aggregating tendency, meanwhile, may restrict its use. Conversely, however, GO is an oxidised version of graphene that dissolves well in water and readily accepts different surface functional groups. GO has shown strong adsorption capacities and exceptional removal efficiencies in the adsorption of organic dyes, heavy metals and medications from wastewater, a topic of considerable research (Thakur et al., 2024). The detailed description of graphene is provided in Sect. 2.3.2.

Carbon Dots

CNMs, or carbon dots (Cdots), are becoming more and more popular. They consist of a graphitic core and functional groups, giving Cdots extraordinary stability, and tunable optical, chemical and electrical properties. Small carbon nanoparticles, known as carbon dots (CDs), are usually less than 10 nm. They have distinct surface, electrical and optical characteristics. Graphene, CNTs and organic molecules are just a few of the carbon sources that can be used to create CDs. CDs are attracting interest for their potential in the treatment of wastewater because of their great dispersibility, high surface area and adjustable surface chemistry. They have demonstrated encouraging adsorption properties for a range of EPs, including chemicals, heavy metals and medications (Thakur et al., 2024).

5.2.2 Oxide-Based Nanoadsorbents

Metal and metal oxide nanoparticles, such as copper oxide, titanium oxide, ferric oxide, AgNPs and so on, are another class of adsorbent nanomaterials that are frequently employed to remove heavy metal ions from wastewater. Both materials have a high surface area and a particular affinity. As a result, nanometal oxides have been used as adsorbents to remove heavy metal ions due to their low environmental impact and poor solubility, which results in secondary contamination (Naser et al., 2021).

5.2.3 nZVI-Based Adsorbents

Despite nZVI's great potential for eliminating contaminants from water, its aggregation to clusters, high surface energy and magnetic performance limits its dispersion and mobility, making remediation more difficult in practice. The three primary elements pertinent to the behaviour of nZVI in water are as follows: Firstly, nZVI exhibits a tendency to agglomerate, primarily influenced by the van der Waals and magnetic dipole moment forces acting between the particles. Secondly, it readily undergoes oxidation, leading to the formation of a passivation layer. Lastly, during the water treatment process, precipitation and corrosion products adhere to the surface of nZVI, serving to encapsulate it and prevent its interaction with pollutants in the water. These characteristics collectively contribute to the complex dynamics of nZVI in water treatment applications (Di et al., 2023).

5.3 *Membrane-Based Nanomaterials*

Because of its advantages over other wastewater treatment techniques, including ease of use, lack of phase switching, easy upscaling and downscaling procedures and seamless integration, membrane technology has recently grown in popularity. Apart from their efficacious ability to cleanse wastewater, membranes provide the potential to be recycled, thereby yielding noteworthy economic, social and environmental advantages. A membrane is a porous covering that permits fluid to pass through it but blocks impurities according to the physical and chemical characteristics of those impurities (Sawunyama et al., 2023).

5.3.1 Polymeric Nanocomposite Membrane

Membranes are created using polymeric materials because they are inexpensive and can be easily combined with other materials to form nanocomposites that are employed in a variety of separation processes (Zielinska et al., 2020). Because of the exceptional qualities of 2D TMDC materials, TMDC nanosheets have been widely used in ultra- and nanofiltration procedures for the treatment of contaminated wastewater as well as the purification of drinking water. Good-quality TMDC nanosheets are combined with a polymeric material to create polymeric nanocomposite membranes (Sahu et al., 2023). These materials can be added to the membrane matrix's support layer, thin top active layer or interface between these two layers.

5.3.2 Ceramic Membrane

Alumina (Al_2O_3), zirconia (ZrO_2), silicon oxide (SiO_2), silicon carbide (SiC) and silicon nitride (Si_3N_4) are used to make ceramic membranes, an inorganic form of membrane. The use of these ceramic membranes is growing because of their structural stiffness, hydrophilic surface that blocks the flow of many solvents and chemical inertness. The expense of the starting materials for the membranes and the high sintering energy requirements makes them far more expensive than organic membranes, even though they have vastly higher selectivity. Usually, these membranes are made up of three layers: a porous support layer, an intermediate layer and a top layer. These layers function as the support layer of the ceramic membrane, a source of mechanical strength and a link between the top layer and the support layer. According to their structural composition, ceramic membranes are classified as dense or porous membranes. Studies on water treatment usually use porous membranes because of their thickness, surface porosity and pore size features (Sawunyama et al., 2023).

6 Challenges and Considerations

A number of difficulties have been noted in the advancement of techniques based on nanomaterials and nanotechnology. The identification and description of nanoparticles pose numerous problems because of their small size. Distinguishing artificial nanoparticles from specific natural materials that are similar has become extremely challenging. Another drawback is the absence of appropriate sensitive and selective procedures needed for exact measurements of the surface area, quantity and size. It is more challenging to identify and properly characterise nanomaterials in nature due to their diverse forms and characteristics. It got harder to monitor nanomaterials in the system and look at their impacts on various systems of organisms exposed to nanomaterials, as they were exposed to more of the natural environment or biological components (El-Kady et al., 2023).

6.1 Toxicity

One of the biggest obstacles facing scientists, engineers and chemists working on the development of nanomaterials is their toxicity. Cytotoxicity, dermal toxicity, pulmonary toxicity, genotoxicity, carcinogenic toxicity, hepatic toxicity, cardiovascular toxicity, haemolytic toxicity and immunological toxicity are among the toxicities involving nanomaterials. The physiochemical characteristics of a nanomaterial, such as its morphology; electrical, magnetic and optical aspects; surface and surface chemistry; charge; size and size distribution; oxidation; composition and crystalline structure; aggregation and concentration; dispersion state; synthesis methods, etc., all impact the toxicology of a material (Mazari et al., 2021).

6.2 Long-Term Stability

Nanomaterials' aggregation, dissolution and interactions with other components in wastewater are influenced by their size, shape, surface charge and porosity. Particles that are smaller and more charged are more likely to aggregate, which decreases their surface area and capacity for adsorption (Sadegh et al., 2017). Certain nanoparticles can dissolve or disintegrate due to factors like pH, salinity and the presence of oxidising chemicals, which can impact the nanomaterials' longevity and functioning (Lu et al., 2016). Wastewater contains microorganisms that can cling to or even break down some nanomaterials, affecting their functionality and possibly posing new hazards.

6.3 Environmental Impact

Nanomaterials infiltrate open spaces and aquatic ecosystems via various sources and pathways. They can originate from both natural and man-made sources, such as burning fossil fuels, traffic, mining and demolition, and the production of nanomaterials and their waste streams. Natural sources of nanomaterials include volcanic activity, forest fires, clay minerals, soil erosion, weathering and dust storms. The resulting nanoparticles are employed in various applications that come into direct or indirect contact with the outside world. There are several ways in which sedimentation of soil occurs, including particle emissions, export of materials from incinerator plants, direct sedimentation through sanitary landfills, sewage treatment plant sludge, etc. (Mazari et al., 2021).

7 Future Perspective

Nanotechnology is a fast-developing field of science with particularly intriguing implications for the treatment of wastewater. Because of their special qualities, nanomaterials present a viable method to address the global water crisis by offering accurate, long-lasting and effective pollution removal solutions. Present-day classifications frequently concentrate on broad categories such as 'catalysts' or 'adsorbents'. It is probable that sub-classifications of contaminants, such as heavy metals, medicines or microplastics, will be the focus of future systems. Along with surface functionalisation and modification approaches, which drastically change how nanomaterials interact with pollutants, the classification will go beyond material type. In order to figure out the best nanomaterials for particular wastewater streams, AI systems will examine enormous databases on nanomaterial properties, pollutant characteristics and treatment procedures. The composition of wastewater will be continuously monitored by sensors and AI, enabling dynamic adjustments to treatment plans and the choice of nanomaterials depending on data collected in real time. The synergistic effects of combining several types of nanomaterials, for example, metal nanoparticles and CNTs, will improve the efficiency and selectivity of the removal of pollutants. Targeted treatment options for complicated wastewater compositions will be made possible by responsive nanomaterials, which alter their properties in response to environmental stimuli such as pH or temperature. The classification of nanoparticles in wastewater treatment has a bright future ahead of it, with precise, sustainable and highly effective solutions for cleaner water in the future. We can use nanotechnology to solve the water issue globally and guarantee clean water for one and all by adopting data-driven strategies, investigating cutting-edge nanomaterials and resolving possible obstacles. Nanomaterial-infused microfluidic devices and miniature robots will make it possible to treat particular contaminants in wastewater streams in a highly targeted and localised manner.

8 Conclusion

This chapter comprehensively reviewed the burgeoning field of nanomaterials for wastewater treatment, unveiling a multidimensional classification system to systematically analyse their diverse functionalities. Our framework, encompassing composition, morphology, functionalisation and application, provides a critical lens through which to assess the strengths and limitations of various nanomaterials.

Understanding the distinct advantages conferred by each compositional category (metal-based nanoparticles, metal oxides, carbon-based nanomaterials and polymeric nanomaterials) enables targeted selection for the removal of a specific contaminant. Similarly, appreciating the diverse morphologies (nanoparticles, nanostructured materials, nanotubes, nanocomposite nanomaterials and nanopolymers) empowers engineers to exploit their unique interaction dynamics with pollutants.

Furthermore, exploring the versatility of functionalisation strategies (chemical, biofunctionalisation and responsive nanomaterials) highlights the immense potential for enhancing selectivity, activity and adaptability of these materials to complex wastewater matrices. Finally, focusing on application-driven classifications (photocatalytic, adsorbent and membrane-based nanomaterials) underscores the critical role of nanomaterials in addressing a multitude of water purification challenges.

This systematic categorisation not only paves the way for future research efforts to optimise design and deployment of nanomaterials in wastewater treatment but also opens doors for novel functionalities. The vast, uncharted territory within each dimension demands further exploration, with promising avenues including exploration of novel compositions, tailored morphologies, bioinspired and stimuli-responsive functionalisation strategies and hybrid multifunctional nanomaterials for synergistic removal of contaminants. By harnessing the power of this intricate classification framework and delving deeper into each dimension, we can unlock the full potential of nanomaterials to revolutionise wastewater treatment and secure a sustainable future for our precious water resources.

References

Amdeha, E. (2023). Biochar-based nanocomposites for industrial wastewater treatment via adsorption and photocatalytic degradation and the parameters affecting these processes. *Biomass Conversion and Biorefinery, 2023*(1), 1–26. https://doi.org/10.1007/S13399-023-04512-2

Anjum, M., Miandad, R., Waqas, M., Gehany, F., & Barakat, M. A. (2019). Remediation of wastewater using various nano-materials. *Arabian Journal of Chemistry, 12*(8), 4897–4919. https://doi.org/10.1016/J.ARABJC.2016.10.004

Aragaw, T. A., Bogale, F. M., & Aragaw, B. A. (2021). Iron-based nanoparticles in wastewater treatment: A review on synthesis methods, applications, and removal mechanisms. *Journal of Saudi Chemical Society, 25*(8), Article 101280. https://doi.org/10.1016/J.JSCS.2021.101280

Asif, M. B., Zhang, Z., Iftekhar, S., & Lehto, V. P. (2021). Polysaccharide-derived biopolymeric nanomaterials for wastewater treatment. *Biopolymeric Nanomaterials: Fundamentals and Applications*, 447–469. https://doi.org/10.1016/B978-0-12-824364-0.00012-5

Baby, R., Hussein, M. Z., Abdullah, A. H., & Zainal, Z. (2022). Nanomaterials for the treatment of heavy metal contaminated water. *Polymers, 14*(3). https://doi.org/10.3390/POLYM14030583

Benamor, A., Nasser, M., El-Naas, M. H., & Qiblawey, H. (2019). Metal-oxide nanotubes functional material tailored for membrane water/wastewater treatment. *IOP Conference Series: Materials Science and Engineering, 634*(1), Article 012048. https://doi.org/10.1088/1757-899X/634/1/012048

Bhatt, P., Joshi, S., Urper Bayram, G. M., Khati, P., & Simsek, H. (2023). Developments and application of chitosan-based adsorbents for wastewater treatments. *Environmental Research, 226*, Article 115530. https://doi.org/10.1016/J.ENVRES.2023.115530

Bhol, P., Yadav, S., Altaee, A., Saxena, M., Misra, P. K., & Samal, A. K. (2021). Graphene-based membranes for water and wastewater treatment: A review. *ACS Applied Nano Materials, 4*(4), 3274–3293. https://doi.org/10.1021/ACSANM.0C03439/ASSET/IMAGES/MEDIUM/AN0C03439_0012.GIF

Birmod, R. P., Lade, V. G., & Shambharkar, R. D. (2021). Responsive membranes for wastewater treatment. Handbook of Nanomaterials for Wastewater Treatment: Fundamentals and Scale up Issues, pp. 673–697. https://doi.org/10.1016/B978-0-12-821496-1.00037-4

Burhan, M., Suhan, K., Al-Mamun, R., Farzana, N., Aishee, S. M., Marwani, H. M., Hasan, M. M., Asiri, A. M., Rahman, M. M., Islam, A., & Rabiul Awual, M. (2023). Sustainable pollutant removal and wastewater remediation using TiO_2-based nanocomposites: A critical review. *Nano-Structures & Nano-Objects, 36*. https://doi.org/10.1016/j.nanoso.2023.101050

Desireddy, S., & Pothanamkandathil Chacko, S. (2021). A review on metal oxide (FeOx/MnOx) mediated nitrogen removal processes and its application in wastewater treatment. *Reviews in Environmental Science and Bio/technology, 20*(3), 697–728. https://doi.org/10.1007/S11157-021-09581-1

Di, L., Chen, X., Lu, J., Zhou, Y., & Zhou, Y. (2023). Removal of heavy metals in water using nano zero-valent iron composites: A review. *Journal of Water Process Engineering, 53*, Article 103919. https://doi.org/10.1016/j.jwpe.2023.103913

Elghamry, I., Gouda, M., & Al-Fayiz, Y. S. S. (2023). Synthesis of chemically modified acid-functionalized multiwall carbon nanotubes with benzimidazole for removal of lead and cadmium ions from wastewater. *Polymers, 15*(6), 1421. https://doi.org/10.3390/POLYM15061421

El-Kady, M. M., Ansari, I., Arora, C., Rai, N., Soni, S., Verma, D. K., Singh, P., & Mahmoud, A. E. D. (2023). Nanomaterials: A comprehensive review of applications, toxicity, impact, and fate to environment. *Journal of Molecular Liquids, 370*, Article 121046. https://doi.org/10.1016/j.molliq.2022.121046

Feliczak-Guzik, A. (2023). Nanomaterials as photocatalysts—synthesis and their potential applications. *Materials, 16*(1). https://doi.org/10.3390/ma16010193

Ghosh, N., Das, S., Biswas, G., & Haldar, P. K. (2022). Review on some metal oxide nanoparticles as effective adsorbent in wastewater treatment. *Water Science and Technology, 85*(12), 3370–3395. https://doi.org/10.2166/WST.2022.153

Gudkov, S. V., Burmistrov, D. E., Serov, D. A., Rebezov, M. B., Semenova, A. A., & Lisitsyn, A. B. (2021). A mini review of antibacterial properties of ZnO nanoparticles. *Frontiers in Physics, 9*, Article 641481. https://doi.org/10.3389/FPHY.2021.641481/BIBTEX

Hudlikar, M., Joglekar, S., Dhaygude, M., & Kodam, K. (2012). Green synthesis of TiO_2 nanoparticles by using aqueous extract of Jatrophacurcas L. latex. *Materials Letters, 75*, 196–199. https://doi.org/10.1016/J.MATLET.2012.02.018

Ighalo, J. O., Sagboye, P. A., Umenweke, G., Ajala, O. J., Omoarukhe, F. O., Adeyanju, C. A., Ogunniyi, S., & Adeniyi, A. G. (2021). CuO nanoparticles (CuO NPs) for water treatment: A review of recent advances. *Environmental Nanotechnology, Monitoring and Management, 15*. https://doi.org/10.1016/J.ENMM.2021.100443

Islam, M. A., Morton, D. W., Johnson, B. B., Mainali, B., & Angove, M. J. (2018). Manganese oxides and their application to metal ion and contaminant removal from wastewater. *Journal of Water Process Engineering, 26*, 264–280. https://doi.org/10.1016/J.JWPE.2018.10.018

Iyyappan, J., Gaddala, B., Gnanasekaran, R., Gopinath, M., Yuvaraj, D., & Kumar, V. (2023). Critical review on wastewater treatment using photo catalytic advanced oxidation process: Role of photocatalytic materials, reactor design and kinetics. *Case Studies in Chemical and Environmental Engineering*, 100599. https://doi.org/10.1016/j.cscee.2023.100599

Javanmirpourshirzadi, Z., Zamani Hargalani, F., Robati, M., & Taghavi, L. (2023). Removal of lead (II) from aqueous solutions using β-cyclodextrin functionalized magnetic graphene oxide nanocomposite, performance, and optimization with response surface methodology. *Journal of Nanoparticle Research, 25*(7), 1–18. https://doi.org/10.1007/S11051-023-05778-3/METRICS

Joshi, N. C., & Gururani, P. (2022). Advances of graphene oxide based nanocomposite materials in the treatment of wastewater containing heavy metal ions and dyes. *Current Research in Green and Sustainable Chemistry, 5*, Article 100306. https://doi.org/10.1016/J.CRGSC.2022.100306

Kausar, A., Ahmad, I., Maaza, M., & Eisa, M. H. (2022). State-of-the-art of polymer/fullerene C60 nanocomposite membranes for water treatment: Conceptions, structural diversity and topographies. *Membranes, 13*(1), 27. https://doi.org/10.3390/MEMBRANES13010027

Kiruba Daniel, S. C. G., Malathi, S., Balasubramanian, S., Sivakumar, M., & Anitha Sironmani, T. (2014). Multifunctional silver, copper and zero valent iron metallic nanoparticles for wastewater treatment. *Application of Nanotechnology in Water Research, 9781118496305*, 435–457. https://doi.org/10.1002/9781118939314.CH15

Kumar, A., Kaladharan, K., & Tseng, F. G. (2021). Nanomaterials: Surface functionalization, modification, and applications. *Springer Series in Biomaterials Science and Engineering, 16*, 405–438. https://doi.org/10.1007/978-981-33-6252-9_14

Kumar, S., Kaur, P., Singh Brar, R., & Nagendra Babu, J. (2017). Nanoscale zerovalent copper (nZVC) catalyzed environmental remediation of organic and inorganic contaminants: A review. *Heliyon*, e10140. https://doi.org/10.1016/j.heliyon.2022.e10140

Kurwadkar, S., Hoang, T. V., Malwade, K., Kanel, S. R., Harper, W. F., & Struckhoff, G. (2019). Application of carbon nanotubes for removal of emerging contaminants of concern in engineered water and wastewater treatment systems. *Nanotechnology for Environmental Engineering, 4*(1), 1–16. https://doi.org/10.1007/S41204-019-0059-1/METRICS

Lakshmi, D., Diana, M. I., Helen, P. A., & Selvin, P. C. (2024). Functionalized nanomaterials, classification, properties, and functionalization techniques. *Materials Horizons: From Nature to Nanomaterials, Part, F1329*, 65–92. https://doi.org/10.1007/978-981-99-3021-0_3/COVER

Liao, Y., Loh, C. H., Tian, M., Wang, R., & Fane, A. G. (2018). Progress in electrospun polymeric nanofibrous membranes for water treatment: Fabrication, modification and applications. *Progress in Polymer Science, 77*, 69–94. https://doi.org/10.1016/J.PROGPOLYMSCI.2017.10.003

Lu, H., Wang, J., Stoller, M., Wang, T., Bao, Y., & Hao, H. (2016). An overview of nanomaterials for water and wastewater treatment. *Advances in Materials Science and Engineering*, 4964828, 10. https://doi.org/10.1155/2016/4964828

Madeła, M. (2020). Effect of copper nanoparticles on biological wastewater treatment. *Desalination and Water Treatment, 199*, 493–498. https://doi.org/10.5004/dwt.2020.26340

Martin, L. M. A., Sheng, J., Zimba, P. V., Zhu, L., Fadare, O. O., Haley, C., Wang, M., Phillips, T. D., Conkle, J., & Xu, W. (2022). Testing an iron oxide nanoparticle-based method for magnetic separation of nanoplastics and microplastics from water. *Nanomaterials, 12*(14), 2348. https://doi.org/10.3390/NANO12142348

Mazari, S. A., Ali, E., Abro, R., Khan, F. S. A., Ahmed, I., Ahmed, M., Nizamuddin, S., Siddiqui, T. H., Hossain, N., Mubarak, N. M., & Shah, A. (2021). Nanomaterials: Applications, waste-handling, environmental toxicities, and future challenges—A review. *Journal of Environmental Chemical Engineering, 9*(2). https://doi.org/10.1016/j.jece.2021.105028

Mishra, S., & Sundaram, B. (2023). Efficacy and challenges of carbon nanotube in wastewater and water treatment. *Environmental Nanotechnology, Monitoring & Management, 19*, Article 100764. https://doi.org/10.1016/J.ENMM.2022.100764

Mittal, H., Alfantazi, A., & Alhassan, S. M. (2024). Recent developments in the adsorption of uranium ions from wastewater/seawater using carbon-based adsorbents. *Journal of Environmental Chemical Engineering, 12*(1), Article 111705. https://doi.org/10.1016/j.jece.2023.111705

Mohan, A., Rout, L., Thomas, A. M., Nagappan, S., Parambadath, S., Park, S. S., & Ha, C. S. (2020). Silver nanoparticles impregnated pH-responsive nanohybrid system for the catalytic reduction of dyes. *Microporous and Mesoporous Materials, 303*, Article 110260. https://doi.org/10.1016/J.MICROMESO.2020.110260

Naser, J. A., Ahmed, Z. W., & Ali, E. H. (2021). Nanomaterials usage as adsorbents for the pollutants removal from wastewater; A review. *Materials Today: Proceedings, 42*, 2590–2595. https://doi.org/10.1016/j.matpr.2020.12.584

Nayl, A. A., Abd-Elhamid, A. I., Aly, A. A., & Bräse, S. (2022a). Recent progress in the applications of silica-based nanoparticles. *RSC Advances, 12*(22), 13706–13726. https://doi.org/10.1039/D2RA01587K

Nayl, A. A., Abd-Elhamid, A. I., Awwad, N. S., Abdelgawad, M. A., Wu, J., Mo, X., Gomha, S. M., Aly, A. A., & Bräse, S. (2022b). Review of the recent advances in electrospun nanofibers applications in water purification. *Polymers, 14*(8), 1594. https://doi.org/10.3390/POLYM14081594

Nguyen, N. T., & Nguyen, V. A. (2020). Synthesis, characterization, and photocatalytic activity of ZnO nanomaterials prepared by a green, nonchemical route. *Journal of Nanomaterials*, 1768371, 8. https://doi.org/10.1155/2020/1768371

Noreen, S., & Abd-Elsalam, K. A. (2021). Biochar-based nanocomposites: A sustainable tool in wastewater bioremediation. *Aquananotechnology: Applications of Nanomaterials for Water Purification*, 185–200. https://doi.org/10.1016/B978-0-12-821141-0.00023-9

Palani, G., Trilaksana, H., Sujatha, R. M., Kannan, K., Rajendran, S., Korniejenko, K., Nykiel, M., & Uthayakumar, M. (2023). Silver nanoparticles for waste water management. *Molecules, 28*(8), 3520. https://doi.org/10.3390/MOLECULES28083520

Pryshchepa, O., Pomastowski, P., & Buszewski, B. (2020). Silver nanoparticles: Synthesis, investigation techniques, and properties. *Advances in Colloid and Interface Science, 284*, Article 102246. https://doi.org/10.1016/J.CIS.2020.102246

Rafeeq, H., Hussain, A., Ambreen, A., Zill-e-Huma, Waqas, M., Bilal, M., & Iqbal, H. M. N. (2022). Functionalized nanoparticles and their environmental remediation potential: A review. *Journal of Nanostructure in Chemistry, 12*(6), 1007–1031. https://doi.org/10.1007/S40097-021-00468-9

Raghav, S., Jain, P., Painuli, R., & Kumar, D. (2021).The role of carbon nanocomposite membranes for water and wastewater treatment. *Green Energy and Technology*, 19–41. https://doi.org/10.1007/978-981-15-6699-8_2/ Rahman, R. O. A., El-Kamash, A. M., & Hung, Y. -T. (2022). Applications of nano-zeolite in wastewater treatment: An overview. *Water, 14*(2), 137. https://doi.org/10.3390/W14020137

Rajendran, S., & Kumar, P. S. (2021).Carbon nanocomposites for wastewater treatment. *The Future of Effluent Treatment Plants: Biological Treatment Systems*, 215–234. https://doi.org/10.1016/B978-0-12-822956-9.00013-1

Rani, U. A., Ng, L. Y., Ng, C. Y., & Mahmoudi, E. (2020). A review of carbon quantum dots and their applications in wastewater treatment. *Advances in Colloid and Interface Science, 278*, Article 102124. https://doi.org/10.1016/J.CIS.2020.102124

Rigoletto, M., Calza, P., Gaggero, E., & Laurenti, E. (2022). Hybrid materials for the removal of emerging pollutants in water: Classification, synthesis, and properties. *Chemical Engineering Journal Advances, 10*, Article 100252. https://doi.org/10.1016/J.CEJA.2022.100252

Sadegh, H., Ali, G. A. M., Gupta, V. K., Makhlouf, A. S. H., Shahryari-ghoshekandi, R., Nadagouda, M. N., Sillanpää, M., & Megiel, E. (2017). The role of nanomaterials as effective adsorbents and their applications in wastewater treatment. *Journal of Nanostructure in Chemistry, 7*(1), 1–14. https://doi.org/10.1007/S40097-017-0219-4

Sahu, A., Dosi, R., Kwiatkowski, C., Schmal, S., & Poler, J. C. (2023). Advanced polymeric nanocomposite membranes for water and wastewater treatment: A comprehensive review. *Polymers, 15*(3), 540. https://doi.org/10.3390/POLYM15030540

Sawunyama, L., Oyewo, O.A., Seheri, N., Onjefu, S. A., Onwudiwe, D. C. (2023). Metal oxide functionalized ceramic membranes for the removal of pharmaceuticals in wastewater. *Surfaces and Interfaces, 38*. https://doi.org/10.1016/j.surfin.2023.102787

Sheoran, K., Kaur, H., Siwal, S. S., Saini, A. K., Vo, D. V. N., & Thakur, V. K. (2022). Recent advances of carbon-based nanomaterials (CBNMs) for wastewater treatment: Synthesis and application. *Chemosphere, 299*, Article 134364. https://doi.org/10.1016/J.CHEMOSPHERE.2022.134364

Shih, Y. J., Chen, Z. S., Chen, C. L., Huang, Y. H., & Huang, C. P. (2022). Enhancing arsenic (III) removal by integrated electrocatalytic oxidation and electrosorption reactions on nano-textured bimetal composite of iron oxyhydroxide and manganese dioxide polymorphs (α-, γ-, β-, and ε-MnxFe1−xO). *Applied Catalysis b: Environmental, 317*, Article 121757. https://doi.org/10.1016/J.APCATB.2022.121757

Thakur, A., Kumar, A., & Singh, A. (2024). Adsorptive removal of heavy metals, dyes, and pharmaceuticals: Carbon-based nanomaterials in focus. *Carbon*, 217. https://doi.org/10.1016/j.carbon.2023.118621

Wang, Y., Li, Z., Wang, J., Li, J., & Lin, Y. (2011). Graphene and graphene oxide: Biofunctionalization and applications in biotechnology. *Trends in Biotechnology, 29*(5), 205–212. https://doi.org/10.1016/J.TIBTECH.2011.01.008

Wang, X., Feng, Y., Dong, P., & Huang, J. (2019). A mini review on carbon quantum dots: Preparation, properties, and electrocatalytic application. *Frontiers in Chemistry, 7*, Article 490939. https://doi.org/10.3389/FCHEM.2019.00671/BIBTEX

Wang, P., Fu, F., & Liu, T. (2021). A review of the new multifunctional nano zero-valent iron composites for wastewater treatment: Emergence, preparation, optimization and mechanism. *Chemosphere, 285*, Article 131435. https://doi.org/10.1016/J.CHEMOSPHERE.2021.131435

Weidner, E., Karbassiyazdi, E., Altaee, A., Jesionowski, T., & Ciesielczyk, F. (2022). Hybrid metal oxide/biochar materials for wastewater treatment technology: A review. *ACS Omega, 7*(31), 27062–27078. https://doi.org/10.1021/ACSOMEGA.2C02909/ASSET/IMAGES/LARGE/AO2C02909_0006.JPEG

Xu, P., Zeng, G. M., Huang, D. L., Feng, C. L., Hu, S., Zhao, M. H., Lai, C., Wei, Z., Huang, C., Xie, G. X., & Liu, Z. F. (2012). Use of iron oxide nanomaterials in wastewater treatment: A review. *Science of the Total Environment, 424*, 1–10. https://doi.org/10.1016/J.SCITOTENV.2012.02.023

Yang, B., Wei, Y., Liu, Q., Luo, Y., Qiu, S., & Shi, Z. (2019). Polyvinylpyrrolidone functionalized magnetic graphene-based composites for highly efficient removal of lead from wastewater. *Colloids and Surfaces a: Physicochemical and Engineering Aspects, 582*, Article 123927. https://doi.org/10.1016/J.COLSURFA.2019.123927

Yin, J., & Deng, B. (2015). Polymer-matrix nanocomposite membranes for water treatment. *Journal of Membrane Science, 479*, 256–275. Elsevier B.V. https://doi.org/10.1016/j.memsci.2014.11.019

Yu, J. G., Yu, L. Y., Yang, H., Liu, Q., Chen, X. H., Jiang, X. Y., Chen, X. Q., & Jiao, F. P. (2015). Graphene nanosheets as novel adsorbents in adsorption, preconcentration and removal of gases, organic compounds and metal ions. *Science of the Total Environment, 502*, 70–79. https://doi.org/10.1016/J.SCITOTENV.2014.08.077

Zahmatkesh, S., Hajiaghaei-Keshteli, M., Bokhari, A., Sundaramurthy, S., Panneerselvam, B., & Rezakhani, Y. (2023). Wastewater treatment with nanomaterials for the future: A state-of-the-art review. *Environmental Research, 216*, Article 114652. https://doi.org/10.1016/J.ENVRES.2022.114652

Zielinska, A., Carreiró, F., Oliveira, A. M., Neves, A., Pires, B., Nagasamy Venkatesh, D., Durazzo, A., Lucarini, M., Eder, P., Silva, A. M., Santini, A., & Souto, E. B. (2020). Polymeric nanoparticles: Production, characterization, toxicology and ecotoxicology. *Molecules, 25*(16), 3731. https://doi.org/10.3390/MOLECULES25163731

Characterisation of Nanomaterials

Onkar Sudhir Ekande and S. Mathava Kumar

1 Introduction

Characterisation of nanomaterials is done to identify the physical, chemical and electrical properties responsible for environmental remediation through the physico-chemical process. Nanomaterials with different morphologies, crystal phases and high specific surface area were employed in different treatment processes such as adsorption, photocatalysis, electrocatalysis and thin-film nanocomposite (TFN) membranes for the treatment of tertiary wastewater aimed at the elimination of nutrients and emerging organic pollutants (EOPs) (Bethi et al., 2016; Ekande & Kumar, 2021a; Miklos et al., 2018). Because of their small size, nanomaterials frequently exhibit unique and favourable features as compared to bulk materials, making them important in a wide range of physico-chemical operations for the treatment of wastewater.

The suitability and performance of a nanomaterial for a particular process can be assessed through different characterisation techniques using spectroscopy and microscopy. The characterisation of nanomaterials is broadly divided into structural characterisation, chemical characterisation, optical and optoelectronic characterisation, electrical characterisation and surface characterisation. These characterisation techniques will help the researchers identify the mechanism of nanomaterial-based physico-chemical treatment technologies for the treatment of wastewater.

O. S. Ekande
Department of Civil and Environmental Engineering, Texas A&M University, College Station, Texas, United States

S. M. Kumar (✉)
Environmental Engineering Division, Department of Civil Engineering, Indian Institute of Technology Madras, Chennai, Tamil Nadu, India
e-mail: mathav@iitm.ac.in

© Capital Publishing Company, New Dehli, India. 2025
R. Neelancherry and S. M. Kumar, *Nanotechnology in Wastewater Treatment*,
https://doi.org/10.1007/978-3-031-94246-4_3

2 Structural Characterisations of Nanomaterials

2.1 Electron Microscopy

Electron microscopes reveal the morphology and structural characterisation of nanomaterials. Scanning electron microscopy (SEM) is a bulk technique used to understand the morphology and composition of nanomaterials, whereas different electron beams reveal the morphology, chemical composition and atomic number contrast (Table 1). SEM combined with energy dispersive X-ray spectroscopy (EDS) helps to know the elemental makeup of nanomaterials. Secondary electrons are generated because of the inelastic collision between the incident electron beam and the loosely bound outer shell electrons. These secondary electrons generated close to the surface ~10 nm region provided the topographic information of the nanomaterial (Fig. 1). However, backscattered electrons penetrate deeper into the sample (~1 μm), resulting in low topographic contrast compared to secondary electron images. Various elements present in the nanomaterial have been detected using characteristic X-rays emitted by the sample, which helps identify the composition of the nanomaterial.

Catalytic treatment and TFN membrane are used to study different morphologies of zinc oxide (ZnO) to improve process efficiency (Rajakumaran et al., 2020; Xu et al., 2009). Nanosize ZnO with a spherical, rod-like and flower-like structure was observed in the backscattered mode of SEM. Different solvents and compositions of precursors were optimised to synthesise ZnO with different morphologies through the analysis of SEM images.

The porous nature of carbonaceous materials such as biochar, graphitic carbon and activated carbon before and after modification is confirmed through SEM images

Table 1 Application of different electron beams for nanomaterial characterisation

Electron source	Application in nanomaterial characterisation
Secondary electrons	Topographic contrast
Backscattered electrons	Atomic number contrast
Characteristic X-ray	Chemical composition

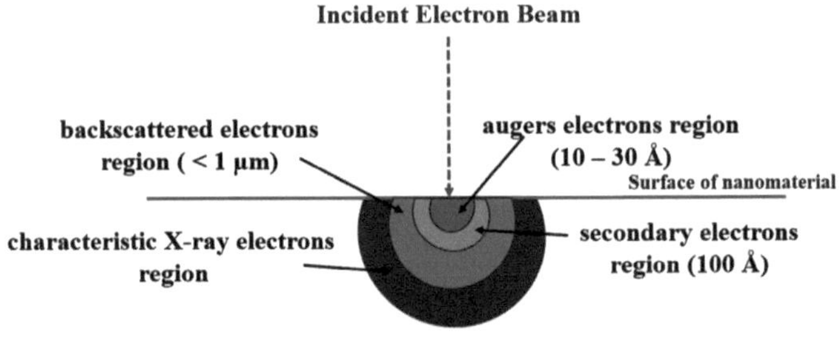

Fig. 1 Interaction between electron beams and sample

Table 2 Differences between SEM and TEM

Parameter	SEM	TEM
Electron source	Scattered electrons	Transmitted electrons
Lowest resolution	Nanometres (nm) to angstroms (Å) 10 nm to 1 nm	Angstroms (Å) (~2 Å)
Image dimension	3D	2D
Information	Topography/surface	Crystallographic and atomic data
Samples	Thick and thin samples	Ultrathin samples

(Ekande & Kumar, 2021a; Mahdi et al., 2019; Zhang et al., 2024a, b). The SEM technique revealed the morphology of the nanomaterials, whereas information on internal structure is not achieved through SEM. This information is acquired by using transmission electron microscopy (TEM), which generates images by capturing transmitted electrons from the ultrathin nanomaterial sample and revealing the crystal structure of the nanomaterial (Table 2). TEM images provide information about the crystal defects such as atomic vacancies, dislocation and grain boundaries in the nanomaterial (Feng et al., 2021; He et al., 2024; Kumar et al., 2024).

Moreover, TEM gives a more accurate size of nanoparticles compared to SEM. HR-TEM images of different morphologies of ZnO revealed the interplanar spacing between two planes. It was observed that ZnO nanowire showed a dominant crystal of the (101) facet exhibiting a 0.247 nm lattice spacing. Similarly, ZnO nanosheets and ZnO nanoflowers showed dominant crystal planes of (110) and (103) facets exhibiting 0.158 nm and 0.147 nm lattice spacings, respectively (Jiang et al., 2024). HR-TEM images confirmed the formation of strong interfacial contact between two different photo/piezocatalysts in heterojunction (Ekande & Kumar, 2023). Further, the deposition of plasmonic Au on oxygen-defective ZnO in heterojunction is responsible for defect healing between Au and ZnO (Mohite et al., 2024). Moreover, annual dark-field imaging for mapping nanomaterial samples in TEM images confirmed the defects in the ZnO crystal structure in heterojunction (Mohite et al., 2024).

2.2 *X-Ray Diffraction (XRD)*

XRD imparts crystallographic information about nanomaterials and reveals the arrangement of atoms, crystal size, unit cell dimensions and residual stresses in nanomaterials. Spacing between crystal lattice planes is determined using Bragg's law, as shown in Eq. 1. The schematic of the unit cell is shown in Fig. 2, and probable types of unit cells depending on the geometric parameters are reported in Table 3.

$$n\lambda = 2d_i \sin\theta \tag{1}$$

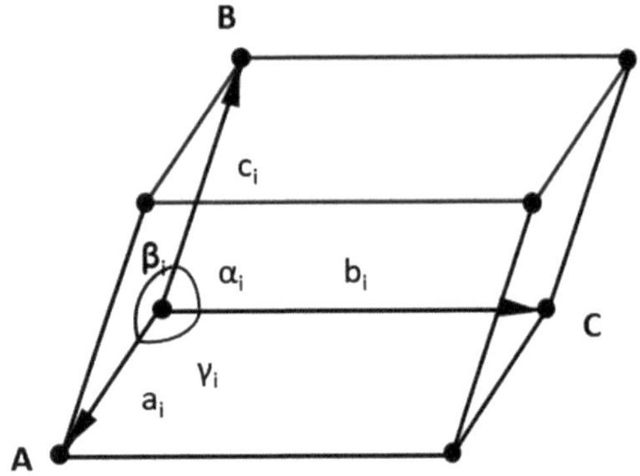

Fig. 2 Schematics of the unit cell

Table 3 Different types of unit cells and geometric parameters

Unit cells	Axial lengths	Angles
Cubic	$a_i = b_i = c_i$	$\alpha_i = \beta_i = \gamma_i = 90°$
Tetragonal	$a_i = b_i \neq c_i$	$\alpha_i = \beta_i = \gamma_i = 90°$
Orthorhombic	$a_i \neq b_i \neq c_i$	$\alpha_i = \beta_i = \gamma_i = 90°$
Rhombohedral	$a_i = b_i = c_i$	$\alpha_i = \beta_i = \gamma_i \neq 90°$
Hexagonal	$a_i = b_i \neq c_i$	$\alpha_i = \beta_i = 90°, \gamma_i = 120°$
Monoclinic	$a_i \neq b_i \neq c_i$	$\alpha_i = \gamma_i = 90° \neq \beta_i$
Triclinic	$a_i \neq b_i \neq c_i$	$\alpha_i \neq \beta_i \neq \gamma_i \neq 90°$

where λ is the X-ray wavelength (Cu = 1.54 Å), d_i is the interplanar spacing and θ is the incidence angle.

$$\frac{K \times \lambda}{B \times \cos\theta_B} \tag{2}$$

where diffraction angle is denoted as θ_B, K is the Scherrer constant and full width at half maximum (FWHM) is denoted by B. Scherrer formula is used to determine the small crystal size up to 100 nm (Eq. 2).

XRD analysis is performed on the different morphologies of ZnO. The characteristic diffraction peaks at 31.8, 34.4, 36.2, 47.5, 56.6, 62.9 and 68° correspond to the (100), (002), (101), (102), (110), (103) and (112) planes of hexagonal wurtzite ZnO (Rajakumaran et al., 2019). All diffraction peaks are present irrespective of ZnO morphology; however, the grain sizes of the ZnO nanoflowers, ZnO nanowires and ZnO nanosheets are 25.3, 32 and 15.9 nm, respectively, calculated using the Scherrer formula (Eq. 2) (Jiang et al., 2024). Broadening of XRD peaks occurs due to the combined effect of crystal size, strain and instrumental broadening (Eq. 3).

$$\beta_{tot} \cos\theta = \varepsilon \sin\theta + \frac{K\lambda}{B} \tag{3}$$

where ε corresponds to the strain, and the crystal size is represented by $\frac{K\lambda}{B}$.

The Williamson–Hall plotconsists of β_{tot} cos θ vs sin θ. The slope of the plot gives the strain in the crystal structure, and the Y-intercept gives the crystal size. In ZnO/CdS and ZnO/CdSe heterojunctions, strain is developed in the ZnO crystal structure during the in situ growth of CdS and CdSe (Eley et al., 2014). This is responsible for the shift in the peaks of ZnO in heterojunction compared to the peak position in pure ZnO. The positive gradient of the Williamson–Hall curve indicates tensile strain, whereas the negative slope indicates compressive stress developed in the crystal structure. In another heterojunction between $NaNbO_3$ and g-C_3N_4, it was observed that the tensile strain generated in the $NaNbO_3$ increased from +1.5% to +2.5% in the $NaNbO_3$/g-C_3N_4 heterojunction (Eq. 3) (Ekande & Kumar, 2023). This was accountable for the transition of the XRD peak of $NaNbO_3$ to a lower angle in heterojunction compared to pure $NaNbO_3$. Overall, XRD analysis provides detailed insights on the crystallographic characterisation of nanomaterials applied in the remediation of wastewater.

3 Chemical Characterisations of Nanomaterials

3.1 Fourier Transform Infrared Spectroscopy (FTIR)

FTIR spectroscopy reveals the presence of functional groups in the nanomaterials used in the different physico-chemical processes. Different chemical bonds absorb infrared (IR) radiation at characteristic frequencies, which help identify the particular functional groups present in nanomaterials. Absorption of IR radiation responsible for the stretching and bending of the bond at different frequencies, that is, absorption of O–H stretching is observed in range of 3200–3700 cm^{-1}, whereas absorption of O–H bending is observed in the range of 1300–1450 cm^{-1} (Table 4).

The FTIR spectrum of the graphitic carbon nitride (g-C_3N_4) shows the existence of C-N and C–C bonds. Also, peaks in the range of 2900–3500 cm^{-1} represent the presence of -NH_2/-NH in the structure of g-C_3N_4 (Ekande & Kumar, 2021a; Tan et al., 2022). In the FTIR spectrum of the ZnO/TiO_2 heterojunction, a peak in the range of 530–550 cm^{-1} was detected due to Ti–O stretching vibration (Upadhaya et al., 2019). The FTIR spectrum is a non-destructive test to confirm the presence of different organic and inorganic bonds in nanomaterials and heterojunctions.

3.2 X-Ray Photoelectron Spectroscopy (XPS)

XPS is used to reveal the elemental composition and chemical states of nanomaterials. X-rays interact with the surface of the sample and are responsible for the emission of electrons from the outermost orbit of the sample. The kinetic energy associated

Table 4 FTIR peak and corresponding functional group

Frequency region (cm_{-1})	Functional group
1700–1800	C=0
1350–1450	C-H
1600–1700	C=C
1000–1250	C-0
1640–1690	C-N
2800–3400	C=N
2220–2260	N-H
2190–2260	C≡N
1180–1200	C≡C
1380–1415	S=O
550–850	C-Cl
1500–1550	N-0
1030	Si-0
1414	NH_4^+

with these electrons reveals the chemical structure of the sample. Information on the different oxidation states of the elements and changes in the composition of nanomaterials and catalysts can be obtained from the XPS spectrum. Carbon correction is performed to account for the background signal due to the presence of adventitious carbon. Carbon correction was performed aligning the C1s spectrum of the sample with 284.8 eV through a horizontal shift without affecting the intensity of the peak. After making the alignment, the background peak intensity is subtracted to get an accurate C1s spectrum in the nanomaterial. Further, it should be taken care of while deconvoluting the peaks for the p, d and f subshells by maintaining the area ratio between spin orbital splitting (Table 5).

XPS of Zn $2p_{3/2}$ and Zn $2p_{1/2}$ in ZnO nanoparticles is detected at 1021 eV and 1044 eV binding energy (BE), respectively, maintaining the ratio of the area under curve Zn $2p_{3/2}$ and Zn $2p_{1/2}$, as shown in Table 5 (Sahai et al., 2014; Zhang & Lü et al., 2024). The O1s broad peak is observed in the range of 529–533 eV. Oxygen vacancies occurring in metal oxides are detected through the XPS of O1s. It is reported that deconvolution of the O1s spectrum to ~530, ~532 and ~533 eV is attributed to oxygen lattice, oxygen vacancies and interstitial oxygen, respectively (Wang et al., 2024).

Table 5 Area ratio for spin orbital splitting

Subshell	Area ratio
$1s_{1/2}$	No
$2p_{1/2}$:$2p_{3/2}$	1:2
$3d_{3/2}$: $3d_{5/2}$	2:3
$4f_{5/2}$: $4f_{7/2}$	3:4

Along with that, peak intensity in XPS is used to calculate the composition of the elements in the nanomaterial.

Ultraviolet photo-electron spectroscopy (UPS) is used to understand the valence band structure and work function of the nanomaterials using ultraviolet light (He I at 21.2 eV). The secondary edge BE was obtained using the UPS, and the tangent drawn at the concavity of the curve to intersect the x-axis gives the BE. Furthermore, the work function (ϕ) is calculated using Eq. 4.

$$\phi = 21.2 - \text{BE} \tag{4}$$

In photocatalytic and piezocatalytic heterojunctions, the ϕ of both catalysts plays a major role in understanding the charge transfer mechanism. Oxidative catalysts such as ZnO and TiO_2 have a lower ϕ than reductive catalysts such as g-C_3N_4, which forms the type-II heterojunction (Ekande & Kumar, 2021b). On the other hand, a higher ϕ of oxidative catalyst compared to a reductive catalyst formed an S-scheme heterojunction (Ekande & Kumar, 2021b).

4 Optical and Optoelectronic Characterisations of Nanomaterials

4.1 Photoluminescence (PL) Spectroscopy

Nanomaterials absorb photons, which excite electrons from VB to CB and again come back to VB (ground state) with the emission of photons with lower energy (Fig. 3). The PL spectrum measures the intensity of emission photons from nanomaterials under specific excitation energies. The PL spectrum is utilised to understand the recombination of charge carriers in the photocatalyst and photocatalytic heterojunctions. Also, a shift in the PL spectrum to the catalyst revealed a change in the band gap of the catalyst (Ekande & Kumar, 2021a).

The PL spectrum of nanosheet g-C_3N_4 showed a low intensity in comparison to layered g-C_3N_4 (Feng et al., 2021). A reduction in the intensity of g-C_3N_4 nanosheets confirmed the reduction in electron–hole recombination, which facilitates efficient interfacial charge transfer at the g-C_3N_4-water interface for formation of reactive species. In g-C_3N_4/ZnO photocatalytic heterojunction, PL spectrum intensity of g-C_3N_4 was reduced significantly in type-II heterojunction between ZnO and g-C_3N_4. The formation of type-II heterojunction transfers the electrons from CB of g-C_3N_4 to ZnO, which inhibits electron–hole recombination in g-C_3N_4.

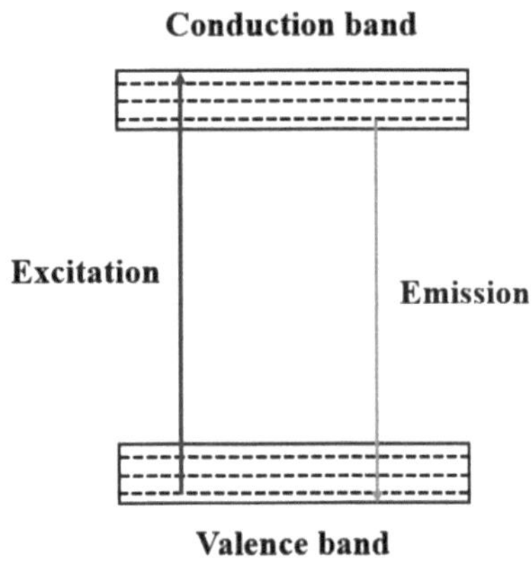

Fig. 3 Excitation and emission processes in the semiconductor catalyst

4.2 *Ultraviolet Diffused Reflectance Spectroscopy (UV-DRS)*

UV-DRS analysis was performed on the nanomaterial to obtain the band gap using the Kubelka–Munk function (Eq. 5). The diffuse reflectance of the solid sample was measured, and it was converted to absorbance using the Kubelka–Munk function.

$$F(R_\infty) = \frac{(1 - R_\infty)^2}{2 \times R_\infty} = \frac{K}{S} \quad (5)$$

For a direct band gap, the highest VB level occurs at the same momentum as the lowest CB level, whereas for an indirect band gap, the lowest energy level of CB occurs at a different momentum than the highest energy of VB. The absorbance of light is directly proportional to the difference between incident light energy and the band gap. Absorption increases with a decrease in the wavelength, that is, an increase in the energy of light.

$$(\alpha h\nu)^{1/n} \propto h\nu - E_g \quad (6)$$

Plot graph between $y = (\alpha h\nu)^{1/n}$, $x = h\nu$, at $y = 0$, x-intercept $= E_g$, direct band gap and indirect band gap has n values of 0.5 and 2, respectively (Eq. 6) (Fig. 4).

UV–visible photocatalyst ZnO has a direct band gap (Preethi et al., 2016). Therefore, the Kubelka–Munk plot for $n = 0.5$ gives a band gap of ~3.2 eV. However, g-C_3N_4 is reported to have an indirect band gap, and the Kubelka–Munk plot for $n = 2$ gives a band gap of ~2.7 eV (Ekande & Kumar, 2023). UV-DRS analysis is important to understand the efficiency of semiconductor photocatalysts before and after structural modification under UV–visible light.

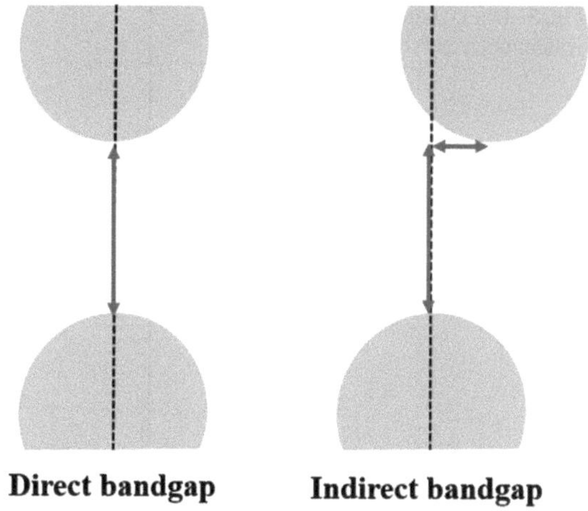

Fig. 4 Direct and indirect band gaps of semiconductor nanomaterial

5 Electrical Characterisations of Nanomaterials

Electrical characterisation of catalytic materials such as photocatalysts, electrocatalysts and piezocatalysts provides a vital role in understanding the type of catalyst (N-type and P-type), transfer of electrons at the interface of catalyst and solution, overpotential, interfacial charge transfer resistance and kinetics of charge carrier transfer.

5.1 Cyclic Voltammetry (CV)

CV analysis is used to understand the electron transfer process (oxidation and reduction) between the catalyst and aqueous system. CV analysis was used to examine the persistence of the photocatalyst, electrocatalyst and the electrochemical quantification of OH in an aqueous system (Elgrishi et al., 2018).

Nerst Equation:

$$E = E^{o} + \frac{RT}{nF} \ln \frac{Ox}{Red} \quad (7)$$

where electrochemical cell potential is denoted by E, standard potential of species is denoted by E^{o}, universal gas constant is denoted by R, F is Faraday constant, *n* is the number of electron transfer and temperature in kelvin is denoted by T.

In a chemically reversible system, all species are present in the reduced form (point *a),* and let's say [Red] is 1, whereas [Ox] is 0 (Fig. 5). It was observed that with an increase in the potential from low to high, Red gets oxidised to Ox and at point *b*, [Red] = [Ox] = 0.5, which further increased and achieved maximum current at point

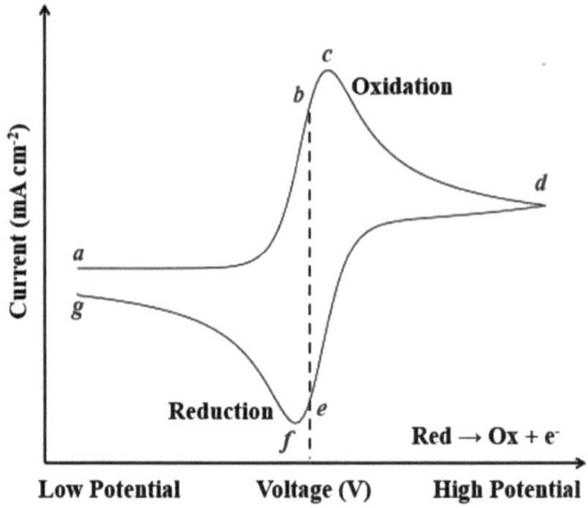

Fig. 5 Duck-shaped cyclic voltammetry curve

c. Current decreased with further increase in potential due to the limiting diffusion of [Red] from bulk to electrode surface with further increase in [Ox]. Finally, at point *d*, all [Red] oxidised to [Ox], [Ox] = 1. At the reverse cycle of potential from high potential to low potential, [Ox] decreased to [Red] and observed [Ox] = [Red] = 0.5 at point *e*. Furthermore, with a decrease in the potential, all [Ox] converted to [Red] (point *g*). From CV analysis, at points *b* and *e, the concentration* of [Red] = [Ox], following the Nerst equation, $E = E_{1/2}$ w.r.t. the reference electrode (Eq. 7).

The stability of different photocatalysts and photocatalytic heterojunctions was studied using CV. CV analysis of ZnO observed the oxidation peak at −1.18 V during the forward scan and −1.33 V in the reverse scan and obtained peaks corresponding to the generation of the zincate ion ($Zn(OH)_4{}^{2-}$) (Qamar et al., 2016). In the presence of light, the oxidation peak is observed at −1.05 V due to the release of Zn^{2+} as a result of self-oxidation. However, ZnO combined with γ-Fe_2O_3 suppresses the CV peaks observed for oxidation and reduction of pure ZnO. CV analysis confirmed the inhibition of the photocorrosion of ZnO in heterojunction with γ-Fe_2O_3.

5.2 Electrochemical Impedance Spectroscopy (EIS)

In an electrochemical cell, the interaction between electroactive species and electrodes is governed by the charge-mass transfer processes from bulk to electrode surface. These processes are represented in the equivalent circuit by incorporating solution resistance in series with the parallel combination of double layer capacitance and electron transfer resistance (Fig. 6) (Magar et al., 2021). The impedance of an electrode system is represented using real and imaginary components, that is, the real component on the x-axis and the imaginary component on the y-axis, called the Nyquist plot. The Nyquist plot can be fitted for the equivalent circuit to calculate the capacitance, solution resistance and electron transfer resistance. EIS was used

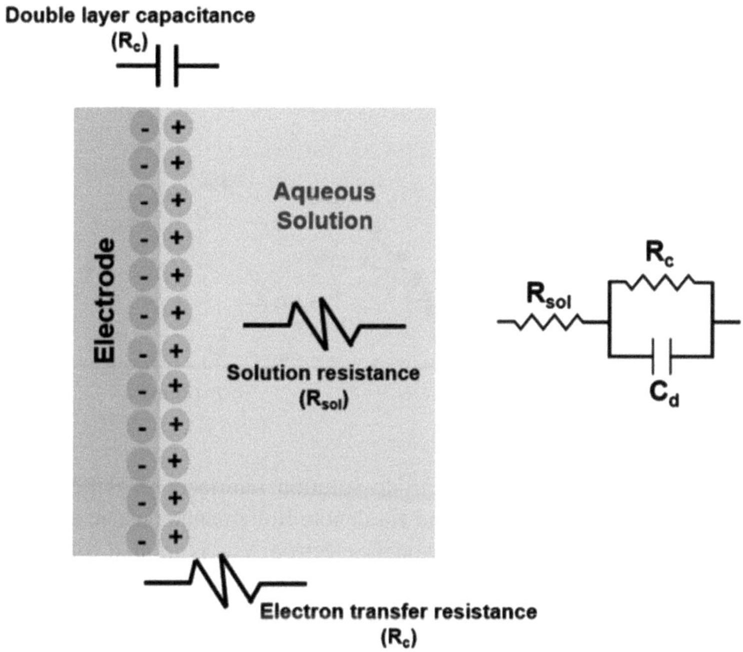

Fig. 6 EIS circuit to set out the redox reactions at the surface of the electrode in electrochemical cells

to obtain the charge transfer resistance in the catalytic processes using the Nyquist plot. Nanomaterial catalyst with a lower radius of the Nyquist plot has low charge transfer resistance and vice versa.

From Fig. 7, the Nyquist plots of ZnO, g-C_3N_4 and ZnO/g-C_3N_4 reveal that the charge transfer resistance of the ZnO is higher than g-C_3N_4. A significant reduction in the Nyquist plot radius of g-C_3N_4/ZnO confirmed that the heterojunction between g-C_3N_4 and ZnO improved the transfer of charge carriers through the heterojunction–water interface.

Mott-Schottky analysis measures the space-charge region capacitance near the electrode–water interface with respect to the applied potential. This was performed to obtain the charge carrier density and flat band potential of the catalyst. From the Mott-Schottky equation (Eq. 8), $1/C^2$ versus V_{app} slope revealed the density of charge carriers, and x-axis intercept gives the flat band potential. For the slope, the Mott-Schottky plot slope is positive and negative for N-type and P-type semiconductors, respectively.

$$\frac{1}{c^2} = \left(\frac{2}{q^{\varepsilon}0^{\varepsilon}_{s}A^2N_d}\right)\left(V_{app} - V_{fb} - \frac{K_BT}{q}\right) \tag{8}$$

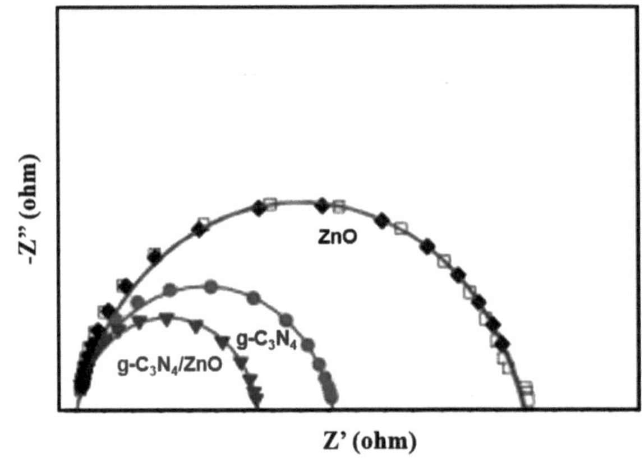

Fig. 7 Nyquist plot of the ZnO, $g\text{-}C_3N_4$ and $g\text{-}C_3N_4/ZnO$

where C is capacitance at the electrode–solution interface; ε_0 is permittivity of vacuum (8.85×10^{-12} F m^{-1}); ε_s and K_B denote Boltzmann constant and dielectric constant, respectively; q is the charge on the electron; V_{app} is applied voltage; density of charge carriers is denoted by N_d (cm^{-3}); and catalyst area at interface is denoted by A (m^2). Mott-Schottky analysis of the ZnO confirmed the N-type semiconductor photocatalyst (Gelderman et al., 2007). The flat band potential of the ZnO was −0.31 V ± 0.033 V vs. SCE in 1 M KCl, and the ND was 2×10^{24} m^{-3} (Gelderman et al., 2007).

6 Surface Characterisations of Nanomaterials

6.1 Atomic Force Microscopy (AFM)

AFM tool acquires high-resolution surface images at an atomic scale. Topographical images, surface roughness and height variations are acquired using AFM. Images are obtained through interaction between a sharp tip made of silicon/silicon carbide and the surface of the sample. Topographical images are formed through the movement of the tip across the surface of the sample. AFM can image nanoparticles and nanostructures with high resolution. It can reveal their size, shape and distribution on surfaces. AFM has been used to study the assembly and aggregation of nanoparticles, as well as their interactions with other molecules. Piezoelectric force microscopy (PFM) is an available technique within AFM to obtain the piezoelectric response of the nanomaterial. The tip is coated with the conducting material to obtain the piezoelectric response of the nanomaterial. For a piezoelectric material, the direction of polarisation was obtained using a phase contrast image, which is useful to understand the domain structure and domain walls in the piezoelectric material. On the other hand, the amplitude contrast image revealed the piezoelectric response at each point on the surface, that is, the magnitude of the piezoelectric response.

Piezoelectric ZnO nanorods are analysed using AFM-PFM analysis (Ma et al., 2024). From the AFM image, the height of nanorods (~100 nm) is clearly identified. Butterfly loop is obtained through PFM analysis, and piezoelectric coefficient calculated is ~18 pm V^{-1}. Similarly, the piezoelectric coefficient of different piezocatalysts, such as $NaNbO_3$, $BaTiO_3$, g-C_3N_4, MoS2, etc. is obtained using the PFM analysis.

6.2 Zeta Potential

In solution, nanoparticles with a specific charge (positive or negative) strongly bound the ions with the opposite charge close to the surface, forming the stern layer. Following the stern layer, a diffuse charge layer is formed, which consists of loosely bound charges. These layers are collectively called the electric double layer, and the net electrostatic potential at the interface of the double layer and bulk solution is called the zeta potential of the nanoparticle (Fig. 8). The zeta potential of a nanoparticle is calculated using the Henry equation (Eq. 9) (McNeil, 2011).

$$U_e = \frac{2\varepsilon Zf(\mathrm{K}a)}{3\eta} \tag{9}$$

where electrophoretic mobility is denoted by U_e, dielectric constant is denoted by ε, h is the absolute zero-shear viscosity of the medium, Henry function is denoted by f(ka) and particle radius to the Debye length ratio is denoted by *κa*.

The zeta potential of nanoparticles depends on their ionic strength, pH, particle size and shape, and temperature (McNeil, 2011). It was found that an increase in the

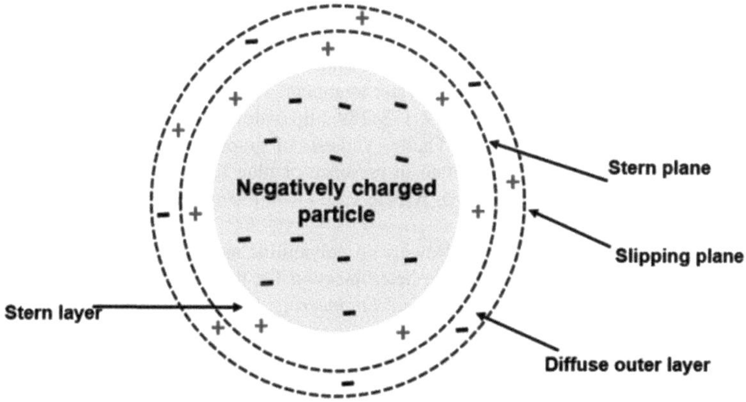

Fig. 8 Distribution of charges near the surface of negatively charged nanoparticles

ionic strength of an aqueous solution increased screening of the surface and reduced the thickness of the double layer, ultimately affecting the zeta potential. Temperature affects the viscosity of the solution, which affects the mobility of the ions responsible for zeta potential.

Nanomaterial adsorbent's zeta potential and catalyst play a major role in determining interaction with the target pollutants. Zeta potential of N-doped porous carbon (NCP) prepared by using hemp stems has an isoelectric point (zeta potential is zero) at pH 7. Therefore, the zeta potential is positive at pH < 7, and the zeta potential is negative at pH > 7 (Tang et al., 2024). Adsorption of the tetracycline (TC) on the NCP depends on the pH, as at a lower pH, the NCP is positively charged, and TC is present in the neutral form (TC^0) and the positive form (TC +), which results in low electrostatic attraction between NCP and TC (Tang et al., 2024). Overall, the zeta potential of nanomaterials has played a vital role in evaluating the efficiency of processes such as adsorption and catalysis through interaction with targeted pollutants.

7 Conclusion

In summary, the characterisation of nanomaterials is crucial for understanding their full capacity and improving the efficiency of the treatment of wastewater. By an in-depth analysis of the structural, chemical and physical characteristics of these materials, scientists can gain insights into the distinct behaviours and features of these materials at a nanoscale. These discoveries have been made possible by the application of sophisticated characterisation techniques like spectroscopy and electron microscopy. Effective characterisation enables the precise tailoring of nanomaterials for the treatment of wastewater.

References

Bethi, B., Sonawane, S. H., Bhanvase, B. A., & Gumfekar, S. P. (2016). Nanomaterials-based advanced oxidation processes for wastewater treatment: A review. *Chemical Engineering and Processing-Process Intensification, 109*, 178–189. https://doi.org/10.1016/j.cep.2016.08.016

Ekande, O. S., & Kumar, M. (2021a). Facile synthesis of graphitic carbon nitride from acetic acid pretreatment to activate persulfate in presence of blue light for photocatalytic removal of metronidazole. *Chemosphere, 276*, Article 130171. https://doi.org/10.1016/j.chemosphere.2021.130171

Ekande, O. S., & Kumar, M. (2021b). Review on polyaniline as reductive photocatalyst for the construction of the visible light active heterojunction for the generation of reactive oxygen species. *Journal of Environmental Chemical Engineering, 9*(4), Article 105725. https://doi.org/10.1016/j.jece.2021.105725

Ekande, O. S., & Kumar, M. (2023). New insight on interfacial charge transfer at graphitic carbon nitride/sodium niobate heterojunction under piezoelectric effect for the generation of reactive oxygen species. *Journal of Colloid and Interface Science, 651*, 477–493. https://doi.org/10.1016/j.jcis.2023.07.189

Eley, C., Li, T., Liao, F., Fairclough, S. M., Smith, J. M., Smith, G., & Tsang, S. C. E. (2014). Nanojunction-mediated photocatalytic enhancement in heterostructured CdS/ZnO, CdSe/ZnO, and CdTe/ZnO nanocrystals. *Angewandte Chemie International Edition, 53*(30), 7838–7842. https://doi.org/10.1002/anie.201404481

Elgrishi, N., Rountree, K. J., McCarthy, B. D., Rountree, E. S., Eisenhart, T. T., & Dempsey, J. L. (2018). A practical beginner's guide to cyclic voltammetry. *Journal of Chemical Education, 95*(2), 197–206. https://doi.org/10.1021/acs.jchemed.7b00361

Feng, X., Lv, B., Lu, L., Feng, X., Wang, H., Xu, B., Yang, Y., & Zhang, F. (2021). Role of surface oxygen vacancies in zinc oxide/graphitic carbon nitride composite for adjusting energy band structure to promote visible-light-driven photocatalytic activity. *Applied Surface Science, 562*, Article 150106. https://doi.org/10.1016/j.apsusc.2021.150106

Gelderman, K., Lee, L., & Donne, S. W. (2007). Flat-band potential of a semiconductor: Using the Mott-Schottky equation. *Journal of Chemical Education, 84*(4), 685. https://doi.org/10.1021/ed084p685

He, S., Mang, E. H., Leff, A. C., Zhou, X., Taheri, M. L., & Marian, J. (2024). Complex dislocation loop networks as natural extensions of the sink efficiency of saturated grain boundaries in irradiated metals. *Science Advances, 10*(18), eadj8395. https://doi.org/10.1126/sciadv.adj8395

Jiang, S., Chen, Y., Cui, X., Sun, Y., Ma, G., Bao, Y., Yao, Y., & Ma, T. (2024). Constructing highly efficient ZnO nanocatalysts with exposed extraordinary (110) facet for CO_2 electroreduction. *ACS Applied Materials & Interfaces, 16*(12), 14965–14973. https://doi.org/10.1021/acsami.4c01797

Kumar, A., Luder, J., Bertotti, M., Araki, K., Montoro, F., Herbst, F., & Gonçalves, J. M. (2024). Deciphering the atomic scale electrocatalytic sites in hierarchically porous gold nanostructures: Implications for ascorbic acid electrooxidation. *ACS Applied Nano Materials, 7*(7), 7213–7225. https://doi.org/10.1021/acsanm.3c06226

Ma, J., Xia, L., Wu, D., Feng, Y., Ban, C., Ruan, L., Guan, J., Zhang, M., Zhou, R., Dai, J., Gan, L., & Zhou, X. (2024). High-efficiency CO_2 conversion via mechano-driven dynamic strain engineering of ZnO nanostructures. *Nano Energy, 121*, Article 109258. https://doi.org/10.1016/j.nanoen.2024.109258

Magar, H. S., Hassan, R. Y., & Mulchandani, A. (2021). Electrochemical impedance spectroscopy (EIS): Principles, construction, and biosensing applications. *Sensors, 21*(19), 6578. https://doi.org/10.3390/s21196578

Mahdi, Z., El Hanandeh, A., & Yu, Q. J. (2019). Preparation, characterization and application of surface modified biochar from date seed for improved lead, copper, and nickel removal from aqueous solutions. *Journal of Environmental Chemical Engineering, 7*(5), Article 103379. https://doi.org/10.1016/j.jece.2019.103379

McNeil, S. E. (Ed.) (2011). Characterization of nanoparticles intended for drug delivery (Vol. 697, pp. 71–82). New York, NY: Humana press. https://doi.org/10.1007/978-1-60327-198-1

Miklos, D. B., Remy, C., Jekel, M., Linden, K. G., Drewes, J. E., & Hübner, U. (2018). Evaluation of advanced oxidation processes for water and wastewater treatment–A critical review. *Water Research, 139*, 118–131. https://doi.org/10.1016/j.watres.2018.03.042

Mohite, S. V., Kim, S., Bae, J., Jeong, H. J., Kim, T. W., Choi, J., & Kim, Y. (2024). Defects healing of the ZnO surface by filling with Au atom catalysts for efficient photocatalytic H_2 production. *Small, 20*(4), 2304393. https://doi.org/10.1002/smll.202470034; https://doi.org/10.1002/smll.202304393

Preethi, S., Anitha, A., & Arulmozhi, M. (2016). A comparative analysis of the properties of zinc oxide (ZnO) nanoparticles synthesized by hydrothermal and sol-gel methods. *Indian Journal of Science and Technology, 9*(40), 1–6. https://doi.org/10.17485/ijst/2016/v9i40/92696

Qamar, M. T., Aslam, M., Ismail, I. M., Salah, N., & Hameed, A. (2016). The assessment of the photocatalytic activity of magnetically retrievable ZnO coated γ-Fe2O3 in sunlight exposure. *Chemical Engineering Journal, 283*, 656–667. https://doi.org/10.1016/j.cej.2015.08.002

Rajakumaran, R., Boddu, V., Kumar, M., Shalaby, M. S., Abdallah, H., & Chetty, R. (2019). Effect of ZnO morphology on GO-ZnO modified polyamide reverse osmosis membranes for desalination. *Desalination, 467*, 245–256. https://doi.org/10.1016/j.desal.2019.06.018

Rajakumaran, R., Kumar, M., & Chetty, R. (2020). Morphological effect of ZnO nanostructures on desalination performance and antibacterial activity of thin-film nanocomposite (TFN) membrane. *Desalination,* 495, 114673. https://doi.org/10.1016/j.desal.2020.114673

Robertson, A. W., & Warner, J. H. (2013). Atomic resolution imaging of graphene by transmission electron microscopy. *Nanoscale, 5*(10), 4079–4093. https://doi.org/10.1039/c3nr00934c

Sahai, A., & Goswami, N. (2014). Probing the dominance of interstitial oxygen defects in ZnO nanoparticles through structural and optical characterizations. *Ceramics International, 40*(9), 14569–14578. https://doi.org/10.1016/j.ceramint.2014.06.041

Tan, J., Li, Z., Li, J., Meng, Y., Yao, X., Wang, Y., Lu, Y., & Zhang, T. (2022). Visible-light-assisted peroxymonosulfate activation by metal-free bifunctional oxygen-doped graphitic carbon nitride for enhanced degradation of imidacloprid: Role of non-photochemical and photocatalytic activation pathway. *Journal of Hazardous Materials, 423,* 127048. https://doi.org/10.1016/j.jhazmat.2021.127048

Tang, C., Ni, Z., Xu, C., Luo, Y., Cai, X., Gao, Q., Fang, Y., Zhong, G., Qui, R., & Zhang, S. (2024). Enhanced adsorption of organic pollutants using N-doped porous carbon derived from hemp stems: Insights into the mechanism. *Separation and Purification Technology, 333*, Article 125878. https://doi.org/10.1016/j.seppur.2023.125878

Upadhaya, D., Kumar, P., & Dhar Purkayastha, D. (2019). Superhydrophobic ZnO/TiO2 heterostructure with significantly enhanced photocatalytic activity. *Journal of Materials Science: Materials in Electronics, 30*(11), 10399–10407. https://doi.org/10.1007/s10854-019-01381-2

Wang, P., Ma, X., Hao, X., Tang, B., Abudula, A., & Guan, G. (2024). Oxygen vacancy defect engineering to promote catalytic activity toward the oxidation of VOCs: A critical review. *Catalysis Reviews, 66*(2), 586–639. https://doi.org/10.1080/01614940.2022.2078555

Xu, L., Hu, Y. L., Pelligra, C., Chen, C. H., Jin, L., Huang, H., Sithambaram, S., Aindow, M., Joesten, R., & Suib, S. L. (2009). ZnO with different morphologies synthesized by solvothermal methods for enhanced photocatalytic activity. *Chemistry of Materials, 21*(13), 2875–2885. https://doi.org/10.1021/cm900608d

Zhang, J., Zheng, Z., Xu, L., Xie, H., Fei, Z., Dyson, P. J., & Yan, N. (2024a). Light-and electro-driven phase change materials derived from activated porous biochar nanosheets and encapsulated polyethylene glycol. *Colloids and Surfaces a: Physicochemical and Engineering Aspects, 690*, Article 133783. https://doi.org/10.1016/j.colsurfa.2024.133783

Zhang, K. N., Lei, Z. L., & Lü, Q. F. (2024b). Flower-like ZnO modified TiO2@ Ti_3C_2Tx composite for efficient adsorption of malachite green and oil-water separation. *Journal of Cleaner Production, 447*, Article 141632. https://doi.org/10.1016/j.jclepro.2024.141632

Development of Polyaniline-Based Calcium Ferrite Nanocomposites for the Treatment of Eriochrome Black T Dye Laden Wastewater

Bibek Saha, Sabin Shaji, and Animesh Debnath

1 Introduction

Water pollution greatly affects human health and is one of the most pervasive environmental threats on a global scale (Zhu et al., 2020). Among the different types of water contaminants, synthetic dye is the most dangerous because of its cancerous and mutagenic properties, which seriously jeopardises human and marine life (Sonal & Mishra, 2021). Dyes are generally highly organic compounds with complex aromatic structures and are not biodegradable (Hou et al., 2019). Most of these dyes are released into the effluent water from textile, food, leather, paper and pulp, rubber, plastic and cosmetic industries, etc. (Mosleh et al., 2016). Even a low concentration of dye (<1.0 ppm) in the water system can block sunlight from entering the water, thereby interfering with the ability of aquatic plants to photosynthesise. It also degrades the amount of dissolved oxygen present in the water system, thereby generating an anaerobic condition in the water system, ultimately reducing the re-oxygenation capacity (Khurana et al., 2017). Azo dye is the most abundant category of dye among the several categories of synthetic dyes used by 70% of the dyeing industries worldwide (De et al., 2022). More than one azo bond (-N = N-) is present in the molecular structure of most of the azo dyes, which function as chromophores (Moeinpour et al., 2014). Eriochrome Black-T (EBT) dye is the most crucial azo dye widely used in several dyeing industries. EBT dye, if present even in a trace amount, will be hard to degrade because of its extreme resistance to heat, water, light and microbial and chemical attacks (Bansal et al., 2020). Gene mutations, cancer and allergic dermatitis are among the negative health impacts of the EBT dye (Akhouairi et al., 2019). Consequently, the effluent from these dye-contaminated firms needs

B. Saha · S. Shaji · A. Debnath (✉)
Department of Civil Engineering, National Institute of Technology Agartala, Jirania, Tripura (W), India
e-mail: debnathanimesh@gmail.com

© Capital Publishing Company, New Dehli, India. 2025
R. Neelancherry and S. M. Kumar, *Nanotechnology in Wastewater Treatment*,
https://doi.org/10.1007/978-3-031-94246-4_4

to be thoroughly treated before being released into the water courses (Modwi et al., 2023).

Currently, to separate dyes from the effluent water, numerous treatment techniques have been investigated and put into practice among which the most widely used techniques are membrane filtration (Liu et al., 2022), coagulation-precipitation (Sonal & Mishra, 2021), advanced oxidation (Zhu et al., 2020), photocatalysis (Mosleh et al., 2016) and adsorption (Modwi et al., 2023). Nevertheless, except the adsorption process, there are a number of drawbacks to these techniques, including insufficient ion removal, excessive energy consumption and the creation of waste products and toxic sludge that require extra treatment before discarding (Agarwal et al., 2016; Tosun Satir et al., 2022). Therefore, the adsorption method is primarily used to separate dyes from the aqua matrix because it is easier to operate, more effective at removing several pollutants and ions and produces sludge that is less harmful overall (Li et al., 2017). For the introduction of the adsorption technology for the separation of dyes from the effluent system, a suitable adsorbent is required (Agarwal et al., 2016). As nanoparticles have a higher specific surface area than standard adsorbents and can increase adsorption efficiency, several types of nano-sized materials have been used as adsorbents recently (Salem et al., 2016). Because magnetic nanoparticles can be easily isolated from treated wastewater using external magnets, researchers have greatly used magnetic nanoparticles as adsorbents (Shahabuddin et al., 2016).

To remove dyes, over the past couple of decades, a variety of nanoscale magnetic metal ferrites have been studied ($CoFe_2O_4$, $CaFe_2O_4$, etc.) (Li et al., 2017; Mukhopadhyay et al., 2021). Extensive surface area, high reactivity, substantial mechanical durability and capacity for regeneration are notable properties of metal ferrite nanoparticles (Mukhopadhyay et al., 2021; Syed et al., 2021). In addition to these advantages, metal ferrite nanoparticles also have a reasonable cost of manufacturing, simplicity in dissociation via the application of an external magnetic field and a reasonably high capacity for adsorption (Das & Debnath, 2021; Deb et al., 2017). According to a few research papers, $CaFe_2O_4$ has better biocompatibility than other metal ferrites (Beyki et al., 2016; Das & Debnath, 2021; Nasiri et al., 2022). Because $CaFe_2O_4$ has a lower threshold of magnetisation, it inhibits binding of particles, which eventually speeds up the diffusion process of aqueous solutions (Seidi & Hedayati, 2020). The adsorbent–adsorbate association is enhanced as a result of this event, which shortens the equilibrium time (Debnath et al., 2016).

On the other hand, the efficiency of metal ferrites' dye uptake is strongly influenced by the pH of the solution. Strongly basic or strongly acidic conditions are more favourable for the adsorption process utilising metal ferrite, in accordance with the kind of dye being employed (Bhowmik et al., 2020). If a very high or very low pH is maintained during the course of wastewater treatment in order to achieve better dye elimination, it could make the procedure more expensive and unfit for secure disposal (Bhowmik et al., 2018). In order to get through this problem, many researchers use a combination of metal ferrite and conducting polymers, which could effectively remove dye from solutions in the pH range that is close to neutral (Nasar et al., 2019). Because of its simple manufacturing process, exceptional redox characteristics, low cost of the monomer, improved durability in the environment and high degree of protonation, polyaniline (PANI), one of the widely used conducting polymers, has been produced in large quantities by combining alongside metal ferrites (Gabal et al., 2019; Mandal et al., 2015). PANI itself has been widely utilised as an adsorbent for eliminating noxious dyes, heavy metals, hazardous medicinal products and other pollutants from discharged industrial wastewater (Tanzifi et al., 2017). However, PANI alone is not a good enough material to employ as an adsorbent because of its weak mechanical strength (Bhaumik et al., 2015; Saad et al., 2017). The problem can be resolved by combining PANI with metal ferrites (Gabal et al., 2019; Mohammadi Nodeh et al., 2018).

As far as we are aware, no research has been done on the development and utilisation of nanocomposite materials made of PANI and $CaFe_2O_4$ for the purpose of removing the EBT dye. Thus, by placing one nanoparticle over the surface of the other, the present investigation aims to enhance the physiochemical characteristics of PANI and $CaFe_2O_4$ nanoparticles (Das et al., 2022; Kharazi et al., 2019). The main aim of this study is to ascertain the ability of a newly fabricated calcium ferrite/polyaniline ($CaFe_2O_4$-PANI) nanocomposite to adsorb the EBT dye from an aqueous medium. The other aims of this study are reducing equilibrium duration and achieving maximal uptake capacity close to neutral pH.

2 Experimental and Analytical Methodology

2.1 *Reagents Used*

Merck (India) has supplied NaOH pellets and 97% pure HCl solution from which 0.1 M NaOH and 0.1 N HCl solutions were made, which were subsequently utilised to adjust the pH of all the testing solutions. Merck (India) also provided analytical grade of calcium chloride dihydrate ($CaCl_2.2H_2O$) (99.5%), ferric chloride anhydrous ($FeCl_3$) (>96%), ammonium peroxydisulphate (APS) ($(NH_4)_2S_2O_8$) (>98%) and aniline monomer ($C_6H_5NH_2$) (>99.5%). Merck (India) has provided > 99% pure EBT dye powder, which was employed without any additional processing. A 500 mg/L stock solution of the EBT dye was produced by accurately measuring out a certain amount of the dye and mixing it in distilled water. Lower concentrations of the necessary EBT dye solutions were produced by progressive dilution of the stock solution using distilled water.

2.2 *Synthesis of $CaFe_2O_4$ Nanoparticles*

$CaFe_2O_4$ nanoparticles were produced using a straightforward chemical precipitation method, in which 30.0 g of NaOH pellets were immersed in 750 mL of distilled water, followed by swirling of the mixture till it became transparent. Subsequently, in two

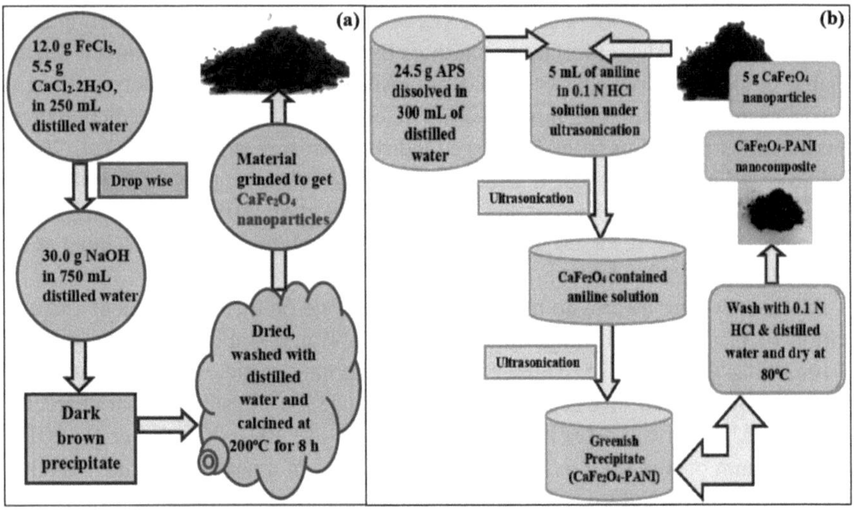

Fig. 1 Diagrammatic presentation illustrating the steps involved in creating **a** $CaFe_2O_4$ nanoparticles and **b** a $CaFe_2O_4$-PANI nanocomposite

distinct beakers, 12.0 g of $FeCl_3$ and 5.5 g of $CaCl_2.2H_2O$ were mixed in 250 mL of distilled water, kept in each beaker and stirred till the liquids became clear. Drop by drop, both metal salt solutions were added to the NaOH solution simultaneously, and the mixture was agitated until a brownish-coloured precipitate was visible. To get rid of any remaining impurities, the brown-coloured precipitate underwent filtering followed by extensive cleaning with distilled water. Ultimately, the rinsed residue had been dried for 9 h at 80 °C in a hot air oven, and then it was calcined for 8 h at 200 °C in a muffle furnace (Vadivel et al., 2016). The brown-coloured $CaFe_2O_4$ composite solid particles were the end product, and they were crushed down even more to create $CaFe_2O_4$ nanoparticles (Debnath et al., 2016). The synthesis process of $CaFe_2O_4$ nanoparticles has been demonstrated in Fig. 1a.

2.3 *Development of the $CaFe_2O_4$-PANI Nanocomposite*

The $CaFe_2O_4$-PANI nanocomposite was developed in two stages: a simple chemical precipitation procedure and an oxidation-based approach of polymerising aniline monomer into PANI. Maintaining the volume to mass ratio of the aniline monomer and $CaFe_2O_4$ at 1:1, the generated calcium ferrite nanoparticle was added to the aniline solution. After that, 0.1 N HCl (150 mL) solution was incorporated to the mixture, and the mixture was sonicated for less than 30 min. The following stage involved combining 24.5 g of APS with 300 mL of distilled water to create the APS solution. Then, drop by drop, it was introduced into the aniline and $CaFe_2O_4$ mixture while being kept under ultrasonication to ensure effective polymerisation by oxidation and the formation of the finished product, or the $CaFe_2O_4$-PANI nanocomposite. After being left uninterrupted for approximately 72 h, for proper precipitation, the resultant precipitate was purified, and any aniline monomer that had not reacted was eliminated by rinsing the substance in 0.1 N HCl solution. The final product was then gently washed using distilled water to reach nearly neutral pH. After drying at 80 °C for 12 h, the composite was effectively blended to form the $CaFe_2O_4$-PANI nanocomposite. The fabrication process of the $CaFe_2O_4$-PANI nanocomposite is shown in Fig. 1b.

2.4 Batch Adsorption Experiment of the EBT Dye

In this investigation, an overhead contact device operating at 350 rpm was used in all of the adsorption experiments. At a particular solution pH level, the predetermined dose of $CaFe_2O_4$-PANI had been added into an EBT dye-containing solution with a given starting EBT dye concentration, and the mixture was swirled for varied amounts of time using the contact device. In light of $CaFe_2O_4$-PANI's magnetic characteristics, the EBT dye was separated from the solution of $CaFe_2O_4$-PANI saturated with EBT dye, using a magnet following the completion of the uptake process. The final concentration of the dye solution at $l_{max} = 540$ nm was then measured using the leftover solution in a UV–VIS spectrophotometer (Thermo Scientific, Orion Aquamate-8000) (De et al., 2022). The EBT dye extraction efficacy and uptake capacity as denoted by R (%) and Q_e (mg/g), respectively, were calculated using Eqs. 1 and 2 (Bhowmik et al., 2020).

$$R(\%) = \frac{\text{Initial } EBT \text{ dye conc.(mg/L)} - \text{ Equilibrium } EBT \text{ dye conc.(mg/L)}}{\text{Initial } EBT \text{ dye conc.(mg/L)}} \times 100 \tag{1}$$

$$Q_e(mg/g) = \frac{Initial\ EBT\ dye\ conc.\ (mg/L) - \ Equilibrium\ EBT\ dye\ conc.\ (mg/L)}{CaFe_2O_4 - \ PANI\ dose\ (g/L)} \tag{2}$$

2.5 Physiochemical Analysis of the $CaFe_2O_4$-PANI Nanocomposite

Using a Bruker D-8 Advance X-ray diffractometer, the crystalline characteristics of the produced $CaFe_2O_4$-PANI nanocomposite were assessed by the X-ray diffraction (XRD) technique. The excellent quality images produced by the JOEL JSM-7600F field emission gun-scanning electron microscope (FEG-SEM) device were used to acquire the data on shape and size of the nanocomposite. The Vertex 80 Fourier transform infrared (FTIR) spectroscopic analysis from Bruker, Germany, was used to identify the operational surface bonds that existed on the adsorbent's surface. The $CaFe_2O_4$-PANI nanocomposite's surface area and pattern of pore sizes were investigated, and data on the N_2 adsorption–desorption isotherm were generated with the use of a surface area analyser device (Quanta Chrome Nova, 1000 series).

2.6 *Experimental Design Based on the Central Composite Design*

RSM is an effective method for maximising or minimising the outcome in any chemical process involving numerous input variables. The most popular design method in RSM, that is, central composite design (CCD), was employed with three variables associated with this study ($CaFe_2O_4$-PANI dose, initial EBT dye concentration and contact period) as process variables to optimise the outcome (EBT dye removal %). Table 1 shows that variables associated with the experiments are encoded at five different levels such as −2, −1, 0, +1 and +2. Twenty test runs in all have been planned using the Design Expert V.11 software to determine the mathematical relationship between the factors under investigation and the EBT dye separation efficacy. The extended quadratic equation that the program suggested for this adsorption research is shown in Eq. 3 (Bhowmik et al., 2018):

$$Y = \beta_0 + \sum_{j=1}^{k} \beta_j X_j + \sum_{j=1}^{k} \beta_{jj} X_j^2 + \sum_{j=1}^{k} \sum_{j=1}^{k} \beta_{ji} X_j X_i \tag{3}$$

where the projected EBT dye extraction efficacy is denoted by Y; X_1,..........X_k are the experimental parameters (in this study, $k = 3$); β_0 represents a constant or zero-order coefficient; and β_j, β_{jj} and β_{ji} present coefficients of the linear, quadratic and interactive terms, respectively.

3 Results and Discussions

3.1 *Characterisation Studies*

Figures 2a and 2b show the XRD patterns of the $CaFe_2O_4$-PANI nanocomposite and $CaFe_2O_4$ nanoparticles, respectively. The miller planes at (310), (201), (221), (400), (340), (331) and (431), respectively, are associated with the diffraction peaks of $CaFe_2O_4$ nanoparticles shown (Fig. 2a) at 2θ values of 29.29°, 35.46°, 39.32°,

Table 1 The adsorption test conditions and their levels undertaken in the CCD optimisation process based on RSM with actual and expected elimination of EBT dye (%) for 20 test trials

Parameters	Levels				
	−2	−1	Central (0)	+1	+2
X_1:$CaFe_2O_4$-PANI dose(g/L)	0.5	0.75	1.0	1.25	1.5
X_2:Initial EBT dye conc	10	30	50	70	90
X_3: Contact time (min)	60	90	120	150	180
Run	***Factors***			***EBT dye removal (%)***	
	X_1	X_2	X_3	***Experimental***	***Predicted***
1	0.75	30	150	82.03	82.79
2	1.5	50	120	93.59	91.47
3	1.25	70	150	89.64	90.27
4	1.0	90	120	78.29	76.90
5	10	50	120	92.02	91.14
6	0.75	70	90	65.69	70.64
7	1.0	50	120	90.68	91.14
8	1.0	50	120	91.58	91.14
9	1.25	30	90	94.48	95.15
10	0.5	50	120	67.19	61.38
11	1.25	30	150	97.86	99.55
12	1.0	50	120	91.93	91.14
13	1.0	50	60	85.55	80.52
14	1.25	70	90	83.30	83.98
15	1.0	10	120	99.46	94.12
16	1.0	50	120	91.21	91.14
17	1.0	50	180	93.52	91.02
18	0.75	30	90	73.79	78.59
19	0.75	70	150	75.98	76.74
20	1.0	50	120	90.88	91.14

43.07°, 47.36°, 48.39° and 56.44°, respectively (Debnath et al., 2016). The diffraction peaks developed appear to be rather comparable to the crystalline structure of $CaFe_2O_4$ nanoparticles developed by Debnath et al. (2016). The (200), (020) and (011) dispersion planes in the XRD pattern of the $CaFe_2O_4$-PANI nanocomposite are represented by the diffraction peaks, which have been detected at 2θ values of 25.59°, 19.41 and 14.77°, respectively (Tanzifi et al., 2017), because of the presence of PANI's emeraldine salt state, as the surface of $CaFe_2O_4$ nanoparticles are almost covered by PANI nanoparticles in the developed nanocomposite. The presence of PANI crystalline planes in the composite is what causes these peaks (Saad et al., 2017).

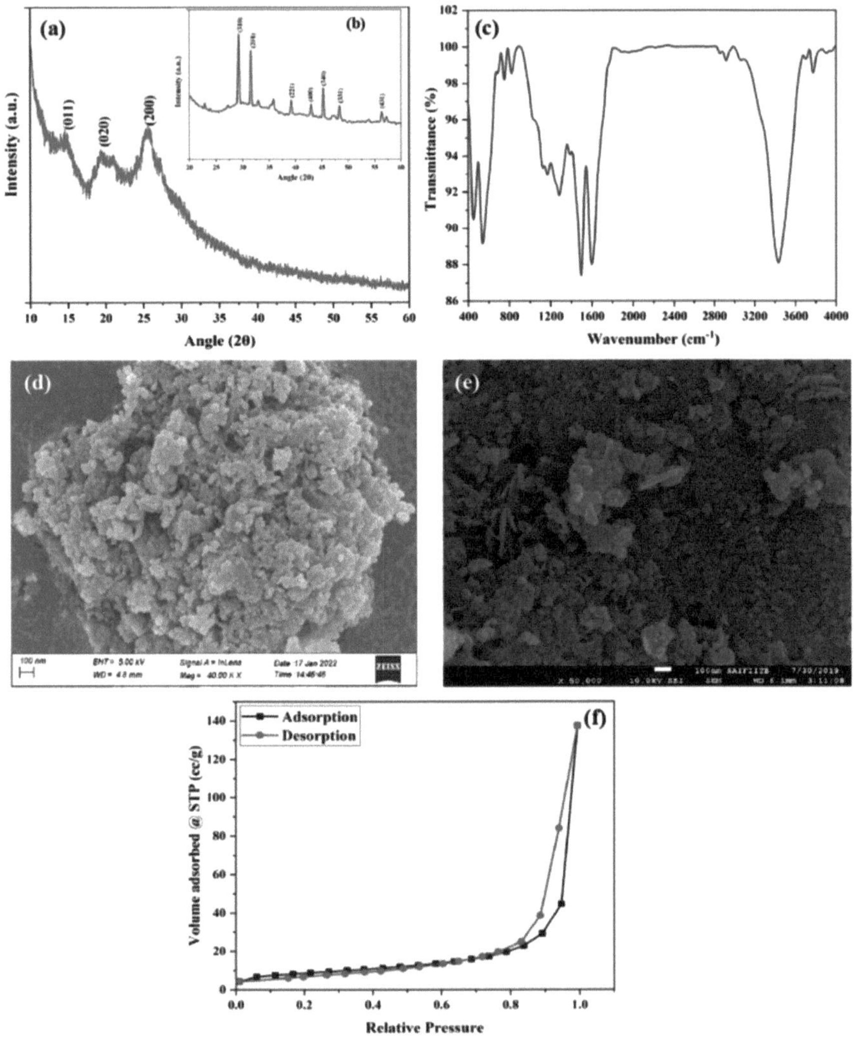

Fig. 2 X-ray diffraction pattern of **a** $CaFe_2O_4$ nanoparticles, **b** $CaFe_2O_4$-PANI nanocomposite, **c** FTIR spectroscopy of $CaFe_2O_4$-PANI nanocomposite, **d**, **e** FEGSEM image of $CaFe_2O_4$-PANI nanocomposite and $CaFe_2O_4$ nanoparticles and **f** N_2 adsorption–desorption isotherm plot of the $CaFe_2O_4$-PANI nanocomposite

$CaFe_2O_4$-PANI's active functional groups were investigated using FTIR spectroscopy. The resulting FTIR spectrum is displayed in Fig. 2c, and it revealed multiple significant peaks that corresponded to the $CaFe_2O_4$ and PANI nanoparticles. Nitrogen quinoid ring structure of PANI and N–H stretching are shown by the bands at 1497 cm^{-1} and 1601 cm^{-1}, respectively (Gabal et al., 2019). In addition, PANI exhibits two notable peaks at 1286 $cm^{-}1$ and 1385 cm^{-1}, which are brought on by the benzenoid and quinoid units' C = C and C = N stretching, respectively (Saad et al., 2017). Moreover, the 1125 cm^{-1} peak was identified as the bending vibration of the benzene ring available in the molecular structure of PANI (Mohammadi Nodeh et al., 2018). O–H bond stretching vibration brought on by water molecules adhering to the samples is responsible for the peak at 3432 cm^{-1} (Bhowmik et al., 2020). The peak values at 412 cm^{-1} and 533 cm^{-1}, respectively, reflect the vibrations caused by stretching of the octahedral groups and the tetrahedral groups of metal oxide surfaces (Debnath et al., 2016).

The shape and structure of the produced $CaFe_2O_4$-PANI nanocomposite and $CaFe_2O_4$ nanoparticles are revealed by the FEGSEM picture, as illustrated by Figs. 2d and e. It is evident that the nanocomposite is composed of aggregated, spherically shaped, non-uniformly sized particles in the nanoscale dimension (Mukhopadhyay et al., 2021). The FEGSEM image also showed several micro-dimension particles, which are the result of magnetic $CaFe_2O_4$ nanoparticles clumping together (Bhowmik et al., 2019). The N_2 adsorption–desorption isotherm (Fig. 2f) plot depicts a type IV isotherm curvature, which is consistent with the composite's mesoporous structure (Gabal et al., 2019). Utilising the Brunauer–Emmett–Teller (BET) method, the specific surface area of $CaFe_2O_4$-PANI nanocomposites has been estimated as 30.086 m^2/g. Using the Barrett–Joyner–Halenda (BJH) method, it was discovered that the $CaFe_2O_4$-PANI nanocomposite had a collective pore volume of 0.217 cc/g.

3.2 *Influence of Solution pH and the $CaFe_2O_4$-PANI Nanocomposite Dose*

Considering that variations in solution pH can affect the amount of accessible charges on the surface of adsorbents and the extent of dye ionisation, solution pH is an essential testing condition in every dye adsorption study (Ivanets et al., 2018). The EBT dye adsorption investigation with regard to the $CaFe_2O_4$–PANI nanocomposite was carried out within the wide pH limit of 2.0–10.0. Figure 3a makes it evident that, with a 1.0 g/L of $CaFe_2O_4$-PANI nanocomposite adsorbent dosage, a maximum EBT elimination rate of 62.87% was accomplished at a solution pH of 3.0 after 30 min of contact. The removal effectiveness declined as the pH of the solution increased, and by the moment it reached a solution pH value of 7.0, it had plummeted to approximately 45%. The ideal pH was chosen as 5.0 for additional research at which approximately 54% of EBT dye removal efficiency was observed, as sustaining an acidic pH in large-scale industrial operations is challenging and costly, and it is potentially challenging to dispose the acidic byproducts (Jamwal et al., 2016).

To assess the impact of the $CaFe_2O_4$-PANI nanocomposite's dosage on the effectiveness of dye elimination, 50 mg/L of initial EBT dye concentration was chosen, and the contact time was kept fixed at 60 min. The pH value of the solution was maintained at 5.0. The dye elimination effectiveness increased from 36.74% to 84.38%, within the $CaFe_2O_4$-PANI dose range of 0.25 g/L to 1.5 g/L, as shown in Fig. 3b. The

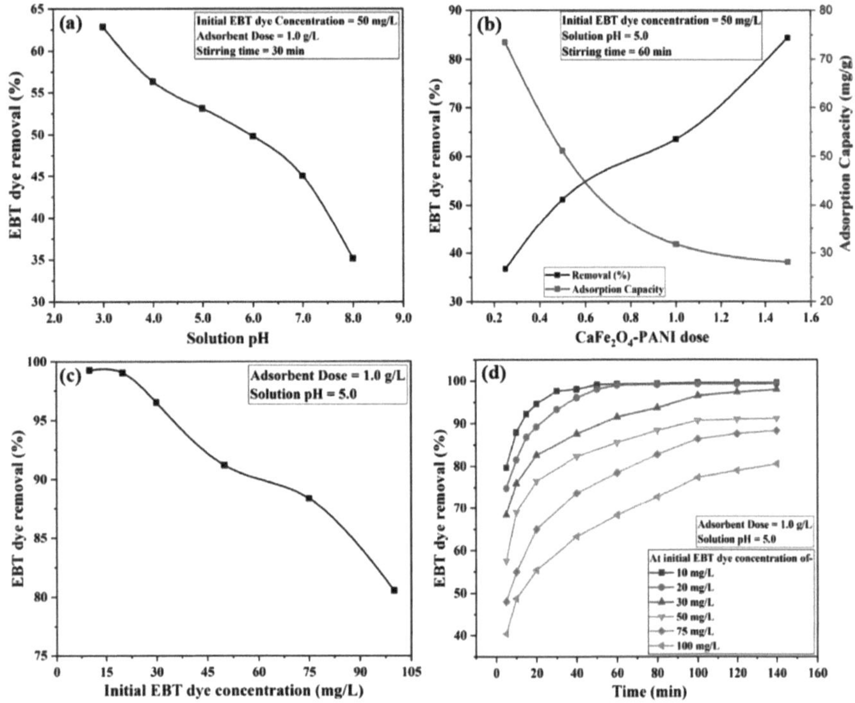

Fig. 3 **a** Changes in EBT dye elimination (%) with variation in solution pH, **b** changes in EBT extraction efficiency with change in the CaFe2O4-PANI nanocomposite dose, **c** influence of initial EBT dye concentration on EBT extraction efficiency and **d** impact of contact time on EBT dye removal (%)

reason for the change is that a higher dose of adsorbent increased the total surface area of materials and, consequently, improved the active sorption sites (Mukhopadhyay et al., 2021). This increased the dye uptake capacity of the $CaFe_2O_4$-PANI nanocomposite. Based on observed data, the optimal dose was selected as 1.0 g/L after accounting for overall feasibility in large-scale applications, at which 65.16% of EBT dye elimination efficiency was obtained (Debnath et al., 2016).

3.3 *Effect of Initial EBT Dye Concentration and Contact Time*

In accordance with the literature survey, it was observed that effluents discharged by many of the dye production and utilising industries contain dye in the concentration range of 10 to 250 mg/L (Li et al., 2017; Saha et al., 2022). Hence, the current adsorption tests were performed within the initial EBT dye concentration range of 10 mg/L to 100 mg/L to analyse the influence of variation in the initial dye concentration and contact time on the effectiveness of dye separation. The $CaFe_2O_4$-PANI dosage and solution pH, which are the other operational variables, were maintained at their optimal values of 1.0 g/L and 5.0, respectively. Figure 3c indicates that the EBT dye elimination (%) decreases as the initial concentration of the EBT dye increases. For a lesser initial EBT dye concentration range of 10–30 mg/L, dye elimination beyond 95% was accomplished. On the contrary, the EBT removal drops to about 90% for higher values of dye concentrations and for 100 mg/L, the removal (%) reduced to 80.56%. EBT dye molecules immediately occupy all the binding sites existing on the surface of $CaFe_2O_4$-PANI nanocomposites because there were enough active sites accessible at a lesser concentration of the EBT dye (Saha et al., 2022). But, with the rise in the initial concentration of the EBT dye, numerous molecules of the EBT dye remain unadsorbed as only limited active sites exist on the surface of the $CaFe_2O_4$-PANI nanocomposites (Li et al., 2017). Thus, based on the above results, the initial concentration was optimised as 50 mg/L for further experiments, which led to the effective elimination of the EBT dye by 92.57%.

The trials to investigate the impact of the duration of contact on the effectiveness of dye elimination were performed within the initial EBT dye concentration range of 10 mg/L to 100 mg/L while tracking variations in extraction percentages at different time intervals. The pH value of the solution was set at 5.0 for the solution, and the adsorbent dose was maintained at 1.0 g/L. Figure 3d indicates that the equilibrium time was reached after approximately 30 min of contact time at the initial EBT dye concentrations of 10 mg/L and 20 mg/L and after around 120 min of contact time for larger dye concentrations (Liang et al., 2018). In light of this, 120 min was determined to be the ideal contact period for maximal adsorption considering larger concentration of dyes in the real scenario industrial effluents (Bhowmik et al., 2020).

3.4 Adsorption Kinetic Analysis

Studies on adsorption kinetics provide crucial details about the sorption process, such as the underlying uptake mechanism and the participation of potential regulating processes. For the development of any adsorption system, these are absolutely necessary. The kinetic study was performed by adapting pseudo-first-order, pseudo-second-order and intra-particle diffusion models as the desired kinetic models. The three kinetic models under consideration have their linear forms displayed by Eqs. 4, 5 and 6, respectively (Saha et al., 2023).

$$\log(Q_e - Q_t) = \log Q_e - \frac{k_f}{2.303}t \tag{4}$$

$$\frac{t}{Q_t} = \frac{1}{K_s Q_e^2} + \frac{1}{Q_e}t \tag{5}$$

$$Q_t = K_{id} t^{\frac{1}{2}} + c \tag{6}$$

The EBT dye adsorption quantity (mg) onto the unit quantity of adsorbents' surface (g) at equilibrium and at contact time t, respectively, are depicted by Q_e and Q_t, given by the pseudo-first-order equation (Eq. 4), where k_{f} represents the first-order rate constant. The constant related to the pseudo-second-order rate is denoted by k_{s} in the pseudo-second-order equation (Eq. 5). c and K_{id} stand for the constant relating to boundary layer thickness and intra-particle diffusion rate, respectively, in the intra-particle diffusion model equation (Eq. 6). Applying the EBT dye uptake data to the three aforementioned linear forms of kinetic models, the key kinetic parameters have been derived, as indicated in Table 2, the results of which are plotted and shown by Figs. 4a, b and c, respectively. It has been noticed that the model's computed values of Q_e were found to be approximately the same as the values determined by the experiment, and the values of R^2 for the pseudo-second-order equation were extremely near to 1.0 (Table 2). It is an implication of the fact that the pseudo-second-order kinetic model can suitably describe the EBT dye uptake onto the $CaFe_2O_4$-PANI nanocomposite. Also, it can be deduced that the chemisorption characteristic of the EBT dye uptake onto the $CaFe_2O_4$-PANI nanocomposite is primarily responsible for the transfer or exchange of electrons that take place between the anionic EBT dye molecules and the surface of the positively charged nanocomposites (Zhou et al., 2014). The experimental findings from the intra-particle diffusion model were put forward by plotting two linear lines, with no linear line crossing the origin (Fig. 4c) (Zhou et al., 2014). This indicated that there may be more than one adsorption process involved in intra-particle diffusion, rather than just one rate-restricting step (Zhou et al., 2014). The first one, of the two linear lines, indicates the migration of the molecules of the EBT dye from the dye-contained mixture to the surface of $CaFe_2O_4$-PANI nanocomposites with the boundary layer dispersion mechanism (Khurana et al., 2017). The EBT dye molecules were adsorbed onto the porous surface of the adsorbent, as indicated by the second line, which shows how the dye molecules diffused from the exterior into the pore spaces of the $CaFe_2O_4$-PANI nanocomposite (De et al., 2022).

3.5 Equilibrium Isotherm Analysis

Adsorption isotherm analysis had been carried out to investigate the chemistry underlying anionic dye sorption onto the synthesised adsorbent, in a manner comparable to kinetic analysis (Saha et al., 2023). The adsorption capabilities determined for the three isotherm models, that is, Langmuir, Freundlich and Temkin models, were investigated under ideal experimental conditions with the objective of determining which equilibrium isotherm model was the most appropriate. Table 3 shows the coefficient values of three isotherm models. Equations 7, 8 and 9 depict the mathematical equations for the Langmuir, Freundlich and Temkin isotherm models (Saha et al., 2023).

Table 2 Pseudo-first-onder, pseudo-second-order and intra-particle diffusion kinetic model variables for the uptake of the EBT dye onto tlie CaFe^04-PANT nana composite (dose of adsorbent 1.0 g/L and initial EBT dye concentration

Kinetic models	Parameters	Initial EBT dye concentration (mg/L)					
		10	20	30	50	75	100
Pseudo-first-order	*kf* (min–1)	10.10	6.75	2.92	4.19	3.22	2.63
	Q_e (mg/g)	3.37	7.21	8.53	21.60	38.36	47.19
	R^2	0.957	0.980	0.990	0.959	0.973	0.973
Pseudo-second-order	$k \times 10^3$ (min–1)	62.55	18.67	9.82	4.88	1.82	1.19
	Q_e(mg/g)	10.19	20.56	29.51	46.95	69.44	84.60
	R^2	0.999	0.999	0.999	0.999	0.998	0.995
Intra-particle diffusion	k_{id1} (mg/g $min^{1/2}$)	0.66	1.32	1.88	4.16	5.70	6.61
	R^2	0.975	0.989	0.981	0.954	0.999	0.974
	C (mg/g)	6.54	12.08	16.52	20.17	23.24	26.43
	$^{k}{}_{id2}$(mg/g $min^{1/2}$)	0.08	0.51	0.71	0.85	2.09	3.25
	R^2	0.929	0.976	0.994	0.927	0.961	0.981
Experimental value		9.93	19.81	28.93	45.61	66.28	80.56

$$\frac{C_e}{Q_e} = \frac{1}{Q_m b} + \frac{C_e}{Q_m} t \quad (7)$$

$$lnC_e = lnK_F + 1/\mathrm{n}lnC_e \quad (8)$$

$$Q_e = \beta_1 \ln K_T + \beta_1 \ln C_e \quad (9)$$

Equation 7 presents the Langmuir equation, wherein the equilibrium adsorption capacity of the $CaFe_2O_4$-PANI nanocomposite to adsorb EBT dye and the EBT dye equilibrium concentration are denoted by Q_e (mg/g) and C_e (mg/L), respectively. The adsorption energy constant is denoted by the symbol b (L/mg), and the maximum EBT dye uptake capability of the $CaFe_2O_4$-PANI nanocomposite is indicated by Q_m (mg/g). The adsorption intensity–related heterogeneity component is represented by n in the Freundlich equation, as shown in Eq. 8. Equation 9 represents the Temkin isotherm equation, where the variables K_T and β_1 represent the constants related to the Temkin isotherm model with respect to the heat of adsorption. The three above-mentioned isotherm models were plotted and represented graphically by Fig. 5a, b and c, respectively, to show the plots of (a) vs. C_e, (b) lnQ_e vs. lnC_e and (c) Q_e against lnC_e. The observations from the isotherm study of EBT dye adsorption depict that the adsorption process follows the Langmuir isotherm model (R^2: 0.9912) over Temkin (R^2: 0.966) and Freundlich (R^2: 0.9626), implying a chemical uptake

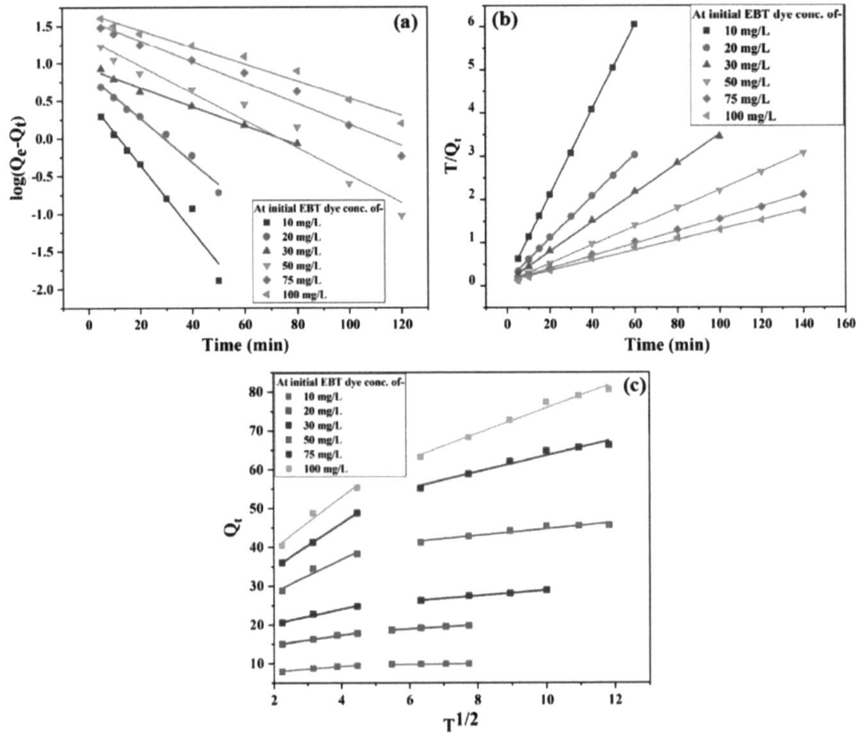

Fig. 4 Adsorption kinetic analysis for the EBT dye. **a** Pseudo-first-order, **b** pseudo-second-order and **c** intra-particle diffusion

Table 3 Langmuir, Freundlich and Temkin isotherm parameters and their values obtained for the adsorption of the EBT dye onto the $CaFe_2O_4$-PANI nanocomposite with a dosage of 1.0 g/L

Models	Values of the isotherm parameters	
Langmuir	Q_m (mg/g)	98.04
	b (L/mg)	0.28
	R^2	0.991
Freundlich	n	3.29
	K_F (mg/g)	27.99
	R^2	0.962
Temkin	B_1	28.72
	K_T (L/mg)	19.52
	R^2	0.966

process which is the monolayer sorption of the EBT dye over the surface of the whole $CaFe_2O_4$-PANI nanocomposite, in which adsorption of the adsorbate happens as the distribution of one layer of molecules of an adsorbate over the surface of an adsorbent (Tanzifi et al., 2017). Thus, the Langmuir isotherm model has been considered

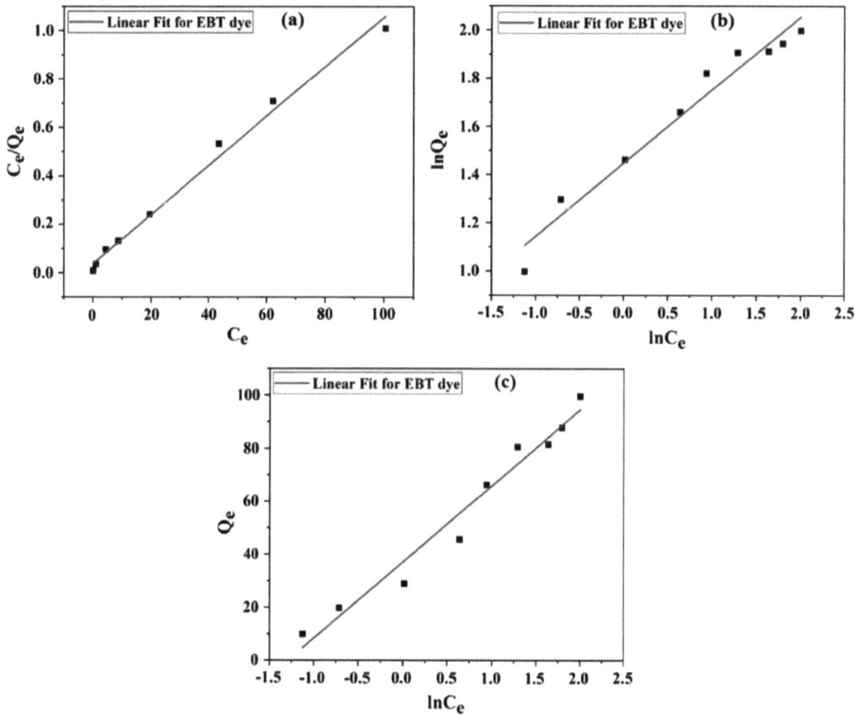

Fig. 5 **a** Langmuir isotherm, **b** Freunlich isotherm and **c** Temkin isotherm models for the sorption of the EBT dye onto the $CaFe_2O_4$-PANI nanocomposite

as the most suitable model for the $CaFe_2O_4$-PANI nanocomposite adsorbent. The Langmuir model estimates that the maximal adsorption capacity of the EBT dye is 98.04 mg/g.

3.6 *Effect of Real Water Sample*

In order to check the real-world application of the $CaFe_2O_4$-PANI nanocomposite adsorbent, the dye solutions were prepared using real water samples which include surface water (from pond), ground water (from tube well) and domestic wastewater (from kitchen sink), and the adsorption process was carried out (Kuang et al., 2020). To determine the extent of interferences caused by the contaminants in the water specimens, sorption tests had been performed using real-world water specimens (Da'na and Sayari, 2012). For this real sample study, the solution pH was fixed at 5.0, the dose of $CaFe_2O_4$-PANI nanocomposite was adjusted to 1.0 g/L, the EBT dye initial concentration was set at 50 mg/L and the contact time was fixed at 120 min. The EBT extraction efficiencies using each of the real water samples are displayed

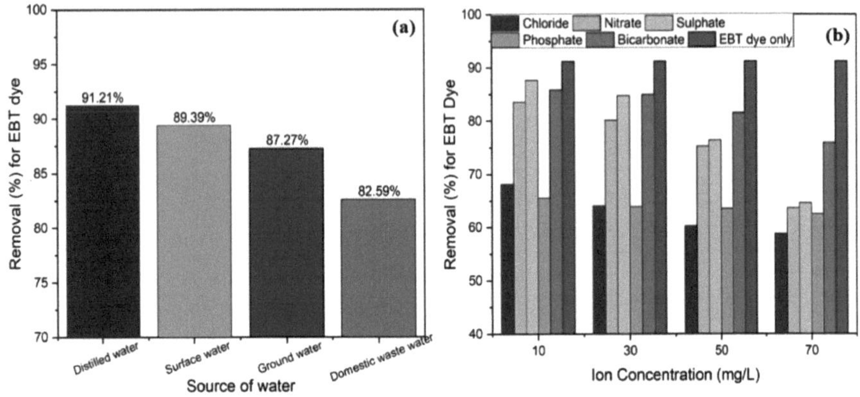

Fig. 6 Influence of **a** real water sample and **b** coexisting ions' EBT dye sequestration efficacy

in Fig. 6a, which indicates that the EBT dye elimination efficiency was hindered the most in experiments using residential wastewater (82.59%), followed by ground water (87.28%) and surface water (89.59%). The distilled water sample exhibited the highest EBT dye elimination efficiency of 92.21%, indicating the absence of any potential interference during the procedure for dye removal. Owing to the prevalence of various other ions as well as other types of pollutants and impediments in the water, raw water samples have lower elimination effectiveness than distilled water, as distilled water is devoid of any kind of impurities. Despite all the interferences, a significant amount of dye elimination was still achieved using the adsorbent, indicating that the $CaFe_2O_4$-PANI nanocomposite performed well in the decontamination method for raw wastewater.

3.7 *Influence of Coexisting Ions*

A major impact can be seen because of the presence of various ions in a raw water/wastewater sample on how different pollutants adsorb onto any appropriate adsorbent (Eltaweil et al., 2023). Thus, studies have been carried out taking into account a number of prevalent ion-containing salts in order to determine the precise hindrance by the existing ions on the sorption process of the EBT dye over the surface of the $CaFe_2O_4$-PANI nanocomposite. In every experiment, the $CaFe_2O_4$-PANI nanocomposite dose, the pH of the solution, the initial EBT dye concentration of the solution and the duration of the contact were maintained fixed at 1.0 g/L, 5.0, 50 mg/L and 120 min, respectively. The required concentrations of salts were introduced into the solution containing dye prior to the start of the sorption process. The effect of the addition of nitrate (NO_3^-), phosphate (PO_4^{3-}), bicarbonate (HCO_3^-), chloride (Cl^-) and sulphate (SO_4^{2-}) ions on the extraction efficiencies of EBT dye are presented in

Fig. 6b. The samples of the preferred salts chosen for this study were present in quantities ranging from 10 to 70 mg/L. Figure 6b implies that, the hindrance created by the introduction of interfering ions keeps increasing as the ionic concentration increases. As a result, the efficiency of the removal of the EBT dye decreases with improved ionic concentration. Existence of Cl^- and PO_4^{3-} ions leads to maximum hindrance or maximum reduction in the EBT dye extraction efficiency, while the presence of HCO_3^- ions causes minimum reduction in the EBT sequestration percentage. The reduction in removal efficiency by the Cl^- ion was approximately between 23 and 32% and that by the PO_4^{3-} ion was approximately between 25 and 28%, whereas by the HCO_3^- ions, the reduction in extraction efficiency is about 6%–17%. The results suggest that Cl^- and PO_4^{3-} ions had a notable impact on the effectiveness of removing the EBT dye, and this effect can be attributed to their substantial attraction to the surface sites of an electrochemically active $CaFe_2O_4$–PANI nanocomposite. This, in turn, reduces the number of potentially accessible sites for the effective uptake of the EBT dye (Eltaweil et al., 2023). $Cl^- > PO_4^{3-} > NO_3^- > SO_4^{2-} > HCO_3^-$ was the decreasing order of effect in the EBT dye elimination efficacy with the use of the $CaFe_2O_4$-PANI nanocomposite.

3.8 RSM-Based Optimisation and Analysis of Variance

RSM was used to predict, optimise and assess the collaborative impact of the test factors on the effectiveness of the elimination of the EBT dye (Mandal et al., 2015). The CCD was adopted for the experimental design (DoE) using RSM, and the data are presented in Table 1. Preliminary experiments were conducted and based on that, each testing variable considered was coded at five levels between –2 and +2. The independent variables chosen were X_1: dose of the $CaFe_2O_4$-PANI nanocomposite (0.5–1.5 g/L), X_2: initial concentration of the EBT dye (10–90 mg/L) and X_3: contact time (60–180 min). Table 4 displays the findings of the model summarised stats and analysis of variance (ANOVA) carried out on the collected information for the uptake of the EBT dye onto the $CaFe_2O_4$-PANI nanocomposite. The developed model was found to be quadratic in nature, which shows the sum of squares' value (1675.77) to be on the higher side, and the accompanying low p-value (<0.0001), further assuring that the developed model is significant (Ghaedi et al., 2016). Maximum number of parameters along with the model indicates a very low p-value (<0.05), indicating the significance of their contribution towards the development of the model (Pan et al., 2020). The quadratic synopsis of statistical values, such as the predicted R^2: 0.9634, adjusted R^2: 0.9898 and correlation coefficients R^2: 0.9946, can also be used to examine the precision of the established model. Additional parameters that demonstrate an effective modelling of the uptake tests' findings include the standard deviation (0.9527), coefficient of variation (1.11%) and an adequate precision of 56.6587 (Tanzifi et al., 2017). Equation 10 is the complete quadratic equation derived from ANOVA that relates each of the independent factors to the outcome (EBT dye extraction efficacy) in terms of real components. The expected EBT dye extraction efficacy for each of the 20 test trials was calculated using this equation, as indicated in Table 4.

$$EBT\ dye\ removal\ efficiency\ (\%) = -42.289 + 120.235X_1 + 0.321X_2 + 0.416X_3 - 0.221X_1X_2 + 0.0087X_1X_3 + 0.0009X_2X_3 - 63.94X_1^2 - 0.0041X_2^2 - 0.0028X_3^3 \quad (10)$$

The plot representing the actual versus projected values of EBT dye extraction (%) had a coefficient of correlation value (R^2: 0.9946), which was very close to 1.0, indicating a good fit (Fig. 7a). Figure 7b illustrates the evolution of residuals for the 20 test trials, and it is evident from the observations that the deviations obtained by subtracting the actual from the projected EBT dye removal values are well below the permissible limit (±2%) (Gupta et al., 2017). The model predicted a maximum EBT dye removal (%) of 99.6% for the $CaFe_2O_4$-PANI nanocomposite dose: 1.28 g/L, contact time: 147 min and initial EBT dye concentration: 29.61 mg/L. Three replica experiments were conducted to validate this output. It was observed that the

Table 4 Findings of the ANOVA and the values of significant statistical parameters related to the developed quadratic model

Source of variation	Sum of squares	DF	Mean square	F-value	P-value	Status
Model	1675.77	9	186.20	205.14	<0.0001	Significant
X_1	905.86	1	905.86	998.00	<0.0001	
X_2	299.61	1	296.61	326.79	<0.0001	
X_3	110.20	1	110.20	121.41	<0.0001	
$X_1 X_2$	5.20	1	5.20	5.73	0.0377	
$X_1 X_3$	0.0190	1	0.0190	0.0209	0.8878	
$X_2 X_3$	1.81	1	1.81	2.00	0.1878	
X_1^2	340.52	1	340.52	375.15	<0.0001	
X_2^2	49.91	1	49.91	54.98	<0.0001	
X_3^2	45.41	1	45.41	50.03	<0.0001	
Residual	9.08	10	0.9077			
Lack of fit	7.56	5	1.51	4.97	0.0516	Not significant
Pure error	1.52	5	0.3042			
Cor total	1684.84	19				
Quadratic summary statistics	R^2	Adj. R^2	P_{red} R^2	Std. Dev	c.v.%	Adeq precision
Response (FRT removal %)	0.9946	0.9898	0.9634	0.9634	1.11	56.6587

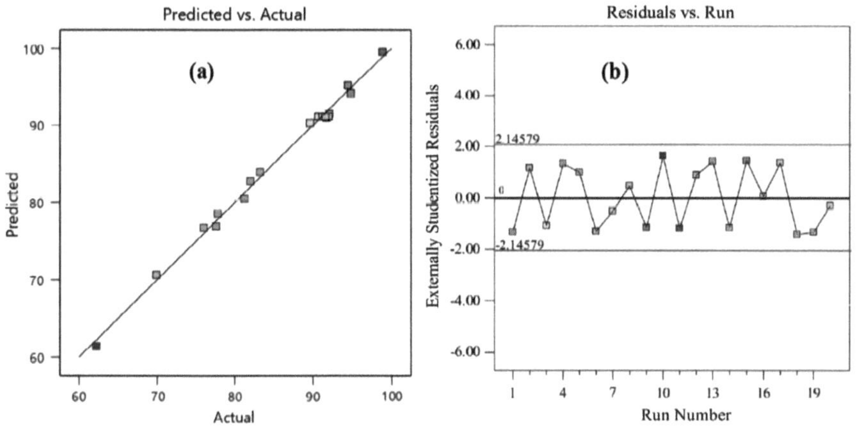

Fig. 7 **a** Plot of absolute versus expected EBT dye abatement percentage and **b** plot of number of runs vs. residuals for EBT dye abatement percentage

average removal from these experiments was 98.93%, which is close to the predicted removal of 99.6%, thus validating the model for the removal of the EBT dye by using the $CaFe_2O_4$-PANI nanocomposite.

3.9 *Regeneration Study of the EBT Dye Uptake*

The EBT dye adsorption regeneration assessment onto the $CaFe_2O_4$-PANI nanocomposite was conducted five times in a row to verify the manufactured adsorbent's potential for reuse. The $CaFe_2O_4$-PANI nanocomposite's dose was kept as 1.0 g/L while the initial EBT dye concentration was maintained at 50 mg/L during the regeneration tests. After setting the solution pH as 5.0, the EBT dye uptake reactions were carried out for 120 min. After the first cycle of the uptake investigation, the EBT dye–impregnated adsorbent was separated using a magnet from the EBT dye solution, and it was then repeatedly cleaned with 0.1 N HCl solution alongside distilled water. After cleaning, the adsorbent was dried at 90 °C for 8 h duration and then ground to make the nanoscale adsorbent ready for the following cycles. The same procedure of the uptake process was followed for the next four cycles. Figure 8 plots the measured EBT elimination (%) throughout the course of five consecutive cycles. The observed results indicated that during the first cycle of the adsorption study, an optimal EBT dye elimination efficiency of 89.958% was reached. As the adsorption study continued, elimination efficiency kept on decreasing. From the first to the third cycle, the EBT dye extraction efficacy decreased at a slower pace (~4% to 6%). However, after the fourth cycle of adsorption, the EBT dye sequestration efficacy was observed to be 77.25%, indicating a substantially higher fall (~11%) in

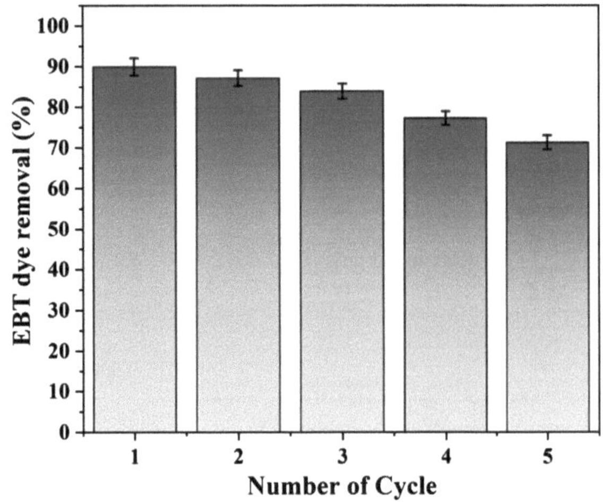

Fig. 8 Regeneration analysis of EBT dye uptake up to five cycles

the removal efficiency. Nevertheless, a significant decrease in the EBT dye extraction efficacy (~18%) was seen following the fifth cycle. Because of the considerable decrease in the dye elimination efficiency, regeneration tests were carried out up to the fifth cycle.

3.10 Evaluation of Performance

The performance evaluation study as shown in Table 5 displays the comparison of the adsorption capability of the $CaFe_2O_4$-PANI nano composite with that of numerous other studies discussing the adsorption of EBT dye, taking into account the highest possible uptake capacity, optimal pH of solution and equilibrium contact time. Several adsorbents, that is, magnetic graphene oxide impregnated with α-Fe_2O_3, PAn/ZnO-PNC, tea waste, $NiFe_2O_4$ magnetic nanoparticles and natural sawdust have been considered in this study, which demonstrated maximum EBT dye uptake capacity of 210.53, 108.69, 151.26, 47.00 and 40.96 mg/g, respectively. $NiFe_2O_4$ magnetic nanoparticles and natural sawdust possess maximum EBT dye uptake capacities which are much lesser than that of the $CaFe_2O_4$-PANI nanocomposite, whereas the time taken to reach equilibrium by natural sawdust is comparable to the $CaFe_2O_4$-PANI nanocomposite (Akhouairi et al., 2019; Moeinpour et al., 2014). On the contrary, although magnetic graphene oxide impregnated with α-Fe_2O_3 and tea waste has got the maximum EBT dye uptake capacity, which is more than that of the $CaFe_2O_4$-PANI nanocomposite used in this study, the solution optimal pH values (2.0 for both the study) were observed to be much acidic leading to a unfeasible and economically untenable large-scale industrial utilisation (Bansal et al., 2020; Khurana et al., 2017). However, PAn/ZnO-PNC has shown improved adsorption performance with adsorption capability on the higher side and reduced equilibrium

Table 5 Performance assessment of EBT dye uptake of the $CaFe_2O_4$-PANI nanocomposite with other adsorbents

Adsorbents	Solution pH	Equilibrium time (mill)	Maximum adsorption capacity (Q J (mg/g)	References
Magnetic graphene oxide impregnated with a-Fe_2O_3	2.0	300	210.53	Khurana et al. (2017)
PAn/ZnO-PNC	6.0	45	108.69	Deb et al. (2021)
Tea waste	2.0	60	151.26	Bansal et al. (2020)
$NiFe_2O_4$ magnetic nanoparticles	6.0	15	47.00	Moeinpour et al. (2014)
Natural sawdust	4.0	120	40.96	Akhouairi et al. (2019)
$CaFe_2O_4$-PANI nanocomposite	5.0	120	98.04	Present study

time. But in this case, separation of the dye-loaded adsorbent from the effluent is still a concerning factor as in this study no magnetic substance is applied (Deb et al., 2021). This shows that the manufactured adsorbent performs better than several other adsorbents investigated in terms of effective functioning close to neutral pH, relatively quick attainment of adsorption equilibrium and achievement of a relatively improved value of optimal dye adsorption capability.

4 Conclusions

This work illustrates the development process of a conducting polymer-coated magnetic metal ferrite nanoadsorbent ($CaFe_2O_4$-PANI) as well as its uses for eliminating the EBT dye. A maximum EBT dye sequestration efficiency of 89.98% was noted at ideal testing conditions, which included solution pH of 5.0, $CaFe_2O_4$-PANI nanocomposite dose of 1.0 g/L, contact period of 120 min and initial EBT dye concentration of 50 mg/L. The Langmuir isotherm model, with $R^2 = 0.991$, suited slightly better than the other two isotherm models, while the pseudo second-order kinetic model was found to be the most appropriate model amongst the three kinetic models. As predicted by the Langmuir isotherm model, an optimal uptake capacity of 98.04 mg/g was found; which is either superior or equivalent to several other specified adsorbents for the extraction of the EBT dye. The effectiveness of the EBT dye elimination was influenced the most (23% to 32%) with the presence of Cl^- and PO_4^{3-} ions in the solution. Introduction of raw water samples indicates that household wastewater causes the most hindrance (10% drop in the effectiveness of the removal of the EBT dye) when compared to the other two raw water samples taken into consideration. The highest possible effectiveness of the removal of the EBT dye of 99.6% was achieved under ideal experimental conditions, which were determined using the RSM model. These parameters included a $CaFe_2O_4$-PANI nanocomposite dosage of 1.28 g/L, a contact period of 147 min and an initial EBT dye concentration of 29.61 mg/L. According to a regeneration study, the $CaFe_2O_4$-PANI nanocomposite containing the EBT dye can be reused up to five times with an EBT dye elimination effectiveness drop of less than 18%.

Declaration of Interest Statement

No potential competing interest was reported by the authors.

References

Agarwal, S., Tyagi, I., Gupta, V. K., Golbaz, F., Golikand, A. N., & Moradi, O. (2016). Synthesis and characteristics of polyaniline/zirconium oxide conductive nanocomposite for dye adsorption application. *Journal of Molecular Liquids, 218*, 494–498. https://doi.org/10.1016/j.molliq.2016.02.040

Akhouairi, S., Ouachtak, H., Addi, A. A., Jada, A., & Douch, J. (2019). Natural sawdust as adsorbent for the eriochrome black T dye removal from aqueous solution. *Water, Air, & Soil Pollution, 230*, 1–15. https://doi.org/10.1007/s11270-019-4234-6

Bansal, M., Patnala, P. K., & Dugmore, T. (2020). Adsorption of Eriochrome Black-T (EBT) using tea waste as a low cost adsorbent by batch studies: A green approach for dye effluent treatments. *Current Research in Green and Sustainable Chemistry, 3*, Article 100036. https://doi.org/10.1016/j.crgsc.2020.100036

Beyki, M. H., Alijani, H., & Fazli, Y. (2016). Poly o-phenylenediamine–MgAl@ $CaFe_2O_4$ nanohybrid for effective removing of lead (II), chromium (III) and anionic azo dye. *Process Safety and Environmental Protection, 102*, 687–699. https://doi.org/10.1016/j.psep.2016.04.027

Bhaumik, M., McCrindle, R. I., & Maity, A. (2015). Enhanced adsorptive degradation of Congo red in aqueous solutions using polyaniline/Fe^0 composite nanofibers. *Chemical Engineering Journal, 260*, 716–729. https://doi.org/10.1016/j.cej.2014.09.014

Bhowmik, M., Debnath, A., & Saha, B. (2018). Fabrication of mixed phase calcium ferrite and zirconia nanocomposite for abatement of methyl orange dye from aqua matrix: Optimization of process parameters. *Applied Organometallic Chemistry, 32*(12), Article e4607. https://doi.org/10.1002/aoc.4607

Bhowmik, M., Kanmani, M., Debnath, A., & Saha, B. (2019). Sono-assisted rapid adsorption of anionic dye onto magnetic $CaFe_2O_4$/$MnFe_2O_4$ nanocomposite from aqua matrix. *Powder Technology, 354*, 496–504. https://doi.org/10.1016/j.powtec.2019.06.009

Bhowmik, M., Debnath, A., & Saha, B. (2020). Fabrication of mixed phase $CaFe_2O_4$ and $MnFe_2O_4$ magnetic nanocomposite for enhanced and rapid adsorption of methyl orange dye: Statistical modeling by neural network and response surface methodology. *Journal of Dispersion Science and Technology, 41*(13), 1937–1948. https://doi.org/10.1080/01932691.2019.1642209

Da'na, E. and Sayari, A. (2012). Adsorption of heavy metals on amine-functionalized SBA-15 prepared by co-condensation: Applications to real water samples. *Desalination, 285*, 62–67. https://doi.org/10.1016/j.desal.2011.09.034

Das, P., & Debnath, A. (2021). Reactive orange 12 dye adsorption onto magnetically separable $CaFe_2O_4$ nanoparticles synthesized by simple chemical route: Kinetic, isotherm and neural network modeling. *Water Practice and Technology, 16*(4), 1141–1158. https://doi.org/10.2166/wpt.2021.064

Das, P., Nisa, S., Debnath, A., & Saha, B. (2022). Enhanced adsorptive removal of toxic anionic dye by novel magnetic polymeric nanocomposite: Optimization of process parameters. *Journal of Dispersion Science and Technology, 43*(6), 880–895. https://doi.org/10.1080/01932691.2020.1845958

De, I., Pahuja, M., ud din Wani, H. M., Dey, A., Dube, T., Ghosh, R., Kankan, N., Mishra, J., Panda, J. J., Maruyama, T., Ghosh, K., & Singh, M. (2022). In-vitro toxicity assessment of a textile dye Eriochrome Black T and its nano-photocatalytic degradation through an innovative approach using Mf-NGr-CNTs-SnO_2 heterostructures. *Ecotoxicology and Environmental Safety, 243*, 113985. https://doi.org/10.1016/j.ecoenv.2022.113985

Deb, A., Kanmani, M., Debnath, A., Bhowmik, K. L., & Saha, B. (2017). Preparation and characterization of magnetic $CaFe_2O_4$ nanoparticles for efficient adsorption of toxic Congo Red dye from aqueous solution: Predictive modeling by artificial neural network. *Desalination and Water Treatment, 89*, 197–209. https://doi.org/10.5004/dwt.2017.21361

Deb, A., Debnath, A., & Saha, B. (2021). Sono-assisted enhanced adsorption of Eriochrome Black-T dye onto a novel polymeric nanocomposite: Kinetic, isotherm, and response surface methodology optimization. *Journal of Dispersion Science and Technology, 42*(11), 1579–1592. https://doi.org/10.1080/01932691.2020.1775093

Debnath, A., Majumder, M., Pal, M., Das, N. S., Chattopadhyay, K. K., & Saha, B. (2016). Enhanced adsorption of hexavalent chromium onto magnetic calcium ferrite nanoparticles: Kinetic, isotherm, and neural network modeling. *Journal of Dispersion Science and Technology, 37*(12), 1806–1818. https://doi.org/10.1080/01932691.2016.1141100

Eltaweil, A. S., Abd El-Monaem, E. M., El-Subruiti, G. M., Ali, B. M., Abd El-Latif, M. M., & Omer, A. M. (2023). Graphene oxide incorporated cellulose acetate beads for efficient removal of methylene blue dye; isotherms, kinetic, mechanism and co-existing ions studies. *Journal of Porous Materials, 30*(2), 607–618. https://doi.org/10.1007/s10934-022-01347-6

Gabal, M. A., Al-Juaid, A. A., El-Rashed, S., Hussein, M. A., Al Angari, Y. M., & Saeed, A. (2019). Structural, thermal, magnetic and electrical properties of polyaniline/$CoFe_2O_4$ nano-composites with special reference to the dye removal capability. *Journal of Inorganic and Organometallic Polymers and Materials, 29*, 2197–2213. https://doi.org/10.1007/s10904-019-01179-z

Ghaedi, M., Khafri, H. Z., Asfaram, A., & Goudarzi, A. (2016). Response surface methodology approach for optimization of adsorption of Janus Green B from aqueous solution onto ZnO/Zn (OH) 2-NP-AC: Kinetic and isotherm study. *Spectrochimica Acta Part a: Molecular and Biomolecular Spectroscopy, 152*, 233–240. https://doi.org/10.1016/j.saa.2015.06.128

Gupta, V. K., Agarwal, S., Asif, M., Fakhri, A., & Sadeghi, N. (2017). Application of response surface methodology to optimize the adsorption performance of a magnetic graphene oxide nanocomposite adsorbent for removal of methadone from the environment. *Journal of Colloid and Interface Science, 497*, 193–200. https://doi.org/10.1016/j.jcis.2017.03.006

Hou, N., Wang, R, Geng, R., Wang, F., Jiao, T., Zhang, L., Zhou, J., Bai, Z., & Peng, Q. (2019). Facile preparation of self-assembled hydrogels constructed from poly-cyclodextrin and poly-adamantane as highly selective adsorbents for wastewater treatment. *Soft Matter*, ***15*(30),** 6097–6106. https://doi.org/10.1039/C9SM00978G

Ivanets, A. I., Srivastava, V., Roshchina, M. Y., Sillanpää, M., Prozorovich, V. G., & Pankov, V. V. (2018). Magnesium ferrite nanoparticles as a magnetic sorbent for the removal of Mn^{2+}, Co^{2+}, Ni^{2+} and Cu^{2+} from aqueous solution. *Ceramics International, 44*(8), 9097–9104. https://doi.org/10.1016/j.ceramint.2018.02.117

Jamwal, P., Lele, S., & Menon, M. (2016). Rethinking water quality standards in the context of urban rivers. In: Urbanization and the environment: Eighth INSEE biennial conference of the Indian society for ecological economics, Bengaluru

Kharazi, P., Rahimi, R., & Rabbani, M. (2019). Copper ferrite-polyaniline nanocomposite: Structural, thermal, magnetic and dye adsorption properties. *Solid State Sciences, 93*, 95–100. https://doi.org/10.1016/j.solidstatesciences.2019.05.007

Khurana, I., Shaw, A. K., Bharti, Khurana, J. M., & Rai, P. K. (2017). Batch and dynamic adsorption of Eriochrome Black T from water on magnetic graphene oxide: Experimental and theoretical studies. *Journal of Environmental Chemical Engineering*, S2213343717306681. https://doi.org/10.1016/j.jece.2017.12.029

Kuang, Y., Zhang, X., & Zhou, S. (2020). Adsorption of methylene blue in water onto activated carbon by surfactant modification. *Water*, **12**(2), 587. https://doi.org/10.3390/w12020587

Li, X., Lu, H., Zhang, Y., & He, F. (2017). Efficient removal of organic pollutants from aqueous media using newly synthesized polypyrrole/CNTs-$CoFe_2O_4$ magnetic nanocomposites. *Chemical Engineering Journal, 316*, 893–902. https://doi.org/10.1016/j.cej.2017.02.037

Liang, Y. D., He, Y. J., & Zhang, Y. H. (2018). Adsorption property of alizarin red S by $NiFe_2O_4$/polyaniline magnetic composite. *Journal of Environmental Chemical Engineering, 6*(1), 416–425. https://doi.org/10.1016/j.jece.2017.12.022

Liu, Y., Du, J., Wu, H., Cong, C., Zhang, H., Wang, J., & Wang, Z. (2022). Antifouling streptomycin-based nanofiltration membrane with high permselectivity for dye/salt separation. *Separation and Purification Technology*, *297*, 121443. https://doi.org/10.1016/j.seppur.2022.121443

Mandal, S., Mahapatra, S. S., & Patel, R. K. (2015). Enhanced removal of Cr (VI) by cerium oxide polyaniline composite: Optimization and modeling approach using response surface methodology and artificial neural networks. *Journal of Environmental Chemical Engineering, 3*(2), 870–885. https://doi.org/10.1016/j.jece.2015.03.028

Modwi, A., Albadri, A., & Taha, K. Ḳ. (2023). High malachite green dye removal by ZrO_2-g-C_3N_4 (ZOCN) meso-sorbent: Characteristics and adsorption mechanism. *Diamond and Related Materials*, 109698. https://doi.org/10.1016/j.diamond.2023.109698

Moeinpour, F., Alimoradi, A., & Kazemi, M. (2014). Efficient removal of Eriochrome black-T from aqueous solution using $NiFe_2O_4$ magnetic nanoparticles. *Journal of Environmental Health Science and Engineering*, *12***,** 1–7. https://doi.org/10.1186/s40201-014-0112-8

Mohammadi Nodeh, M. K., Soltani, S., Shahabuddin, S., Rashidi Nodeh, H., & Sereshti, H. (2018). Equilibrium, kinetic and thermodynamic study of magnetic polyaniline/graphene oxide based nanocomposites for ciprofloxacin removal from water. *Journal of Inorganic and Organometallic Polymers and Materials, 28*, 1226–1234. https://doi.org/10.1007/s10904-018-0782-2

Mosleh, S., Rahimi, M. R., Ghaedi, M., Dashtian, K., & Hajati, S. (2016). $BiPO_4/Bi_2S_3$-HKUST-1-MOF as a novel blue light-driven photocatalyst for simultaneous degradation of toluidine blue and auramine-O dyes in a new rotating packed bed reactor: Optimization and comparison to a conventional reactor. *RSC Advances, 6*(68), 63667–63680. https://doi.org/10.1039/C6RA10385E

Mukhopadhyay, A., Tripathy, B. K., Debnath, A., & Kumar, M. (2021). Enhanced persulfate activated sono-catalytic degradation of brilliant green dye by magnetic $CaFe_2O_4$ nanoparticles: Degradation pathway study, assessment of bio-toxicity and cost analysis. *Surfaces and Interfaces, 26*, Article 101412. https://doi.org/10.1016/j.surfin.2021.101412

Nasar, A., & Mashkoor, F. (2019). Application of polyaniline-based adsorbents for dye removal from water and wastewater—a review. *Environmental Science and Pollution Research, 26*, 5333–5356. https://doi.org/10.1007/s11356-018-3990-y

Nasiri, A., Rajabi, S., & Hashemi, M. (2022). $CoFe_2O_4$@ Methylcellulose/AC as a new, green, and eco-friendly nano-magnetic adsorbent for removal of reactive Red 198 from aqueous solution. *Arabian Journal of Chemistry, 15*(5), Article 103745. https://doi.org/10.1016/j.arabjc.2022.103745

Pan, S., Liu, Y. H., Wang, Z., Huang, W., Song, L., Liu, R. J. (2020). Optimization on adsorption process of Congo red onto magnetic Ni^0. $5Cu^0$. $5Fe_2O_4/SiO_2$ nanocomposites and their adsorption mechanism. *Journal of Nanoscience and Nanotechnology, 20*(2), 789–801. https://doi.org/10.1166/jnn.2020.16916

Saad, M., Tahir, H., Khan, J., Hameed, U., & Saud, A. (2017). Synthesis of polyaniline nanoparticles and their application for the removal of crystal violet dye by ultrasonicated adsorption process based on response surface methodology. *Ultrasonics Sonochemistry, 34*, 600–608. https://doi.org/10.1016/j.ultsonch.2016.06.022

Saha, B., Debnath, A., & Saha, B. (2022). Fabrication of PANI@ Fe–Mn–Zr hybrid material and assessments in sono-assisted adsorption of methyl red dye: Uptake performance and response surface optimization. *Journal of the Indian Chemical Society, 99*(9), Article 100635. https://doi.org/10.1016/j.jics.2022.100635

Saha, B., Gayen, S., & Debnath, A. (2023). Sequestration of paracetamol from aqueous solution using zinc oxide/polypyrrole nanocomposite: Cost analysis, scale-up design, and optimization of process parameters. *Journal of Hazardous, Toxic, and Radioactive Waste, 27*(4), 04023032. https://doi.org/10.1061/JHTRBP.HZENG-1213

Salem, M. A., Elsharkawy, R. G., & Hablas, M. F. (2016). Adsorption of brilliant green dye by polyaniline/silver nanocomposite: Kinetic, equilibrium, and thermodynamic studies. *European Polymer Journal, 75*, 577–590. https://doi.org/10.1016/j.eurpolymj.2015.12.027

Seidi, F., & Hedayati, K. (2020). A facile synthesis and study of photocatalytic properties of magnetic $CaFe_2O_4$-CeO_2 nanocomposites applicable for separation of toxic azo dyes. *Journal of Nanostructures, 10*(3), 497–508.

Shahabuddin, S., Sarih, N. M., Mohamad, S., & Baharin, S. N. A. (2016). Synthesis and characterization of Co_3O_4 nanocube-doped polyaniline nanocomposites with enhanced methyl orange adsorption from aqueous solution. *RSC Advances, 6*(49), 43388–43400. https://doi.org/10.1039/C6RA04757B

Sonal, S., & Mishra, B. K. (2021). Role of coagulation/flocculation technology for the treatment of dye wastewater: Trend and future aspects. *Water Pollution and Management Practices*, 303–331. https://doi.org/10.1007/978-981-15-8358-2_13

Syed, A., Elgorban, A. M., Bahkali, A. H., & Sillanpää, M. (2021). Visible-light sensitization and recombination delay through coupling $CaFe_2O_4$ on Bi_2O_3 nanocomposite for high performance photocatalytic and antibacterial applications. *Surfaces and Interfaces, 26*, Article 101336. https://doi.org/10.1016/j.surfin.2021.101336

Tanzifi, M., Hosseini, S. H., Kiadehi, A. D., Olazar, M., Karimipour, K., Rezaiemehr, R., & Ali, I. (2017). Artificial neural network optimization for methyl orange adsorption onto polyaniline nano-adsorbent: Kinetic, isotherm and thermodynamic studies. *Journal of Molecular Liquids, 244*, 189–200. https://doi.org/10.1016/j.molliq.2017.08.122

Tosun Satir, I., Ozdemir, N., & Donmez Gungunes, C. (2022). Bone meal and chitosan biocomposite: A new biosorbent for the removal of Victoria Blue R from wastewater. *Chemical Engineering Communications, 209*(9), 1231–1245. https://doi.org/10.1080/00986445.2021.1957850

Vadivel, S., Maruthamani, D., Habibi-Yangjeh, A., Paul, B., Dhar, S. S., & Selvam, K. (2016). Facile synthesis of novel $CaFe_2O_4$/g-C_3N_4 nanocomposites for degradation of methylene blue under visible-light irradiation. *Journal of Colloid and Interface Science, 480*, 126–136. https://doi.org/10.1016/j.jcis.2016.07.012

Zhou, C., Wu, Q., Lei, T., & Negulescu, I. I. (2014). Adsorption kinetic and equilibrium studies for methylene blue dye by partially hydrolyzed polyacrylamide/cellulose nanocrystal nanocomposite hydrogels. *Chemical Engineering Journal, 251*, 17–24. https://doi.org/10.1016/j.cej.2014.04.034

Zhu, G., Fang, H., Xiao, Y., & Hursthouse, A. S. (2020). The application of fluorescence spectroscopy for the investigation of dye degradation by chemical oxidation. *Journal of Fluorescence*, 30, 1271–1279.https://doi.org/10.1007/s10895-020-02591-2

Nanomaterials–Based Strategies for Addressing Arsenic Pollution in Water and Wastewater: Recent Developments

Rutuja T. Kamble, Bishnupriya Pradhan, Rashmi Rekha Majhi, and Sujana M. Gude

1 Introduction

Promising applications of science and technology involving nanoscience to address the challenges faced in nature are truly inspiring. 'Nanoscience' is the study of materials sized in the range of 1–100 nm (nanoscale) that show extraordinary properties, functions and phenomena because of their small size. The characteristic qualities of nanotechnology are mostly determined by the size of the nanoparticles (NP_S). To enhance certain special properties, nanomaterials are 'engineered deliberately' at the nanoscale range. Currently, nanotechnology is emerging as a solution to numerous challenges in food production, agriculture, medicines, industrial applications and other sectors (Beyene et al., 2017). 'Nanotechnology' can be defined as the designing and development of devices and systems using nanoscale materials at an atomic and molecular level in the fields of material science and engineering, involving the principles of physics, chemistry and biology. Nanomaterials assume significance because of their small dimension, better accuracy and high reactivity. Adding to the viability, they are cost–effective and can be produced by using certain environmentally friendly techniques. This technology has also shown great efficiency in resolving water–related issues that have been a global concern. In recent years, the development of new instruments and methods made possible the use of nanotechnology, particularly

R. T. Kamble
Academy of Scientific and Innovative Research (AcSIR), Ghaziabad, Uttar Pradesh, India

B. Pradhan · R. R. Majhi
Hydro & Electrometallurgy Department, CSIR–Institute of Minerals and Materials Technology, Bhubaneswar, Odisha, India

R. T. Kamble · S. M. Gude (✉)
Environment & Sustainability Department, CSIR–Institute of Minerals and Materials Technology, Bhubaneswar, Odisha, India
e-mail: sujana@immt.res.in

© Capital Publishing Company, New Dehli, India. 2025
R. Neelancherry and S. M. Kumar, *Nanotechnology in Wastewater Treatment*,
https://doi.org/10.1007/978-3-031-94246-4_5

in the domain of water and wastewater treatment, by providing viable new options for more economical and efficient purification techniques. (Beyene et al., 2017; Bora & Dutta, 2014; Jain et al., 2021).

Arsenic, a heavy metal, mostly known for its hazardous carcinogenic properties, is naturally found in water bodies. Arsenic exposure has proven to be very fatal to the environment and humans, causing life–threatening health issues. Long–term exposure to this heavy metal causes many cancerous and non–cancerous diseases (Habuda-Stanić & Nujić, 2015). The World Health Organization (WHO), BIS and US EPA have set a revised standard guideline of $10\,\mu g/L$ as the acceptable limit of arsenic in drinking water. The contaminated groundwater is a principal source of arsenic exposure in humans (Shaji et al., 2021). Apart from groundwater, arsenic is also abundantly found in wastewater. The maximum contamination level (MCL) set by the US EPA for arsenic is 0.01 mg/L.

Of the several cost–efficient treatment techniques to eliminate arsenic from water, nanotechnology is the most potent technique. This chapter focuses on the application of various NP_S and nano–techniques for the removal of arsenic from water and wastewater. The chapter also summarises metals, metal oxides and mixed metal NP_S, in addition to a few commercially accessible and cost–effective nanoparticle–infused sorbents, nano–filters and photocatalysts that are tested and studied for the removal of arsenic from various aqueous solutions (Bora & Dutta, 2014; Habuda-Stanić & Nujić, 2015; Jain et al., 2021).

Scientifically, the term 'heavy metal' indicates elements having a density higher than 5 g/cm^3, but it is generally used for arsenic, lead, cadmium, mercury, chromium and copper elements (Tahoon et al., 2020). Of these, arsenic contaminants are a great concern, because their carcinogenic and bio–accumulative nature can seriously threaten human health, as well as the aquatic ecosystems (Mudhoo et al., 2011). Arsenic is present abundantly (ranks 20th) in the earth's crust. It is a significant element of about 200 minerals, for example, realgar (As_4S_4), arsenolite (As_2O_3), arsenopyrite (FeAsS), orpiment (As_2S_3), etc. Weathering of these minerals releases a major portion of arsenic into water bodies; other natural sources are volcanic activities and geothermal waters (Siddique et al., 2020). Various recently published reports have documented arsenic pollution in the USA, Mexico, New Zealand, Taiwan, Canada, Argentina, Poland, Chile, China, Hungary, Japan, India and Bangladesh. Among these nations, citizens of Bangladesh and the state of West Bengal in India are severely affected. Some of the tubewells in Bangladesh have arsenic concentrations of more than 4 mg/L (Alka et al., 2021). Mining and smelting of ores, purification of petroleum, pharmaceuticals, production of glass and cement, combustion of fossil fuels, production of pulp and paper, application of arsenic–containing agro–chemicals, landfill leaching and wood preservation can also produce arsenic (Hao et al., 2018).

Generally, humans are exposed to arsenic through ingestion of contaminated water and food. Prolonged contact with arsenic may cause cardio–vascular and neurological disorders, as well as cancers of the bladder, lungs, skin, etc. Also, acute poisoning by arsenic can cause diarrhoea, abdominal pain, vomiting, etc. (Jain, 2022). The toxicity of arsenic to the human body changes with its forms. It is reported that inorganic arsenicals, generally found in aqueous solutions, are more toxic than organic arsenicals, which specifically enter the human body through marine food. An inorganic arsenite enters cells through channels that typically transport small solutes like glycerol. It can inhibit enzyme function and cellular respiration by binding to sulfhydryl groups in proteins. Arsenate, due to its structural similarity with phosphate, enters cells via phosphate transporters. After entering the cell, the arsenate gets reduced to arsenite by the action of reducing agents present. The arsenite species cause pyruvic acid accumulation in the blood, which affects the various cellular functions (Alka et al., 2021; Siddique et al., 2020).

2 Arsenic Chemistry in Aqueous Solutions

In a natural environment, arsenic may occur in various oxidation states (−3, 0, + 3 and +5). The inorganic forms of arsenic are arsenites, arsenates, arsenic acids, arsenious acids and arsine, while methyl arsenic acid and dimethyl arsenic acid are the organic species (Fig. 1). In aqueous systems, inorganic arsenic is mostly found as oxyanions, for example, H_3AsO_3, $H_2AsO_3^-$ and $HAsO_3^{2-}$ (As III) and H_3AsO_4, $H_2AsO_4^-$, $HAsO_4^{2-}$ and AsO_4^{3-} (As *V*) (Hao et al., 2018).

The existence of a specific arsenic oxidation state in water is principally influenced by two environmental factors, namely, reduction–oxidation reactions and pH conditions. Other parameters, for example, general hydrochemistry, microbial activity and other ionic distribution can also influence arsenic speciation (Siddique et al., 2020). Arsenite (As III) species are prevalent under anoxic reducing conditions and predominantly exist in the uncharged form, H_3AsO_3 at pH values below 9.2. Arsenate (As V) species are stable under aerobic conditions and at a higher oxidation potential.

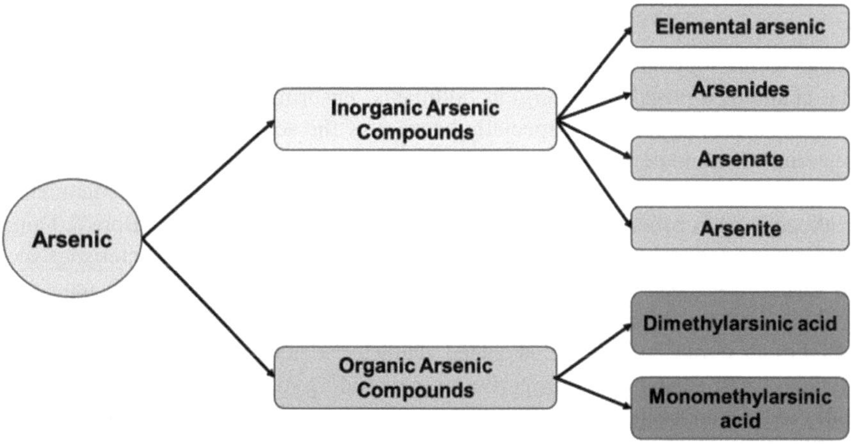

Fig. 1 Arsenic species in aqueous solutions

The $H_2AsO_4^-$ species is dominant in the pH range of 2–6.8, while the $HAsO_4^{2-}$ species are dominant when the solution pH is between 6.8 and 11.5. The other two species are H_3AsO_4, found in extremely acidic conditions (pH value below 2), and AsO_4^{3-}, found at higher alkaline conditions (pH value above 11.5) (Altowayti et al., 2021). The dissolution reactions for both As (III) and As (V) are given below (Hao et al., 2018).

Arsenite:

$$H_3AsO_3 \rightleftharpoons H_2AsO + H^+pK_{a1} : 9.2$$

$$H_2AsO \rightleftharpoons HAsO + H^+pK_{a2} : 12.1$$

$$HAsO \rightleftharpoons AsO + H^+pK_{a3} : 13.4$$

Arsenate:

$$H_3AsO_4 \rightleftharpoons H_2AsO + H^+pK_{a1} : 2.3$$

$$H_2AsO \rightleftharpoons HAsO + H^+pK_{a2} : 6.8$$

$$HAsO \rightleftharpoons AsO + H^+pK_{a3} : 11.5$$

3 Nano–technology for Water and Wastewater Purification

Nano–technology has enabled the development of innovative strategies that address the challenges posed by emerging pollutants, ensuring safe potable water for a growing world population, especially because of the advanced properties of nanomaterials. Nanomaterials, such as nano–catalysts, nano–adsorbents and nanomembranes, have shown remarkable capabilities in eliminating organic contaminants, pathogens, heavy metals and microplastics present in water sources (Bora & Dutta, 2014; Nicomel et al., 2016). These nano–materials have unique characteristics such as tuneable surface charges, a large surface–to–volume ratio and high reactivity, making them ideal for improving water treatment processes and treatment efficiency (Taghipour et al., 2019). The advanced characteristics of nanomaterials are owing to changes in fundamental properties corresponding to their particle size, specifically to the nanoscale dimensions of the material, which result in a higher atom to surface ratio, increased surface energy, spatial confinement and decreased defects; these features are found absent in bulk materials. The higher surface area of nanomaterials offers additional adsorption sites, leading to more efficiency with regard to the removal of contaminants. The use of nanomaterials in the purification of water and wastewater can be a substantial step forward for environmental protection and sustainable water resource management (Jain, 2022).

4 Types of Nanomaterials for Water and Wastewater Purification Processes

Various categories of nanomaterials have been examined for the elimination of arsenic from aqueous solutions. The recent advancement in nanomaterials is because of their chemical and ecological stability, exceptional removal efficiencies and rapid kinetics. These characteristics of NP_S significantly depend on their size. Due to their larger specific surface area and high surface–to–volume ratio, the NP_S show higher absorption capabilities and a high reaction rate for absorption (Nicomel et al., 2016).

4.1 Metal–Based NPs

Metal NP_S are synthesised by taking metal salts as precursors. These are ideal for arsenic elimination because of their high removal capacity and outstanding selectivity. A huge number of metal oxide nanomaterials, for example, iron (Fe), zinc (Zn), titanium (Ti), magnesium (Mg), manganese (Mn), cerium (Ce) and zirconium (Zr) oxides in their pure or composite form are being studied for the elimination of arsenic from water and wastewater (Altowayti et al., 2021). Generally, iron oxides are considered the most feasible metal oxides for the removal of arsenic. In addition, the NP_S of iron oxides exhibit superior performance as compared to micron–sized iron particles. Many studies have explored the fast and efficient removal of arsenic species using various types of iron nanomaterials; for example, Mamindy–Pajany et al. (2011) comprehensively examined the arsenic elimination capacity of goethite, hematite, magnetite and ZVI among which ZVI showed maximum adsorption efficiency. Singh et al. (2021) synthesised sulphide–modified nZVI, which showed maximum adsorption efficiencies of 79.37 mg/g for arsenite and 89.29 mg/g for arsenate. Here, the layer of FeS covering the adsorbent surface provides the passivation to the surface and extends the lifetime to S–nZVI.

Magnetite nanomaterials are extensively utilised in water and wastewater purification, especially to eliminate heavy metals, because of their superparamagnetic characteristics that make separation easy from aqueous systems by the application of a magnetic field externally (Bora & Dutta, 2014; Jain, 2022). Feng et al. (2012) investigated NP_S of Fe_3O_4 coated with ascorbic acid for the removal of arsenic. The material was synthesised via the hydrothermal route without using any templates, as it possesses superparamagnetic properties and a large surface area (179 m^2/g). It showed a maximum elimination efficiency of 46 mg/g for arsenite. Other than iron oxides, aluminium oxides are also widely used, as they show consistent results over a wide range of pH (Ghosh et al., 2019). Numerous mixed–metal oxides have also been investigated for the adsorption of arsenic species; for example, Babaee et al. (2018) synthesised Fe/Cu bi–metallic NP_S by the coprecipitation method. The adsorbent showed excellent arsenite and arsenate removal efficiencies of 19.68 and 21.32 mg/g, individually, even at lower arsenic concentrations. Further, Yu et al. (2018) have developed a mixed–metal oxide composite of Yttrium with magnetite. As compared to pure magnetite NP_S reported in the literature, the material achieved a higher elimination efficiency for arsenate (170.48 mg/g) and arsenite (84.22 mg/g) at a pH of 7.0.

4.2 *Photocatalytic NPs*

In recent years, titanium dioxide (TiO_2) has received attention because of its robust photocatalytic reactivity, low cost and photo–chemical and biological stability. Deedar et al. (2009) investigated the TiO_2 NP_S for photocatalytic removal of arsenic by conducting experiments in various conditions, that is, air–light, air–sunlight, nitrogen–dark and air–dark. Other than TiO_2, ZnO NP_S have emerged as a promising photocatalyst as they possess similar band gap energies. Massoudinejad et al. (2020) investigated the arsenic elimination capacity of ZnO NP_S by combining the adsorption with UV/H_2O_2–facilitated photocatalytic oxidation, where a maximum removal of 85% was achieved. Further, a metal–metal nanocomposite, CuO/ZnO, was examined for photocatalytic oxidation and subsequent elimination of arsenic. In controlled experiments carried out under dark conditions and UV radiation, the removal of arsenite by photocatalytic treatment was higher by 6.5 times as compared to the adsorption carried out in the dark (Samad et al., 2016).

4.3 *Carbon Nanomaterials*

Metal oxide functionalized carbon nanomaterials, such as activated carbon, carbon nanotubes and graphene, are emerging as promising materials for the removal of toxic metals (Tahoon et al., 2020). Chandra et al. (2010) investigated the magnetite–reduced GO composite for the elimination of As (V); outcomes showed that the adsorbent can reduce the As (V) content by over 99.9% from initial concentrations of 5 mg/L. Similarly, Purwajanti et al. (2017) synthesised $Mg(OH)_2$ –MgO @ RGO nano–composite, which achieved a maximum efficiency of 681.3 mg/g for the elimination of arsenite from aqueous solutions.

4.4 Composite Nanomaterials

Nano–composites are matrix materials in which the matrix is embedded with NP_S. Matrix materials are generally polymers, ceramics, metals or any organic phases (Taghipour et al., 2019). A nano–composite, porous MIL–53(Al) MOF, was explored for the elimination of arsenate from water solutions, with a maximum elimination capacity of 105.6 mg/g. It is believed that the Lewis acid site in MOF facilitates the reaction between the contaminant molecule and surface, thereby possessing high adsorption efficiency (Li et al., 2014). According to the literature, polymers embedded with metal oxide NP_S have high efficiency for the removal of arsenic. Martínez–Cabanas et al. (2016) synthesised a magnetite nanoparticle via a green chemistry route using eucalyptus extract; further, these NP_S were encapsulated into chitosan polymer matrix beads and investigated for their As (V) elimination capability. The magnetite–chitosan composite showed considerable regeneration ability, that is, adsorption efficiency was persistent even after four regeneration cycles.

4.5 Nano–Infused Sensors

Recently, arsenic detection in aqueous solutions has gained a tremendous amount of interest in the field of research and technology. Studies have shown that arsenic in water can be detected in the parts per billion range by using certain analytical techniques like ICP–MS, Atomic Absorption Spectroscopy (AAS), Atomic Emission Spectroscopy (AES), and Atomic Fluorescence Spectroscopy (AFS), respectively. However, these techniques have some drawbacks and hence require more convenient and sensitive methods of detection. Considerable research efforts have been made to develop electrochemical sensing systems with advantageous qualities like high sensitivity, easy use, selectivity, time efficiency and simple apparatus to determine the content of inorganic arsenic in the aqueous solutions (Kempahanumakkagari et al., 2017).

The nanoparticle–modified electrodes in electrochemical sensor technology have proved to be very effective and efficient. Novel and ultra–sensitive nano–infused sensors have been developed for water quality screening as well as water purification. Some of the NP_S used for the modification of the electrode–based sensors using nanotechnology for the detection of arsenic include metal–oxide NP_S (metals like: Fe, Ti, Sn, Mn, etc.), bimetallic NP_S, carbon–based NP_S and noble–metal NP_S. Carbon–based NP_S mainly comprise CNTs and different forms of graphene oxide (Kempahanumakkagari et al., 2017). Currently, the electrodes modified using gold (Au) NP_S have gained great attention for the detection of heavy metals including arsenic. The detection of As (III) using the AuNP–modified electrodes was found to show better sensitivity and selectivity than the bulk Au electrode because of the excellent electrical conductivity, higher surface area and catalytic properties of the Au metal. Therefore, by using nanotechnology, the electrochemical sensors infused with nanomaterials have offered new and advantageous opportunities for the detection of arsenic in water and wastewater (Kempahanumakkagari et al., 2017; Thakkar et al., 2021).

5 Arsenic Removal Technologies

As the composition of wastewater or effluent released from each of these sources is different and complex, specific strategies for the treatment of wastewater should be planned (Altowayti et al., 2021). Various techniques, including phytoremediation, ion–exchange, chemical precipitation, catalytic degradation, membrane filtration, adsorption, etc., have been investigated to eliminate the arsenic present in contaminated water and wastewater. These techniques can be employed directly for the treatment of water (Siddique et al., 2020); however, the treatment of wastewater or industrial effluent requires a multistep process. It involves preliminary screening followed by primary treatment, in which suspended solids and organic matter are removed in a sedimentation tank. The secondary treatment is used to degrade the microbial content, for example, trickling filters and finally, the pre–treated water free from suspended materials and microbial agents undergo tertiary treatments (adsorption, ion exchange, etc.), which can remove dissolved contaminants such as arsenic (Taghipour et al., 2019).

Some of the available arsenic removal techniques are given in Fig. 2.

5.1 Oxidation

The chemical composition of polluted waters or wastewater influences the removal of arsenic. This aspect should be considered while determining the treatment method to be employed (Altowayti et al., 2021). Many of the available decontamination methods are more efficient for the removal of arsenate than that of un–charged arsenite. Therefore, if a two–step approach is employed, the treatment technologies can work more effectively. This involves a primary oxidation from arsenite to arsenate and the subsequent use of other techniques to remove the arsenate formed. Although the oxidation process alone cannot remove arsenic from an aqueous solution, it can facilitate the removal of the total arsenic. Some of the common oxidants used can be listed as natural atmospheric oxygen, permanganate chlorine, ozone, hypochlorite, etc. Sorlini et al. (2010) examined the As (III) oxidation efficiencies of various chemicals, among which chlorine dioxide oxidised 86% of As (III) in the initial contact time of 1 h, while potassium permanganate could achieve complete oxidation within a few minutes.

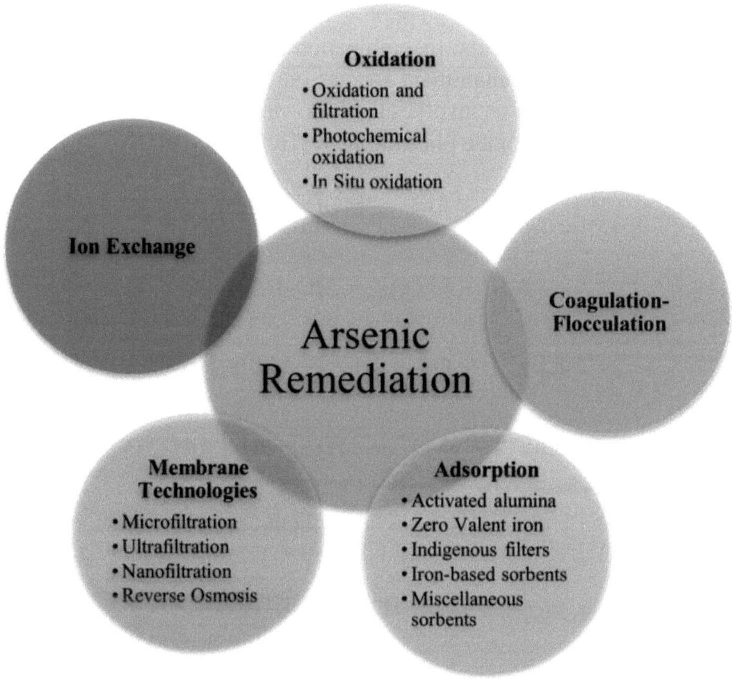

Fig. 2 Arsenic treatment techniques available for water and wastewater

5.2 *Photocatalysis*

Nanocatalysts have unique features because of the structural and electrical changes that distinguish them from bulk materials. In recent decades, several materials like oxides of titanium, aluminium, iron, and zinc have been used as nanoscale catalysts (Somwanshi et al., 2022). In nanocatalysis, when light energy interacts with metallic NP_S, it results in high and widespread photocatalytic activity. This makes it a popular treatment. Photocatalysts alter the rate of reactions without actively participating in the chemical transformation process. Nano–photocatalysts are comprehensively utilised for the purification of water and wastewater, as they have improved catalytic reactivity because of their higher surface area and other exclusive size–dependent characteristics. It has been established that nano–photocatalysts enhance oxidation capabilities by generating oxidising species at the surface of the material; this facilitates the breakdown of contaminant species (Gómez–Pastora et al., 2017). Photocatalysis by the NP_S of metal oxides such as TiO_2, WO_3, ZnO and Zn_2SnO_4 can effectively purify water by eliminating chemical and biological pollutants.

One captivating feature of photocatalysis can be the use of sunlight, making it energy–efficient; this facilitates its application in remote areas. The heterogeneous photocatalysis, as semiconductors, possess a wide band gap that produces electron–hole pairs (excitons) when exposed to light with energy higher than that of the band gap energy of the material. The photogenerated electrons go up to the conduction band, some of which escape recombination and start the reduction–oxidation reactions in the absorbed contaminant species on the surface of the photocatalyst, oxidising them. Photocatalysis involves the following fundamental process (Baruah et al., 2016). {MO—metal oxide photocatalyst (TiO_2, ZnO, etc.)}

$$MO + h\nu 4 \rightarrow MO\left(e^- + h^+\right) \quad (1)$$

$$MO\left(e^-\right) + O_2 \rightarrow MO + \cdot O_2- \quad (2)$$

$$MO\left(e^-\right) + \cdot O_2^- + 2H+ \rightarrow MO + H_2O_2 \quad (3)$$

$$MO\left(e^-\right) + H_2O_2 \rightarrow MO + \cdot OH + OH^- \quad (4)$$

$$\cdot O_2^- + H_2O_2 \rightarrow \cdot OH + OH^- + O_2 \quad (5)$$

$$\cdot O_2^- + H^+ \rightarrow \cdot OH_2 \quad (6)$$

$$MO\left(e^-\right) + \cdot OH_2 \rightarrow MO + HO_2^- \quad (7)$$

$$HO_2^- + H^+ \rightarrow H_2O_2 \quad (8)$$

$$2 \cdot OH_2 \rightarrow O_2 + H_2O_2 \quad (9)$$

Oxidation reactions initiated by generated holes are:

$$MO\left(h^+\right) + H_2O \rightarrow MO + \cdot OH + H^+ \quad (10)$$

$$MO\left(h^+\right) + H_2O \rightarrow MO + \cdot OH + H^+ \quad (11)$$

$$MO\ \left(h^+\right) + OH^- \rightarrow MO + \cdot OH \quad (12)$$

Most nano–photocatalysts have merits such as being non–poisonous, economical, conveniently accessible, chemically stable and having outstanding photoactive characteristics at the nano–scale (ranging from 1 to 100 nm) (Ciambelli et al., 2020). Photocatalysis is an effective and advanced method for the elimination of arsenic from aqueous solutions, involving the use of catalysts to convert arsenic into less toxic or more easily removable forms. It primarily takes place through the oxidation of As (III) to As (V) and then adsorption of the generated As (V) on the catalyst surface. Various available metal oxides, including titanium dioxide (TiO_2), zinc oxides (ZnO), etc., are in use as catalysts. Among them, TiO_2 has strong photostability, and many other nano–photocatalysts, such as copper–based materials, metal sulphides, zinc oxide, etc., show comparatively low chemical stability due to photo–corrosion (Weng et al., 2019). Therefore, scientific research is currently focusing on the development of additional nano–composite materials that have lower toxicity and can be safely used in water treatment processes, as well as the development of novel photocatalysts that can work over a wide range of visible light as well as for long–term (Yaqoob et al., 2020).

5.3 *Adsorption*

Adsorption is a well–recognised technology employed for arsenic decontamination of water and wastewater. The adsorption process offers many advantages over other techniques, including low energy requirements, secure operational handling, cost–effectiveness, regeneration ability and being free from the generation of sludge. Furthermore, this method can completely remove contaminants from the aqueous solution. This technique is particularly advantageous for developing countries, which may have limited access to skilled personnel (Alka et al., 2021; Siddique et al., 2020). Adsorption is considered a surface phenomenon that includes the accumulation of adsorbate species from a solution on the surface of the solid adsorbent material. The elimination of contaminants can occur through physical adsorption using van der Waal forces, electrostatic interaction, etc., which involves the formation of outer–sphere surface–complexes (Fig. 3). On the other path, chemi–sorption took place through the development of chemical bonds and inner–sphere surface–complexes between the adsorbate species and the adsorbent surface (Hao et al., 2018). Following are the reactions that may take place at the adsorbent surface,

$$MOH_{(surface)} + H_2AsO_{(aq)} \rightarrow MOH + \ldots\ldots H_2AsO_{(surface)} \quad (13)$$

$$MOH_{(surface)} + HAsO_{(aq)} \rightarrow MOH + \ldots\ldots H_2AsO_{(surface)} \quad (14)$$

$$MOH_{(surface)} + H_2AsO_{(aq)} \rightarrow MOH + \ldots\ldots H_2AsO_{(surface)} \quad (15)$$

Different kinds of materials have already been examined for their efficiency in the elimination of arsenic, including activated carbon; biochar; minerals; clays; industrial byproducts, such as red mud and fly ash; metal oxides, that is, iron oxides and aluminium oxides; and biopolymers like alginate and chitosan (Hao et al., 2018). Many of the conventional adsorbents have limitations including lower surface area and adsorption capacity; slow kinetics; lesser chemical stability, etc. The employment of nanomaterials for the purification of water and wastewater can provide promising opportunities. With the size–dependent properties of nanomaterials, including porosity, surface area and specific surface charge, it is possible

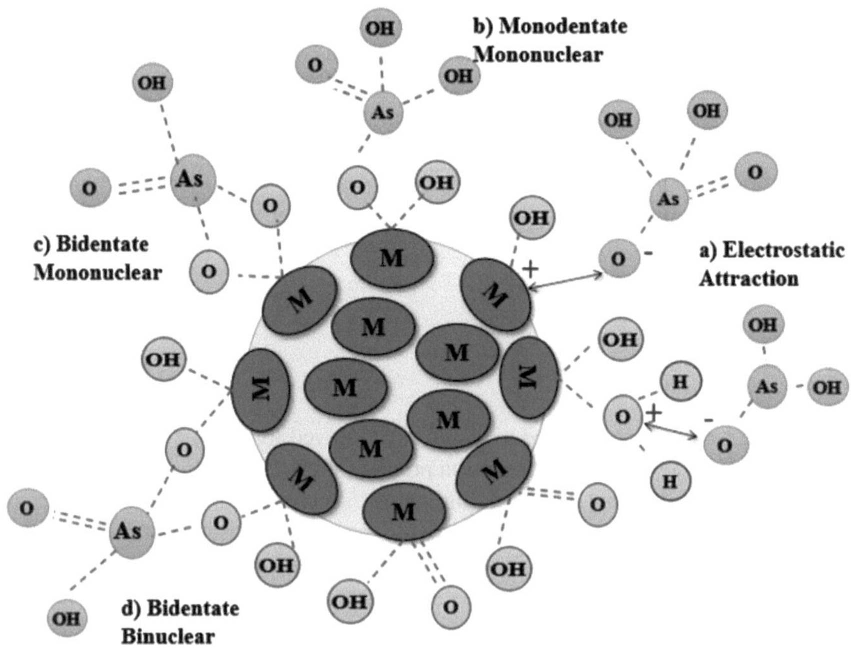

Fig. 3 The mechanism of arsenic adsorption

to achieve higher efficiency, selectivity, chemical stability, the presence of abundant functional groups and reusability, which offer a more sustainable approach for arsenic decontamination. Among the hundreds of adsorbents available, iron–based materials were comprehensively investigated and found to be promising for arsenic adsorption because of their high removal efficacy, non–toxicity, cost–effectiveness and high availability of precursors (Jain, 2022).

Also, it must be considered that the efficacy of this method can be influenced by different factors, that is, contact period, material dose, solution pH, temperature and initial arsenic levels. The existence of other ionic species, such as silicates, phosphates, sulphates and organic matter, can influence the elimination of arsenic through either the competition with arsenic species for adsorption sites or because of the blockage of adsorption sites by macromolecules (Altowayti et al., 2021). Further, the regeneration ability of a saturated adsorbent is one of the important aspects that need to be examined comprehensively. For physical adsorption, the weak bonds can be broken down by simple thermo–mechanical forces, while for chemisorption, saturated adsorbents should be treated with eluent solutions, which increase the surface charges of the adsorbent and thereby release the adsorbed contaminant (Jain, 2022). This technique also suffers from some disadvantages, that is, the requirement for the replacement of adsorbent after exhaustion, lack of self–monitoring and efficiency for low arsenic concentrations (Alka et al., 2021).

5.4 Chemical Precipitation

Chemical precipitation is a widely employed technique for the treatment of water and wastewater because of its advantages, such as simplicity, effectiveness and the pre—treatment step not being necessary. It involves the transformation of dissolved contaminant species into insoluble solid particles, facilitating their sedimentation, which can then be removed from the water. Specific reagents such as lime (calcium hydroxide), alum (aluminium sulphate) and ferric chloride are added to the water, which act as precipitants and react with contaminants. Finally, the precipitates can be sedimented or filtered from the water (Nicomel et al., 2016).

Many researchers have investigated metal salts of iron, magnesium, aluminium, calcium, titanium, zirconium, etc. for the effective removal of arsenic from aqueous solutions by the precipitation method, among which ferric salts showed high efficiencies. These salts in aqueous systems dissociate into positively charged electrolytic ions (Al^{3+}, Fe^{3+}), which reduce the concentrations of negatively charged arsenic oxyanion species by forming larger colloid particles of insoluble solids. These colloids ultimately get precipitated, for example, ferric chloride reacts with arsenate and forms an insoluble ferric arsenate ($FeAsO_4$) compound, aluminium sulphate reacts to form a precipitate of aluminium arsenate ($AlAsO_4$) and lime (calcium hydroxide) reacts to give a precipitate of calcium arsenate ($Ca_3(AsO_4)_2$) (Alka et al., 2021).

$$Fe^{3+} + AsO \leftrightarrow FeAsO_4(s) \tag{16}$$

$$Al^{3+} + AsO \leftrightarrow AlAsO_4(s) \tag{17}$$

$$3Ca^{2+} + 2AsO \leftrightarrow Ca_3(AsO_4)_2(s) \tag{18}$$

Chemical precipitation techniques are effective for reducing the levels of arsenic in waters or effluents containing higher concentrations of arsenic. However, lowering the content of arsenic below the permissible limit of $10\mu g/L$ is generally difficult to achieve through this method (Hao et al., 2018).

Though this technique is widely used, it suffers from some drawbacks, such as the production of huge volumes of unwanted sludge, leading to disposal issues; the removal reaction being highly pH dependent, which needs to be controlled; in addition, the precipitation process could be inhibited due to the existence of complex forming agents (Altowayti et al., 2021). To overcome this limitation, the approach of combining chemical precipitation with other methods such as filtration, ion–exchange or adsorption can enhance the efficiency of the removal of arsenic. Nano composites of iron, aluminium and calcium oxyhydroxides/chlorides/sulphates were also used for the improved precipitation of arsenic. It is also observed from the literature that the efficiency of the removal of arsenic by chemical precipitation relies on the solution pH, arsenic concentrations, type of precipitating reagent used, etc.

As mentioned earlier, As (III) is difficult to remove by chemical precipitation as its un–charged species is present dominantly till pH 9.2. Therefore, it should be oxidised first to arsenate and then employed for precipitation reactions (Nicomel et al., 2016).

5.5 *Membrane Filtration*

In the process of water and wastewater purification, membrane filtration is a well–known technology for the efficient elimination of arsenic as well as other pollutants (Hao et al., 2018). The membranes employed in this technique are highly porous and can act as barriers to selective contaminants. Considering their pore size and the applied pressure, membranes can be categorised into micro–filtration, ultra–filtration, nano–filtration and reverse osmosis (RO). Among these, due to their larger pore sizes, ultrafiltration (10–100 nm) and microfiltration (100–10,000 nm) membranes alone cannot be employed to eliminate the arsenic species dissolved in water or wastewater (Siddique et al., 2020). Hence, pre–treatments, such as coagulation–flocculation or precipitation, need to be employed before filtration to increase the size of arsenic–containing colloids. Han et al. (2002) investigated the feasibility of flocculation followed by microfiltration for the removal of arsenic. Where $Fe_2(SO_4)_3$ and $FeCl_3$ salts were used as coagulants, results showed that the flocculation process significantly increased the rate of removal. However, the efficiency was influenced by solution pH and the membrane pore size.

The high–pressure membrane techniques, namely, nanofiltration (NF) and RO, were found to be promising for eliminating the dissolved arsenic species from water, especially for household and community–level applications where arsenic concentrations can be reduced below permissible limits by using these processes (Alka et al., 2021). The RO membrane technique possesses the highest efficiency for arsenic elimination because of its small pore size (<1 nm), but it also suffers from some significant disadvantages that limit its use. This process is highly energy–intensive and requires skilled personnel for operation and maintenance. Further, it also generates large volumes of residual water, which needs to be managed properly; all these issues make this process non–feasible for rural applications (Altowayti et al., 2021). Schmidt et al. (2016) studied the elimination of arsenic from real groundwater at a pilot scale using the RO technique, where the arsenic–contaminated water after pre–treatment was fed to the RO membrane. It was observed that arsenic concentrations lowered by an average of 89% from 480μg/L.

Nanofiltration (NF) is considered as an efficient and comparatively low–priced technique than RO. It uses membranes with pore sizes of 1–10 nm and operates at a pressure of 2–10 bar. It removes dissolved arsenic species through size–based sieving and Donnan exclusion. The disposal of waste from NF is less challenging compared to low–pressure membranes (such as microfiltration and ultrafiltration) because the waste stream from NF is smaller and contains fewer dissolved ions. However, because of their dependency on the mechanism of charge–based exclusion, the NF membranes may not be effective for uncharged arsenic species. It is observed in the literature that arsenate can be effectively removed by nano–filtration membranes, while for arsenite, a pre–oxidation step is required (Siddique et al., 2020). Other than the arsenic speciation, the presence of suspended solids can significantly influence the removal of arsenic by NF and RO membranes (Nicomel et al., 2016). Despite the high efficiencies of these techniques, issues like membrane fouling, reduced flux and concentrate discharge need to be addressed.

The sensitivity of these organic polymeric membranes to extreme water or wastewater conditions may lead to their degradation. To overcome this limitation, research has been focusing on the development of membranes with higher chemical and mechanical strength, and antifouling properties. In this regard, various other options, namely, CNT, zeolites, ceramic membranes, etc. have been investigated properly. It is observed that these membranes offer merits such as durability, high mechanical strength, resistance to harsh water conditions and energy–efficiency. Further, the efficiency of the elimination of arsenic of these membranes can be enhanced by modifying them with metal oxides (Taghipour et al., 2019; Zaspalis et al., 2007).

6 Advantages and Limitations

Though the use of nanomaterials offers substantial benefits for the treatment of water and wastewater, their use can also have negative effects, including technical, environmental and health concerns. Due to their higher surface charges, NP_S tend to agglomerate, reducing their efficiency in the treatment of water. Further, the difficulty in the separation of nanomaterials from treated water can pose a technical problem. If they get mobilised into a water system, nanomaterials can penetrate cell membranes because of their small size, potentially causing toxicity to humans and aquatic animals. Therefore, the usage of nanomaterials in the purification of water and wastewater needs advanced infrastructure and skilled personnel, reducing the feasibility of the treatment process. Considering all these limitations, research should be carried out for the development of environment–friendly, non–toxic and biodegradable nanomaterials. Also, attention should be given to improving the practical applications of nanomaterials in large–scale water and wastewater purification systems (Taghipour et al., 2019; Tahoon et al., 2020).

7 Summary and Conclusions

Nanotechnology is a promising approach for addressing arsenic contamination of water and wastewater. This chapter reviewed the significant properties of nanomaterials and their application in available purification technologies for water and wastewater, focusing on the removal of arsenic. The respective removal mechanisms of these technologies and the case studies are also discussed in detail. It was observed from the literature that NP_S of metal oxides, composite nanomaterials and functionalised carbon nanostructures possess outstanding properties, that is, large adsorption efficiencies, selectivity and regeneration abilities. In conclusion, the employment of nanotechnology for the removal of arsenic has been providing efficient and sustainable solutions for the purification of water and wastewater. Apart from the adsorbents, nanomaterials have also been employed as sensors for the detection of arsenic, which demonstrated promising performance.

The application of the NP_S in different available water and wastewater purification techniques, such as oxidation, adsorption, photocatalysis and membrane filtration, has significantly improved their feasibility, making them viable alternatives to conventional materials. Despite many advantages, the practical application of nanomaterials requires further thorough investigation. In addition, the economic synthesis and selective properties of nanomaterials, along with their feasible reusability and disposal after exhaustion, are some of the approaches that require further thorough evaluation.

While many of the nanotechnologies discussed in this chapter are in the laboratory research stage, some have progressed to pilot testing or even commercialisation. Nanoadsorbents, nanotechnology–enabled membranes and nano–photocatalysts have shown potential for full–scale application in the near future based on their research and development stages, commercial availability, cost of nanomaterials and compatibility with the existing infrastructure. Although the abovementioned nanotechnologies have commercial products, they have not yet been applied in large–scale treatment of water or wastewater. In addition, numerous other water treatment nanotechnologies have found better use in point–of–use (POU) systems. The challenges faced during the application of different nanotechnologies, such as technical hurdles, high costs and potential environmental and human risks, in water and wastewater treatment systems are noteworthy, but they can be mitigated in the near future. More collaboration is needed between R&D organisations and industry partners or stakeholders to overcome these barriers and enhance nanotechnology, which can continuously provide robust solutions to water and wastewater treatment challenges, both incremental and revolutionary.

Acknowledgements The authors are thankful to the Director, CSIR–IMMT, for permitting to publish this work. Rutuja T. Kamble, UGC–SRF, one of the authors, acknowledges UGC, New Delhi, for financial support and extends her thanks to AcSIR, Ghaziabad.

References

Alka, S., Shahir, S., Ibrahim, N., Ndejiko, M. J., Vo, D. V. N., & Abd Manan, F. (2021). Arsenic removal technologies and future trends: A mini review. *Journal of Cleaner Production, 278*, Article 123805. https://doi.org/10.1016/j.jclepro.2020.123805

Altowayti, W. A. H., Othman, N., Shahir, S., Alshalif, A. F., Al–Gheethi, A. A., Altowayti, F. A. H., Saleh, Z. M., & Haris, S. A. (2021). Removal of arsenic from wastewater by using different technologies and adsorbents: A review. *International Journal of Environmental Science and Technology*, 1–24. https://doi.org/10.1007/s13762-021-03660-0

Babaee, Y., Mulligan, C. N., & Rahaman, M. S. (2018). Removal of arsenic (III) and arsenic (V) from aqueous solutions through adsorption by Fe/Cu nanoparticles. *Journal of Chemical Technology & Biotechnology, 93*(1), 63–71. https://doi.org/10.1002/jctb.5320

Baruah, S., Najam Khan, M., & Dutta, J. (2016). Perspectives and applications of nanotechnology in water treatment. *Environmental Chemistry Letters, 14*, 1–14. https://doi.org/10.1007/s10311-015-0542-2

Beyene, H. D., Werknch, A. A., Bezabh, H. K., & Ambaye, T. G. (2017). Synthesis paradigm and applications of silver nanoparticles (AgNPs), a review. *Sustainable Materials and Technologies, 13*, 18–23. https://doi.org/10.1016/j.susmat.2017.08.001

Bora, T., & Dutta, J. (2014). Applications of nanotechnology in wastewater treatment—A review. *Journal of Nanoscience and Nanotechnology, 14*(1), 613–626. https://doi.org/10.1166/jnn.2014.8898

Chandra, V., Park, J., Chun, Y., Lee, J. W., Hwang, I. C., & Kim, K. S. (2010). Water–dispersible magnetite–reduced graphene oxide composites for arsenic removal. *ACS Nano, 4*(7), 3979–3986. https://pubs.acs.org/doi/abs/https://doi.org/10.1021/nn1008897

Ciambelli, P., La Guardia, G., & Vitale, L. (2020). Nanotechnology for green materials and processes. *Studies in Surface Science and Catalysis, 179*, 97–116. https://doi.org/10.1016/B978-0-444-64337-7.00007-0

Deedar, N. A. B. I., & Aslam, I. (2009). Evaluation of the adsorption potential of titanium dioxide nanoparticles for arsenic removal. *Journal of Environmental Sciences, 21*(3), 402–408. https://doi.org/10.1016/S1001-0742(08)62283-4

Feng, L., Cao, M., Ma, X., Zhu, Y., & Hu, C. (2012). Superparamagnetic high–surface–area Fe_3O_4 nanoparticles as adsorbents for arsenic removal. *Journal of Hazardous Materials, 217*, 439–446. https://doi.org/10.1016/j.jhazmat.2012.03.073

Ghosh, S., Prabhakar, R., & Samadder, S. R. (2019). Performance of γ–aluminium oxide nanoparticles for arsenic removal from groundwater. *Clean Technologies and Environmental Policy, 21*, 121–138. https://doi.org/10.1007/s10098-018-1622-3

Gómez-Pastora, J., Dominguez, S., Bringas, E., Rivero, M. J., Ortiz, I., & Dionysiou, D. D. (2017). Review and perspectives on the use of magnetic nanophotocatalysts (MNPCs) in water treatment. *Chemical Engineering Journal, 310*, 407–427. https://doi.org/10.1016/j.cej.2016.04.140

Habuda-Stanić, M., & Nujić, M. (2015). Arsenic removal by nanoparticles: A review. *Environmental Science and Pollution Research, 22*, 8094–8123. https://doi.org/10.1007/s11356-015-4307-z

Han, B., Runnells, T., Zimbron, J., & Wickramasinghe, R. (2002). Arsenic removal from drinking water by flocculation and microfiltration. *Desalination, 145*(1–3), 293–298. https://doi.org/10.1016/S0011-9164(02)00425-3

Hao, L., Liu, M., Wang, N., & Li, G. (2018). A critical review on arsenic removal from water using iron–based adsorbents. *RSC Advances, 8*, 39545–39560. https://doi.org/10.1039/c8ra08512a

Jain, R. (2022). Recent advances of magnetite nanomaterials to remove arsenic from water. *RSC Advances, 12*(50), 32197–32209. https://doi.org/10.1039/D2RA05832D

Jain, K., Patel, A. S., Pardhi, V. P., & Flora, S. J. S. (2021). Nanotechnology in wastewater management: A new paradigm towards wastewater treatment. *Molecules, 26*(6), 1797. https://doi.org/10.3390/molecules26061797

Kempahanumakkagari, S., Deep, A., Kim, K. H., Kailasa, S. K., & Yoon, H. O. (2017). Nanomaterial–based electrochemical sensors for arsenic—A review. *Biosensors and Bioelectronics, 95*, 106–116. https://doi.org/10.1016/j.bios.2017.04.013

Li, J., Wu, Y. N., Li, Z., Zhu, M., & Li, F. (2014). Characteristics of arsenate removal from water by metal–organic frameworks (MOFs). *Water Science and Technology, 70*(8), 1391–1397. https://doi.org/10.2166/wst.2014.390

Mamindy-Pajany, Y., Hurel, C., Marmier, N., & Roméo, M. (2011). Arsenic (V) adsorption from aqueous solution onto goethite, hematite, magnetite and zero–valent iron: Effects of pH, concentration and reversibility. *Desalination, 281*, 93–99. https://doi.org/10.1016/j.desal.2011.07.046

Martínez-Cabanas, M., López-García, M., Barriada, J. L., Herrero, R., & de Vicente, M. E. S. (2016). Green synthesis of iron oxide nanoparticles. Development of magnetic hybrid materials for efficient As (V) removal. *Chemical Engineering Journal, 301*, 83–91. https://doi.org/10.1016/j.cej.2016.04.149

Massoudinejad, M., Keramati, H., & Ghaderpoori, M. (2020). Investigation of photo–catalytic removal of arsenic from aqueous solutions using UV/H2O2 in the presence of ZnO nanoparticles. *Chemical Engineering Communications, 207*(11), 1605–1615. https://doi.org/10.1080/00986445.2019.1674813

Mudhoo, A., Sharma, S. K., Garg, V. K., & Tseng, C. H. (2011). Arsenic: An overview of applications, health, and environmental concerns and removal processes. *Critical Reviews in Environmental Science and Technology, 41*(5), 435–519. https://doi.org/10.1080/10643380902945771

Nicomel, N. R., Leus, K., Folens, K., Van Der Voort, P., & Du Laing, G. (2016). Technologies for arsenic removal from water: Current status and future perspectives. *International Journal of Environmental Research and Public Health, 13*(1), 62. https://doi.org/10.3390/ijerph13010062

Purwajanti, S., Huang, X., Liu, Y., Yang, Y., Noonan, O., Song, H., & Yu, C. (2017). Mg (OH) 2–MgO@ reduced graphene oxide nanocomposites: The roles of composition and nanostructure in arsenite sorption. *Journal of Materials Chemistry A, 5*(46), 24484–24492. https://pubs.rsc.org/en/content/articlehtml/2017/ta/c7ta07629k

Samad, A., Furukawa, M., Katsumata, H., Suzuki, T., & Kaneco, S. (2016). Photocatalytic oxidation and simultaneous removal of arsenite with CuO/ZnO photocatalyst. *Journal of Photochemistry and Photobiology A: Chemistry, 325*, 97–103. https://doi.org/10.1016/j.jphotochem.2016.03.035

Schmidt, S. A., Gukelberger, E., Hermann, M., Fiedler, F., Großmann, B., Hoinkis, J., Ghosh, A., Chatterjee, D., & Bundschuh, J. (2016). Pilot study on arsenic removal from groundwater using a small–scale reverse osmosis system—Towards sustainable drinking water production. *Journal of Hazardous Materials, 318*, 671–678. https://doi.org/10.1016/j.jhazmat.2016.06.005

Shaji, E., Santosh, M., Sarath, K. V., Prakash, P., Deepchand, V., & Divya, B. V. (2021). Arsenic contamination of groundwater: A global synopsis with focus on the Indian Peninsula. *Geoscience Frontiers, 12*(3), Article 101079. https://doi.org/10.1016/j.gsf.2020.08.015

Siddique, T. A., Dutta, N. K., & Roy Choudhury, N. (2020). Nanofiltration for arsenic removal: Challenges, recent developments, and perspectives. *Nanomaterials, 10*(7), 1323. https://doi.org/10.3390/nano10071323

Kinetics and mechanism of arsenic removal using sulfide–modified nanoscale zerovalent iron. *Chemical Engineering Journal*, *412***,** 128667. https://doi.org/10.1016/j.cej.2021.128667

Somwanshi, S. B., Somvanshi, S. B., & Kharat, P. B. (2020). Nanocatalyst: A brief review on synthesis to applications. *Journal of Physics: Conference Series, 1644*(1), Article 012046. https://doi.org/10.1088/1742-6596/1644/1/012046

Sorlini, S., & Gialdini, F. (2010). Conventional oxidation treatments for the removal of arsenic with chlorine dioxide, hypochlorite, potassium permanganate and monochloramine. *Water Research, 44*(19), 5653–5659. https://doi.org/10.1016/j.watres.2010.06.032

Taghipour, S., Hosseini, S. M., & Ataie-Ashtiani, B. (2019). Engineering nanomaterials for water and wastewater treatment: Review of classifications, properties and applications. *New Journal of Chemistry, 43*(21), 7902–7927. https://doi.org/10.1039/c9nj00157c

Tahoon, M. A., Siddeeg, S. M., Salem Alsaiari, N., Mnif, W., & Ben Rebah, F. (2020). Effective heavy metals removal from water using nanomaterials: A review. *Processes, 8*(6), 645. https://doi.org/10.3390/pr8060645

Thakkar, S., Dumée, L. F., Gupta, M., Singh, B. R., & Yang, W. (2021). Nano–enabled sensors for detection of arsenic in water. *Water Research, 188*, Article 116538. https://doi.org/10.1016/j.watres.2020.116538

Weng, B., Qi, M. Y., Han, C., Tang, Z. R., & Xu, Y. J. (2019). Photocorrosion inhibition of semiconductor–based photocatalysts: Basic principle, current development, and future perspective. *ACS Catalysis, 9*(5), 4642–4687. https://doi.org/10.1021/acscatal.9b00313

Yaqoob, A. A., Parveen, T., Umar, K., & Mohamad Ibrahim, M. N. (2020). Role of nanomaterials in the treatment of wastewater: A review. *Water, 12*(2), 495. https://doi.org/10.3390/w12020495
Yu, Y., Yu, L., Shih, K., & Chen, J. P. (2018). Yttrium–doped iron oxide magnetic adsorbent for enhancement in arsenic removal and ease in separation after applications. *Journal of Colloid and Interface Science, 521*, 252–260. https://doi.org/10.1016/j.jcis.2018.02.046
Zaspalis, V., Pagana, A., & Sklari, S. (2007). Arsenic removal from contaminated water by iron oxide sorbents and porous ceramic membranes. *Desalination, 217*(1–3), 167–180. https://doi.org/10.1016/j.desal.2007.02.011

Silver Nanoparticles Production from Plant Extracts Using Green Synthesis Route for Wastewater Remediation

Upasana Priyadarshini and Remya Neelancherry

1 Introduction

Nanotechnology is a rapidly developing field, and it plays a crucial role in applied sciences. Unlike macroscale materials, nanomaterials exhibit distinct behaviour characterised by their size restriction, the predominance of interfacial phenomena owing to their augmented surface area and quantum effects (Danish et al., 2022). A new avenue has been created within the scientific community to explore the most secure approaches to synthesise, utilise and dispose of nanomaterials, while also mitigating any potential adverse effects on the environment because of the promising interest in nanoparticle (NP) research and its applications in medicines, environmental science and material science. Purification techniques for wastewater based on nanomaterials are exceedingly beneficial, and the utilisation of engineered nanomaterials as catalysts for the degradation of toxic substances is an area of research that is gaining prominence at present. Over the years, researchers have been using physical and chemical methods to develop uniform, evenly distributed and spherical metal NPs. Among these approaches, the chemical approach, which includes chemical reduction, electrochemical procedures and photochemical procedures, is most often used. However, these procedures are unsustainable, as they involve the use of dangerous chemicals, solvents and heat energy. Researchers worldwide are developing a novel approach for synthesising nanomaterials that adheres to the principles of green chemistry. In recent years, the utilisation of microorganisms and plant material to convert bulk materials into nanomaterials has evolved as a feasible replacement to conventional physical and chemical processes because of its simplicity, rapidity, environmental friendliness and cost-effectiveness (Narayanaswamy et al., 2023). A variety

U. Priyadarshini · R. Neelancherry (✉)
School of Infrastructure, Indian Institute of Technology, Bhubaneswar, Odisha, India
e-mail: remya@iitbbs.ac.in

U. Priyadarshini
e-mail: corina.vandergiessen@springer.com

© Capital Publishing Company, New Dehli, India. 2025
R. Neelancherry and S. M. Kumar, *Nanotechnology in Wastewater Treatment*,
https://doi.org/10.1007/978-3-031-94246-4_6

of valuable metals, including copper, silver, gold and palladium, were used in the process of NP synthesis.

Among the numerous types of NPs, silver nanoparticles (AgNPs) are very relevant because of their wide range of applications in fields such as medical, laboratories, pharmaceutical industry and cosmetics. They also possess therapeutic properties (Kaur et al., 2023). The remarkable characteristics of AgNPs, including their optical behaviour, catalytic action, electrical conductivity and chemical stability, have garnered considerable interest. Multiple approaches, including physicochemical and biological processes, are used to produce the AgNP. Because AgNP is safe for the environment and does not produce any harmful by-products, the biological process of synthesis is much more desirable. This chapter focusses on the extraction of plant parts and further green synthesis of AgNPs from plant extracts. Also, the application of AgNPs in the remediation of wastewater is emphasised.

2 Synthesis of AgNPs

Nanomaterials are produced through biological, chemical or physical processes. Depending on the material used for their production, there are two types of procedures for synthesising AgNPs (Fig. 1). Firstly, the top-down synthesis method is utilised when the size of the primary material is larger than nanoscale, necessitating particle reduction via lithographic techniques, grinding, thermal ablation or milling. This technique for reducing particle size necessitates an immense load of energy and can bring to surface defects in NPs, which can have major impacts on their physical and chemical properties. Secondly, the bottom-up production process, referred to as the 'self-assembly approach', employs chemical and biological techniques whereby atoms aggregate at nucleation centres to create NPs. The process of biosynthesis of AgNPs is an economical bottom-up approach that enables the rapid formation of a significant quantity of NPs with minimal impurities and a uniform chemical composition (Khan et al., 2022b).

Physicochemical techniques of AgNP synthesis include chemical reduction, photochemical reduction, microwave-assisted procedure, electrochemical procedures and laser ablation. However, the biological technique is preferable to the physical and chemical procedures for the synthesis of AgNPs because of the inherent limitations of the latter. Both physically and chemically synthesised AgNPs have short lives of up to 20 min, and the methods produce contaminants as residues and also release sweet-smelling amines. In addition, the synthesis process is energy-intensive, expensive and it requires extensive maintenance (Danish et al., 2022).

Conversely, bacteria, fungi, algae, plants or plant extracts are the primary substrates used in the biological synthesis of AgNPs. These substrates function as reducing agents, stabilising agents or capping agents for reducing Ag^+ to Ag^0. Plants or plant extracts are inherently safe, readily available and non-toxic sources of a large number of metabolites required for AgNPs synthesis. Considering the drawbacks of physical and chemical procedures, the biological route of synthesising AgNPs has

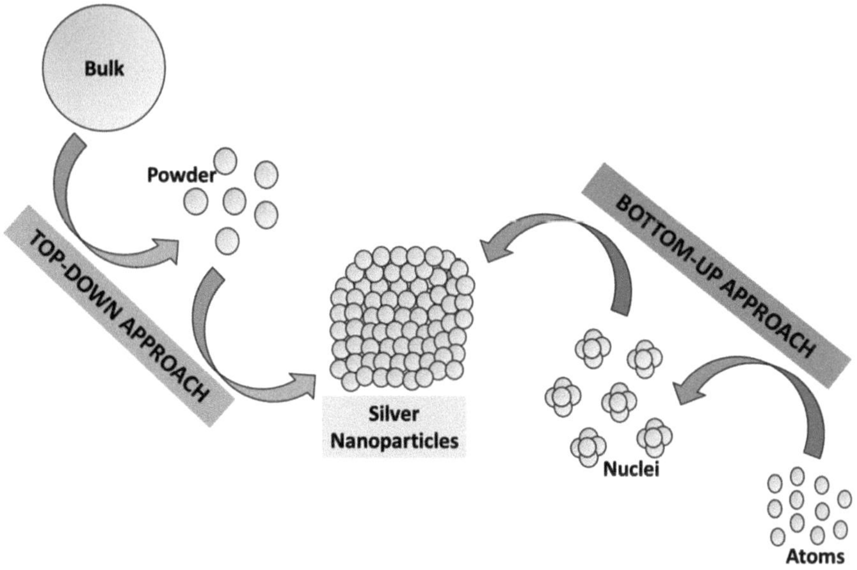

Fig. 1 Synthesis of AgNPs utilising top-down and bottom-up approaches

been viewed as potentially safer, non-toxic and cost-effective (Narayanaswamy et al., 2023).

3 Extraction Methods of Plants for AgNPs Synthesis

Multiple plant parts or products, such as leaves, stems, bark, seeds, secondary metabolites, latex, roots, peels, branches, fruits, tissues, seedlings and essential oils can yield extracts. The extraction methods have a substantial impact on the overall quality of the plant extract. The selection of solvents and the plant part intended for extraction has a profound influence on the success of extracting specific components. Several primary extraction techniques are used in the synthesis of AgNPs from plants. These methods include maceration extraction, extraction from solvent, microwave-assisted extraction and ultrasound-assisted extraction. Figure 2 illustrates the various techniques of plant extraction. Key attributes of the optimal extraction method are economic viability, simplicity, time efficiency and universal applicability across laboratory environments (Ritu et al., 2023).

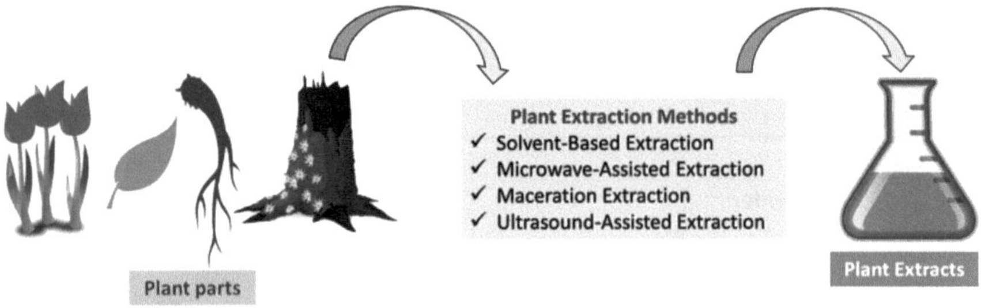

Fig. 2 Extraction of plant parts for the synthesis of AgNPs

3.1 Maceration Extraction

Maceration extraction is a conventional and simple extraction method involving three essential stages: (i) the plant is broken down into small particles, (ii) a suitable solvent is introduced into a sealed container, the type of compound to be extracted is identified and the vessel is left to soak for 3 days under room temperature and (iii) the liquid phase is separated by pressing or straining by filtration (Oroian et al., 2020). However, in terms of extraction efficiency and time consumption, this conventional extraction method is substandard.

3.2 Solvent-Based Extraction

Solvent-based extraction is the main technique for obtaining natural products. It involves the separation of soluble metabolites from plants through the use of a suitable solvent, while insoluble substances are discarded. In order to select an appropriate solvent for extraction, it is critical to have a detailed understanding of the properties of the plant, including the particulate size of the raw material and components of the constituents, prior to initiation. Extraction temperature, duration and solvent-to-solid ratio are all equally crucial parameters. Improving the processes of solute diffusion and solubilisation can substantially increase the experimental efficiency (Dekebo, 2019). The extraction technique comprises several phases, commencing with the solvent infiltrating the solid portion to initiate interactions with the surface area available. Solutes must subsequently diffuse away from the solid portion in order to be collected. There has been an increased recognition and application of environmentally friendly solvents, including ionic liquids (ILs) and ethyl lactate. These solvents are considered biocompatible, biodegradable and non-toxic alternatives to conventional ones. Moreover, these solvents require less effort to prepare and are economical. Notable progress in green solvent technology encompasses bio-derived solvents, ILs, deep eutectic solvents (DESs) and natural deep eutectic solvents (NDESs) (Sánchez-Camargo et al., 2019). Bio-based solvents are non-toxic and biodegradable solvents derived from renewable sources. ILs are non-crystallising compounds at ambient temperature composed primarily of quaternary nitrogen cations. DESs are solid or liquid mixtures of two or more organic compounds whose combined melting point is lower than that of either component used alone. NDES refers to DES that are derived from natural substances, including alcohols, choline derivatives, organic acids and sugars.

3.3 *Microwave-Assisted Extraction*

The efficient utilisation of microwave energy in the process of extracting from plants can be achieved by the rapid heating of both polar and polarisable compounds, including water and those found in the plant matrix sample. This process facilitates the transformation of plants into the necessary energy levels through the conduction of heat. The simultaneous occurrence of heat and mass transfer generates a synergistic effect that enhances and speeds up the recovery of the plant extract. Therefore, microwave-assisted extraction needs less solvent volume and extraction time compared to alternative techniques (Lux et al., 2019). The benefits are enhanced extraction yield and targeted heating of plant material. Two variations exist for this approach, namely, solvent-based and solvent-free extractions. Typically, the former is recommended for non-volatile compounds, while the latter is employed for volatile molecules. However, there are two important considerations in this regard. Firstly, precautions are necessary to prevent thermal degradation of samples, and secondly, this method is primarily suited for small-molecule phenolic compounds (Rao et al., 2020).

3.4 *Ultrasound-Assisted Extraction*

By utilising ultrasonic wave energy in the frequency range of 20–2000 kHz, ultrasound-assisted extraction increases the surface contact between the precursor and the plant parts, thereby extracting natural compounds from plant parts. Plant extract is prepared by immersing finely ground plant parts in an ultrasonic bath containing an appropriate solvent such as ethanol, acetone or methanol. This method offers numerous advantages compared to other techniques, including diminished demand for solvents, decreased consumption of energy, shorter duration of extraction and requirement for lower temperature (Altemimi et al., 2017).

4 AgNPs Production from Plant Extracts Using the Green Synthesis Route

Plant extracts provide an abundant source of enzymes, polyphenols, proteins, flavonoids and carbohydrates. These phytochemicals are isolated and used as stabilising and reducing agents in the extracellular production of metallic NPs, substituting potentially toxic substances such as sodium borohydride ($NaBH_4$). Due to the wide diversity of phytoconstituents present in the extracts, the specific mechanism underlying this phenomenon is not yet defined. Although proteins, polyphenols and organic acids are believed to be the major reducing agents, it is expected that the various phytochemicals will operate simultaneously (Ritu et al., 2023). Thus, green synthesis of AgNPs has the advantage of manufacturing huge amounts of contaminant-free NPs that have a more precisely defined size and morphology. The various green synthesis routes of AgNPs from different phytochemicals are shown in Fig. 3.

4.1 Polysaccharide Method

In this technique, polysaccharides are used as a capping agent and water as an ecologically friendly solvent; in other instances, polysaccharides are used as both capping and reducing agents. For synthesis of starch-AgNPs in a mildly heated environment, starch was used as a capping agent and β-D-glucose as a reducing agent. The presence of starch in the sample solution negates the need for using hazardous organic solvents. Furthermore, the intermolecular connections between starch and AgNPs are feeble and may be reversed at elevated temperatures, enabling the isolation of

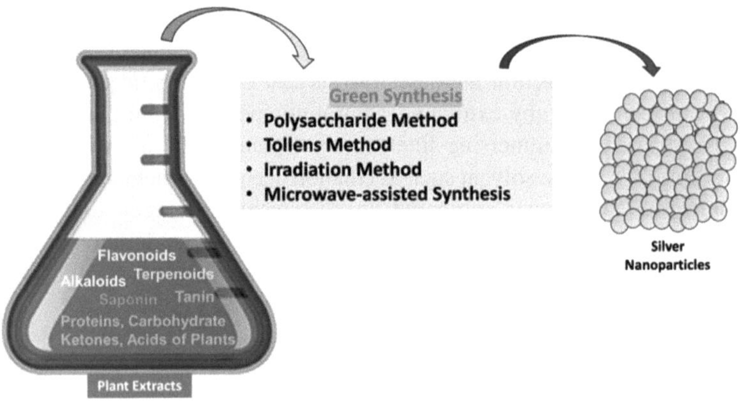

Fig. 3 Green synthesis of AgNPs from various phytochemicals

the synthesised particles. Moreover, starch-AgNPs are readily included into systems intended for biological and pharmacological applications (Sharma et al., 2009).

4.2 Tollens Method

The Tollens synthesis method produces size-controlled AgNPs in a single phase. The Tollens reaction is a chemical process in which the aldehyde reduces the Tollens reagent, $Ag(NH_3)_2^+$ (aq), as shown in Eq. (1).

$$Ag(NH_3)_2^+ \text{ (aq)} + RCHO \text{ (aq)} \rightarrow Ag(s) + RCOOH \text{ (aq)} \quad (1)$$

The Tollens synthesis method demonstrated that the reduction of Ag^+ ions by saccharides resulted in the formation of NPs at the lowest concentration of ammonia (Sharma et al., 2009). In another study, AgNPs coated with D-Maltose were generated using a modified Tollen's technique at ambient temperature. D-Maltose was chosen as both a reducing agent and a surface stabiliser. The presence of D-Maltose on the surface of AgNPs enabled easy functionalisation with different antibiotics via strong hydrogen bonding (Tippayawat et al., 2017).

4.3 Irradiation Method

AgNPs are synthesised effectively using a range of irradiation techniques. AgNPs with distinct shapes and size distribution may be produced by exposing an aqueous solution of Ag salt and surfactant to laser irradiation. This approach does not need a reducing agent. The gamma irradiation method is relatively useful for the synthesis of smaller metallic NPs. It can be performed at room temperature without reducing agents. The gamma irradiation method used a templating approach that confined the reaction centre within simulated 'nanoreactors' or 'nanomoulds', rather than using a particle dimension and a shape-controlled or template-free paradigm. The formation of NPs necessitates varying irradiation doses based on the circumstances and metal ion, influencing both the quantity and size of the NPs. By augmenting the irradiation dose, the quantity of NPs is elevated while their dimensions are reduced (Flores-Rojas et al., 2020).

4.4 Microwave-Assisted Synthesis

This method utilises microwave heating instead of conventional heating during synthesis. The plant extract was mainly added to a precursor such as $AgNO_3$ and after stirring kept in a microwave oven until a colour change was observed confirming synthesis of AgNPs (Anjana et al., 2021).

5 Characterisation of AgNPs

Various characterisation approaches have been used to examine the characteristics of the AgNPs produced. The primary qualitative assessment of the effective synthesis was conducted by observing the colour change in the AgNPs solution. The colour of AgNPs varies in intensity from yellow to dark brown due to the phenomenon of surface plasmon resonance (Khan et al., 2022b). The prevailing methods employed by most researchers to characterise NPs include UV-vis spectroscopy, Fourier transform infrared spectroscopy (FT-IR), X-ray powder diffraction (XRD), energy dispersive X-ray spectroscopy (EDX), scanning electron microscopy (SEM), transmission electron microscopy (TEM), dynamic light scattering (DLS), atomic force microscopy (AFM) and zeta potential measurement.

5.1 UV-Vis Spectroscopy Characterisation

This is used to determine the highest level of light absorption by the NP, which allows for the identification of its quantity and the measurement of the amount of light being captured by the particle. The synthesised NPs will have a distinct peak, displaying high absorption in the visible area. It has been proposed that the absorption band would fall between the wavelength range of 200 and 800 nm. The chemical environment, the dielectric medium and the particle size all affect how well AgNPs absorb light. For example, when moisture is absorbed by the dielectric medium and particles are treated at high temperatures, the dielectric constant changes and the deposited films are lost due to water loss (Pirtarighat et al., 2019).

5.2 FTIR

It is possible to identify the particular functional group that is present in a compound through the utilisation of FTIR spectroscopy. In addition, it can be utilised for the assessment of a variety of capping agents. It is possible to analyse the FTIR spectra of the AgNPs that were manufactured in order to determine the extent to which the plant extract participated in the process of producing AgNPs. In the majority of cases, the FTIR spectra of both the plant extract and the generated AgNPs were displayed. This made it possible to make a useful comparison between the spectra of the two different entities. The peaks also contribute to the identification of phytochemicals in the plant extracts (Kaur et al., 2023).

5.3 XRD

For analysing the crystalline structure of AgNPs, XRD procedure that deeply permeates the material with X-rays was used. The crystalline structure may be seen in the diffraction pattern that was produced as a result. The presence of broad and intense peaks in the XRD spectra of AgNPs suggests that they have a crystalline structure and are composed of tiny particles (Mishra et al., 2020).

5.4 SEM

Analysis of the representation and morphology of AgNPs is conducted using the SEM approach, which is used to determine the micro- and nanoscale size of the NPs. A distinct, spherical AgNP with no agglomeration was detected using the SEM (Mishra et al., 2020). The results of SEM examinations revealed that AgNPs created during the synthesis using the extract from leaves of *Parkia speciosa* had a spherical form and were sized between 26 and 39 nm (Khan et al., 2022a).

5.5 TEM

TEM has been utilised for determining the morphological characteristics, composition and dimension distribution of the AgNPs and is an excellent analytical instrument for determining the size of manufactured AgNPs. Study findings indicate that green AgNPs produced by biosynthesis had a roughly spherical shape, with dimensions ranging between 10 and 50 nm (Mishra et al., 2020).

5.6 DLS

DLS helps to identify the size of particles and their distribution within a liquid phase. This method is frequently used to characterise AgNPs, with the measurement being dependent on the size of the core of the particle, the concentration of particles and the various ions present in the medium. It is extensively utilised to modify magnetic AgNPs in a liquid state. Therefore, it is used to ascertain the mean dimensions of AgNPs in liquids (Kaur et al., 2023).

6 Factors Affecting AgNP Synthesis

In the context of green synthesis, the morphological characteristics of NPs can be modified by adjusting variables like temperature, reactant concentration, pH and reaction duration. The significance of these parameters in relation to environmental influences on NPs synthesis has been widely acknowledged. Furthermore, these elements may have a crucial role in optimising the synthesis of AgNPs (Vijayaram et al., 2024).

6.1 *Temperature*

Various degrees of research studies are being conducted globally to analyse the regulation of temperature on AgNPs. The temperature significantly influences the dimension, morphology and production of AgNPs, making it the primary determinant in their alteration. The synthesis of AgNPs may be customised based on temperature to produce various forms such as triangles, octahedral platelets, spheres and rods. Because of the rise in temperature, the reaction rate increases, leading to the development of more nucleation centres (Rana et al., 2020). The kinetic energy of the reaction is substantially impacted by temperature, resulting in an increased demand for silver ions and, as a consequence, a diminished probability of growth in particle size. The optimal temperature for the synthesis of AgNPs was determined to be 40 °C, which promotes the creation of AgNPs with smaller sizes (Sikdar & Sikdar, 2023).

6.2 *pH*

The pH has a crucial role in determining the structure of AgNPs. Along with temperature, pH also controls the development of nucleation centres. By increasing the pH, the nucleation centres are enlarged (Vijayaram et al., 2024). The optimal production of AgNPs occurs at alkaline pH levels, whereas little or no AgNPs generation is seen under acidic pH conditions. Instead of the accumulation of NPs, higher pH facilitates a significant rise in nuclei formation, leading to the production of a greater number of NPs with reduced diameters. The optimal pH for achieving the highest yield of AgNPs from clove buds' extract was determined to be at pH 10 (Sikdar & Sikdar, 2023). In another study, pH 11 was found to be the optimum pH where olive leaf and green tea leaf extracts were used to synthesise AgNPs. At low pH, the reduction potential of functional groups is diminished. However, as alkalinity increases and pH ranges expand, the reduction potential of all functional groups also increases, facilitating the construction of stable structures (Bergal et al., 2022).

6.3 *Reaction Time*

The impact of reaction time on the green synthesis of AgNPs using plant extracts is significant and can affect several aspects of the synthesis of NPs. The impact can be observed across various parameters including NP size, yield, stability and overall process efficiency. Reaction time can influence the size and morphology of the synthesised AgNPs. Generally, longer reaction times can lead to larger NPs due to prolonged reduction and nucleation processes. Shorter reaction times might yield smaller NPs. Prolonged reaction times might introduce impurities or oxidation of NPs, whereas shorter times could result in incomplete reduction or aggregation (Sikdar & Sikdar, 2023). After 60 min, there was no substantial change seen, indicating the end of the response. Increasing the duration of the reaction leads to a greater reduction of Ag^+ by biomolecules in the tea extract. However, a rapid reaction is more prone to causing the aggregation of AgNPs (Peng et al., 2021).

7 Applications of AgNPs in the Degradation of Pollutants

7.1 *AgNPs for Disinfection*

Disinfectants are used to eradicate microorganisms so as to ensure the provision of sanitary water. The most prevalent conventional approach to disinfecting water and wastewater has been through the utilisation of free chlorine. However, this method generates detrimental disinfection by-products, including trichloroacetaldehyde, chlorophenols, acetic acid (chlorinated), trihalomethanes, chlorite, halogenated acetonitrile and chloropicrin. Upon ingestion, these disinfection by-products may adversely affect human health, because of their cancerous nature, particularly colon and urinary bladder cancer. An additional potential disinfectant against protozoan cysts, viruses and bacteria is ozone; however, the method is expensive and generates reactive by-products including bromate and brominated by-products (Yang et al., 2024). The emergence of nanotechnology has facilitated the implementation of NPs in the disinfection of water and effluents. The AgNPs exhibit powerful and extensive antibacterial activity against both bacteria (gram-positive and -negative), as well as fungi and viruses. Consequently, they are promising substitutes for disinfecting effluent and treating potable water (Cao et al., 2020). The various AgNPs and their composites utilised as disinfection agents and the contaminants targeted are compiled in Table 1. AgNPs as well as modified AgNPs work efficiently in disinfecting various microbial contaminants in an aqueous medium.

Table 1 AgNPs for the disinfection of wastewater for the elimination of microbial contamination

AgNPs and composites of AgNPs	Synthesis method of AgNPs	Microbial contaminants	Remarks	References
Ultra-small AgNPs-loaded random coil lichenan nanocomposite (L-AgNP)	One-step reduction strategy	*Escherichia coli*, *Staphylococcus aureus*	>98% disinfection ratio	Yang et al. (2024)
Ag–Ru/ZnO bimetallic nanocomposite	Biosynthesis using the green method	*Escherichia coli*	85.15% of control of the antibacterial rate	Jha et al. (2024)
AgNPs	Biosynthesis from fresh leaves of Moringa oleifera extract	*Klebsiella pneumoniae*, *E. coli*, *Acinetobacter baumannii*, *Pseudomonas Aeruginosa*	Minimum bacterial concentration (MBC) = 500, 250, 500 and 250 μg/L, respectively	Younas et al. (2023)
AgNPs/ crystalline jute cellulose	Green synthetic bioreduction method	*S. aureus*, *Shigella dysenteriae*, *E. coli*, *Shigella boydii*	Zone of inhibitions: 7.5–18 mm	Rabbi et al. (2020)
Chitosan-AgNPs	Biosynthesis from grape pomace extract followed by dispersion and mixing of chitosan	*E. coli*	Total count of bacteria reduced by 47%	Raota et al. (2019)

7.2 AgNPs for the Degradation of Emerging Contaminants (ECs) in Aqueous Solutions

ECs that have gained widespread recognition include pharmaceutical residues, dyes, parabens, phthalates, phenols, pesticides and bisphenols (Khan et al., 2022a). These ECs are discharged directly from industrial and domestic sources. Conventional chemical, biological and physical methods have been employed for treating wastewater and addressing these concerns. These techniques include filtration, flotation, chemical oxidation, adsorption and solvent extraction. AgNPs are widely used because of their exceptional chemical, biological and physical capabilities, making them very popular among the many types of NPs (Sankar Sana et al., 2022). Table 2 enlists various AgNPs and their composites utilised in the degradation of ECs in an aqueous medium. The AgNPs and their different composites are efficient in degrading various ECs under optimum operating conditions in various wastewater sources.

Table 2 AgNPs and their composites for the degradation of dye from an aqueous medium

AgNPs and composites of AgNPs	Synthesis method of AgNPs	Target ECs	Optimum operating conditions	Degradation efficiency (%)	References
AgNPs	Green synthesis with *Ophiorrhiza mungos*	Methylene blue (MB)	pH = 7.4–8 Optimum dosage = 600 mg/L Initial EC concentration = 10 mg/mL Reaction time = 60 min	88.1 ± 1.74	Akhi et al. (2024)
AgNPs@chitosan/PVA	Green synthesis with chitosan/PVA as support material	MB, methylene red, methyl orange, safranin, crystal violet	Optimum dosage = 10 mg/L Initial EC concentration = 1 mg/L Reaction time = 90 min	77.41, 85.44, 81.69, 54.50 and 9.13, respectively	Meydan et al. (2023)
AgNPs/metal-organic frameworks (MOF)	*Argemone mexicana* L. leaf extract AgNPs coated with MOF	Metronidazole	pH = 5 Optimum dosage = 0.5 g/L Initial EC concentration = 100 mg/L Reaction time = 100 min	95.8	Oni et al. (2023)

(continued)

Table 2 (continued)

AgNPs and composites of AgNPs	Synthesis method of AgNPs	Target ECs	Optimum operating conditions	Degradation efficiency (%)	References
AgNPs/cellulose aerogel	Biosynthesis from Jasminum subtriplinerve blume leaves extract	p-Nitrophenol (PNP)/ $NaBH_4$	Optimum dosage = 1 g/L Initial EC concentration = 1/200 M ratio of PNP/ $NaBH_4$ Reaction time = 10 min	98.6	Gia-Thien Ho et al. (2023)
Graphene oxide/ AgNPs	Graphene oxide and silver ions are reduced concurrently via an aqueous extract of cinnamon plant bark	MB	pH = 2 Optimum dosage = 80 mg Initial EC concentration = 10 mg/L Reaction time = 50 min	99.98	Mamo and Zeleke (2022)
Cellulose-based AgNPs	In situ green synthesis of AgNPs from *Laurus nobilis* extract	MB	pH = 7 Optimum dosage = 10 mg Initial EC concentration = 10 mg/L Reaction time = 60 min	100	Kucukcobanoglu et al. (2022)

(continued)

Table 2 (continued)

AgNPs and composites of AgNPs	Synthesis method of AgNPs	Target ECs	Optimum operating conditions	Degradation efficiency (%)	References
AgNPs/wheat straw biochar	Green synthesis of *Ligustrum lucidum* leaf extract	Phenol	pH = 3 Optimum dosage = 0.05 g/L Initial EC concentration = 10 mg/L Reaction time = 90 min	78	Khan et al. (2022a)
AgNPs/titanium nanotube array	Green synthesis from *Camellia sinensis* leaves and *Coffea* powder	Ibuprofen	pH = 7 Optimum dosage = 1 g/L Initial EC concentration = 5 mg/L Reaction time = 20–30 min	49.6	Peng et al. (2021)
AgNPs	Green synthesis with extract of *Daucus carota* leaves	Rhodamine 6G dye	Optimum dosage = 0.1 mL Initial EC concentration = 10 μM Reaction time = 30 min	89.4	Joseph Kirubaharan et al. (2020)

(continued)

Table 2 (continued)

AgNPs and composites of AgNPs	Synthesis method of AgNPs	Target ECs	Optimum operating conditions	Degradation efficiency (%)	References
AgNPs	Biosynthesis with *M. piperita* extract	Crystal violet dye	pH = 9 Optimum dosage = 2 g/L Initial EC concentration = 1000 mg/L Reaction time = 20 min	Adsorption capacity = 704.7 mg/g	Sharbaf Moghadas et al. (2020)
AgNPs	*Orthosiphon stamineus* leaves extract	2,4-Dichlorophenoxyacetic acid	pH = 3.24 Optimum dosage = 0.009 g/L Initial EC concentration = 8.15 mg/L	97.8	Kamarudin et al. (2020)
Biogenic AgNPs	Green synthesis with aqueous extract of pineapple peel waste	MB	pH = 9.96 Optimum dosage = 0.5 mM Initial EC concentration = 40 mg/L Reaction time = 173 min	98.04	Agnihotri et al. (2018)

8 Conclusion

Nanotechnology, especially the use of AgNPs, offers significant potential because of their stability and low reactivity. However, traditional methods of production of AgNPs often involve harmful chemicals, highlighting the need for green synthesis techniques. Efforts towards various plant-based extraction methods like solvent-based, maceration, microwave and ultrasound-assisted extraction represent a promising method for extracting phytochemicals. The exploration of various green synthesis routes such as polysaccharide, Tollens, irradiation and microwave-assisted methods showcases the diversity of the available approaches. Further characterisation of AgNPs and optimisation of various parameters having impact on the green synthesis showed the potential application of AgNPs from plant extracts. For efficient applications of wastewater remediation, especially microbial disinfection and emerging contaminant treatment, AgNPs are one of the promising NPs.

References

Agnihotri, S., Sillu, D., Sharma, G., & Arya, R. K. (2018). Photocatalytic and antibacterial potential of silver nanoparticles derived from pineapple waste: Process optimization and modeling kinetics for dye removal. *Applied Nanoscience (Switzerland), 8*(8), 2077–2092. https://doi.org/10.1007/s13204-018-0883-9

Akhi, A. A., Hasan, A., Saha, N., Howlader, S., Bhattacharjee, S., Dey, K., & Ganguli, S. (2024). Ophiorrhiza mungos-mediated silver nanoparticles as effective and reusable adsorbents for the removal of methylene blue from water. *ACS Omega, 9*(4), 4324–4338. https://doi.org/10.1021/acsomega.3c05992

Altemimi, A., Lakhssassi, N., Baharlouei, A., Watson, D. G., & Lightfoot, D. A. (2017). Phytochemicals: Extraction, isolation, and identification of bioactive compounds from plant extracts. *Plants, 6*, 42. https://doi.org/10.3390/plants6040042

Anjana, V. N., Joseph, M., Francis, S., Joseph, A., Koshy, E. P., & Mathew, B. (2021). Microwave assisted green synthesis of silver nanoparticles for optical, catalytic, biological and electrochemical applications. *Artificial Cells, Nanomedicine and Biotechnology, 49*(1), 438–449. https://doi.org/10.1080/21691401.2021.1925678

Bergal, A., Matar, G. H., & Andaç, M. (2022). Olive and green tea leaf extracts mediated green synthesis of silver nanoparticles (AgNPs): Comparison investigation on characterizations and antibacterial activity. *BioNanoScience, 12*(2), 307–321. https://doi.org/10.1007/s12668-022-00958-2

Cao, Y., Naseri, M., He, Y., Xu, C., Walsh, L. J., & Ziora, Z. M. (2020). Journal of global antimicrobial resistance non-antibiotic antimicrobial agents to combat biofilm-forming bacteria. *Integrative Medicine Research, 21*, 445–451. https://doi.org/10.1016/j.jgar.2019.11.012

Danish, M. S. S., Estrella-pajulas, L. L., Alemaida, I. M., Grilli, M. L., Mikhaylov, A., & Senjyu, T. (2022). Green synthesis of silver oxide nanoparticles for photocatalytic environmental remediation and biomedical applications. *Metals, 12*(5), 769. https://doi.org/10.3390/met12050769

Dekebo, A. (2019). Introductory chapter: Plant extracts. *Plant Extracts*. https://doi.org/10.5772/INTECHOPEN.85493

del Pilar Sánchez-Camargo, A., Bueno, M., Parada-Alfonso, F., Cifuentes, A., & Ibáñez, E. (2019). Hansen solubility parameters for selection of green extraction solvents. *TrAC Trends in Analytical Chemistry*, **118:** 227–237. https://doi.org/10.1016/J.TRAC.2019.05.046

Flores-Rojas, G. G., López-Saucedo, F., & Bucio, E. (2020). Gamma-irradiation applied in the synthesis of metallic and organic nanoparticles: A short review. *Radiation Physics and Chemistry, 169*, 107962. https://doi.org/10.1016/J.RADPHYSCHEM.2018.08.011

Gia-Thien Ho, T., Thao Truong, D. P., Nguyen, H. B., Long Do, B., Dinh, T. A., Ton-That, P., & Tri, N. (2023). Green synthesized nano-silver/cellulose aerogel as a robust active and recyclable catalyst towards nitrophenol hydrogenation. *Chemical Engineering Journal*, 471, 144604. https://doi.org/10.1016/j.cej.2023.144604

Jha, P. K., Jaidumrong, T., Rokaya, D., & Ovatlarnporn, C. (2024). Callistemon viminalis leaf extract phytochemicals modified silver-ruthenium bimetallic zinc oxide nanocomposite biosynthesis: Application on nanocoating photocatalytic *Escherichia coli* disinfection. *RSC Advances, 14*(16), 11017–11026. https://doi.org/10.1039/d4ra01355g

Joseph Kirubaharan, C., Fang, Z., Sha, C., & Yong, Y. C. (2020). Green synthesis of Ag and Pd nanoparticles for water pollutants treatment. *Water Science and Technology, 82*(11), 2344–2352. https://doi.org/10.2166/wst.2020.498

Kamarudin, N. S., Jusoh, R., Jalil, A. A., Setiabudi, H. D., & Sukor, N. F. (2020). Synthesis of silver nanoparticles in green binary solvent for degradation of 2,4-D herbicide: Optimization and kinetic studies. *Chemical Engineering Research and Design, 159*, 300–314. https://doi.org/10.1016/j.cherd.2020.03.025

Kaur, N., Singh, A., & Ahmad, W. (2023). Microwave assisted green synthesis of silver nanoparticles and its application: A review. *Journal of Inorganic and Organometallic Polymers and Materials, 33*(3), 663–672. https://doi.org/10.1007/s10904-022-02470-2

Khan, M. N., Siddique, M., Mirza, N., Khan, R., Bilal, M., Riaz, N., & Khan, A. (2022a). Synthesis, characterization, and application of Ag-biochar composite for sono-adsorption of phenol. *Frontiers in Environmental Science, 10*(February), 1–13. https://doi.org/10.3389/fenvs.2022.823656

Khan, S. A., Jain, M., Pandey, A., Pant, K. K., Ziora, Z. M., Blaskovich, M. A. T., & Aminabhavi, T. M. (2022b). Leveraging the potential of silver nanoparticles-based materials towards sustainable water treatment. *Journal of Environmental Management*, **319**(July), 115675. https://doi.org/10.1016/j.jenvman.2022.115675

Kucukcobanoglu, Y., Ayisigi, M., Haseki, S., & Aktas, L. Y. (2022). In situ green synthesis of cellulose based silver nanocomposite and its catalytic dye removal potential against methylene blue. *Journal of Cluster Science, 33*(4), 1623–1633. https://doi.org/10.1007/s10876-021-02093-6

Lux, C., Lubio, A., Ruediger, A., Robert, S., & Muehlethaler, C. (2019). Optimizing the analysis of dyes by surface-enhanced Raman spectroscopy (SERS) using a conventional-microwave silver nanoparticles synthesis. *Forensic Chemistry, 16*, 100186. https://doi.org/10.1016/J.FORC.2019.100186

Mamo, B. Y., Zeleke, T. D. (2022). Green synthesis of rGO/Ag nanocomposite using extracts of Cinnamomum verum plant bark: Characterization and evaluation of its application for methylene blue dye removal from aqueous solutions. *International Journal of Nano Dimension, 13*(4), 414–434. https://doi.org/10.22034/IJND.2022.1963053.2159

Meydan, I., Aygun, A., Tiri, R. N. E., Gur, T., Kocak, Y., Seckin, H., & Sen, F. (2023). Chitosan/PVA-supported silver nanoparticles for azo dyes removal: Fabrication, characterization, and assessment of antioxidant activity. *Environmental Science: Advances, 3*(1), 28–35. https://doi.org/10.1039/d3va00224a

Mishra, A. K., Tiwari, K. N., Saini, R., Kumar, P., Mishra, S. K., Yadav, V. B., & Nath, G. (2020). Green synthesis of silver nanoparticles from leaf extract of Nyctanthes arbortristis L. and assessment of its antioxidant, antimicrobial response. *Journal of Inorganic and Organometallic Polymers and Materials, 30*(6), 2266–2278. https://doi.org/10.1007/s10904-019-01392-w

Narayanaswamy, S., Bhaskar, R., Jayadevappa, R. K. K., Ramachandran, S. K. M., & Prasad, M. A. (2023). A comprehensive review on the antimicrobial and photocatalytic properties of green synthesized silver nanoparticles. *Letters in Applied NanoBioScience, 12*(4), 1–18. https://doi.org/10.33263/LIANBS124.140

Oni, B. A., Sanni, S. E., Agu, K. C., & Tomomewo, O. S. (2023). Green synthesis of Ag nanoparticles from Argemone mexicana L. leaf extract coated with MOF-5 for the removal of metronidazole antibiotics from aqueous solution. *Journal of Environmental Management, 342*, 118161. https://doi.org/10.1016/j.jenvman.2023.118161

Oroian, M., Dranca, F., & Ursachi, F. (2020). Comparative evaluation of maceration, microwave and ultrasonic-assisted extraction of phenolic compounds from propolis. *Journal of Food Science and Technology, 57*(1), 70–78. https://doi.org/10.1007/S13197-019-04031-X/FIGURES/2

Peng, Y. P., Liu, C. C., Chen, K. F., Huang, C. P., & Chen, C. H. (2021). Green synthesis of nano-silver–titanium nanotube array (Ag/TNA) composite for concurrent ibuprofen degradation and hydrogen generation. *Chemosphere*, 264. https://doi.org/10.1016/j.chemosphere.2020.128407

Pirtarighat, S., Ghannadnia, M., & Baghshahi, S. (2019). Green synthesis of silver nanoparticles using the plant extract of Salvia spinosa grown in vitro and their antibacterial activity assessment. *Journal of Nanostructure in Chemistry, 9*(1), 1–9. https://doi.org/10.1007/s40097-018-0291-4

Rabbi, M. A., Rahman, M. M., Minami, H., Habib, M. R., & Ahmad, H. (2020). Ag impregnated sub-micrometer crystalline jute cellulose particles: Catalytic and antibacterial properties. *Carbohydrate Polymers, 233.* https://doi.org/10.1016/j.carbpol.2020.115842

Rana, A., Yadav, K., & Jagadevan, S. (2020). A comprehensive review on green synthesis of nature-inspired metal nanoparticles: Mechanism, application and toxicity. *Journal of Cleaner Production*, 272. https://doi.org/10.1016/j.jclepro.2020.122880

Rao, S. S., Saptami, K., Venkatesan, J., & Rekha, P. D. (2020). Microwave-assisted rapid synthesis of silver nanoparticles using fucoidan: Characterization with assessment of biocompatibility and antimicrobial activity. *International Journal of Biological Macromolecules, 163*, 745–755. https://doi.org/10.1016/J.IJBIOMAC.2020.06.230

Raota, C. S., Cerbaro, A. F., Salvador, M., Delamare, A. P. L., Echeverrigaray, S., Da Silva Crespo, J., da Silva, T. B., & Giovanela, M. (2019). Green synthesis of silver nanoparticles using an extract of Ives cultivar (Vitis labrusca) pomace: Characterization and application in wastewater disinfection. *Journal of Environmental Chemical Engineering*, *7*(5), 1–9. https://doi.org/10.1016/j.jece.2019.103383

Ritu Verma, K. K., Das, A., & Chandra, P. (2023). Phytochemical-based synthesis of silver nanoparticle: Mechanism and potential applications. *BioNanoScience, 13*(3), 1359–1380. https://doi.org/10.1007/s12668-023-01125-x

Sankar Sana, S., Haldhar, R., Parameswaranpillai, J., Chavali, M., & Kim, S. C. (2022). Silver nanoparticles-based composite for dye removal: A comprehensive review. *Cleaner Materials, 6*, 100161. https://doi.org/10.1016/j.clema.2022.100161

Sharbaf Moghadas, M. R., Motamedi, E., Nasiri, J., Naghavi, M. R., & Sabokdast, M. (2020). Proficient dye removal from water using biogenic silver nanoparticles prepared through solid-state synthetic route. *Heliyon, 6*(8), e04730. https://doi.org/10.1016/j.heliyon.2020.e04730

Sharma, V. K., Yngard, R. A., & Lin, Y. (2009). Silver nanoparticles: Green synthesis and their antimicrobial activities. *Advances in Colloid and Interface Science, 145*(1–2), 83–96. https://doi.org/10.1016/j.cis.2008.09.002

Sikdar, S., & Sikdar, M. (2023). Green synthesis, optimization and analyzing of silver nanoparticles encapsulated with Syzygium aromaticum extract: Evaluating antibacterial and photocatalytic properties. *Bioresource Technology Reports, 24.* https://doi.org/10.1016/j.biteb.2023.101669

Tippayawat, P., Sapa, V., Srijampa, S., Boueroy, P., & Chompoosor, A. (2017). D-Maltose coated silver nanoparticles and their synergistic effect in combination with ampicillin. *Monatshefte fur Chemie, 148*(7), 1197–1203. https://doi.org/10.1007/s00706-017-2004-y

Vijayaram, S., Razafindralambo, H., Sun, Y. Z., Vasantharaj, S., Ghafarifarsani, H., Hoseinifar, S. H., & Raeeszadeh, M. (2024). Applications of green synthesized metal nanoparticles—A review. *Biological Trace Element Research, 202*(1), 360–386. https://doi.org/10.1007/s12011-023-03645-9

Yang, Z., Gong, X., Hu, Y., Yue, P., Lü, B., & Peng, F. (2024). Green synthesis of lichenan-decorated silver nanoparticles for catalytic hydrogenation of organic dyes and bacterial disinfection. *Chemical Engineering Journal*, *487*. https://doi.org/10.1016/j.cej.2024.150516

Younas, M., Rasool, M. H., Khurshid, M., Khan, A., Nawaz, M. Z., Ahmad, I., & Lakhan, M. N. (2023). Moringa oleifera leaf extract mediated green synthesis of silver nanoparticles and their antibacterial effect against selected gram-negative strains. *Biochemical Systematics and Ecology*, *107*. https://doi.org/10.1016/j.bse.2023.104605

Advances in Nanofiltration

Swathi Desireddy and P. C. Sabumon

1 Introduction

Over the last few years, environmental pollution has deteriorated the quality of water. The ever-increasing pollution and global population have increased the demand for clean water. Conventional physical/chemical/biological treatment techniques such as precipitation, adsorption, coagulation-flocculation, electrocoagulation, ion-exchange, photocatalysis, oxidation, biological filters and microbial degradation have been extensively used for the elimination of several pollutants such as nutrients, organic dyes, radioactive materials and heavy metals (Santos et al., 2024; Siddique et al., 2020). Nevertheless, these techniques are cost-intensive and also target only a narrow spectrum of emerging pollutants (Feng et al., 2019). Recently, various classes of emerging pollutants are being detected in the treated effluents from pharmaceutical and food industries and agricultural units. This calls for the increased use of high-pressure membranes in the treatment and reuse of wastewater (Park & Snyder, 2020). In this context, membrane technologies including microfiltration (MF), ultrafiltration (UF), nanofiltration (NF) and reverse osmosis (RO) have received much interest in the treatment of industrial wastewaters (Koyuncu et al., 2015; Li et al., 2024b).

These membranes act as selective barriers between homogenous phases and carry out separation processes without chemically or physically altering the components as in distillation, extraction, fractionation and adsorption (Purkait et al., 2018). These membrane processes have several advantages like: (i) give good-quality water

S. Desireddy
School of Civil Engineering, Vellore Institute of Technology, Chennai, India

P. C. Sabumon (✉)
Centre for Climate Change and Environment, Vellore Institute of Technology, Chennai, India
e-mail: pcsabumon@vit.ac.in

© Capital Publishing Company, New Dehli, India. 2025

R. Neelancherry and S. M. Kumar, *Nanotechnology in Wastewater Treatment*,
https://doi.org/10.1007/978-3-031-94246-4_7

because of the efficient removal of dissolved solids and low molecular weight contaminants, (ii) simple and low-cost technique, (iii) no requirement of additives or chemicals as they work on the principle of molecular size, (iv) less energy requirement, (v) can be easily up-scaled or down-scaled, (vi) no generation of sludge, (vii) continuous flow operation, (viii) high efficiency and (ix) less footprint (Gupta et al., 2024; Tang et al., 2018; Wu, 2019).

Reliance on the pore flow model, which permits multivalent ions to pass through its membrane pores makes MF an inefficient low-pressure filtration technique. On the contrary, RO membranes excel at removing contaminants but are energy-intensive, have higher operational costs and are prone to significant scaling (Sengur-Tasdemir et al., 2020). NF is of great interest to researchers because of its high efficiency and lower operational costs (de Almeida et al., 2024). NF membranes separate substances through both size exclusion and Donnan exclusion, efficiently rejecting multivalent dissolved ions from water (Sarkar et al., 2021). These membranes exhibit high permeability to monovalent ions, low permeability to divalent ions and offer higher flux compared to RO membranes (Yang et al., 2024). Furthermore, NF membranes are known for their increased reliability and integrity, which result in longer cycle times and reduced costs (García Doménech et al., 2020). These characteristics make NF an excellent separation technology. As a result, it has been widely adopted in various industries, including pharmaceuticals, food, agriculture and biotechnology, for producing high-quality products and treating wastewater.

Despite the advantages of NF, it still has several shortcomings such as membrane fouling, advanced handling of concentrates, membrane resistance to certain chemicals and inadequate rejection in wastewater treatment. Furthermore, extensive industrial usage requires budget-friendly nanomembranes to bring down the operational costs compared to that of RO, though it is still expensive compared to UF and MF (Shahkaramipour et al., 2017). Thus, further advancement in NF technology is of prime importance and has prompted substantial research interest. This review focusses on the latest advancements in NF, exploring innovative membrane materials and their newly identified potential applications, building upon previously published reviews on the topic (Arola et al., 2019; Bashambu et al., 2021; García Doménech et al., 2020; Wang et al., 2018b; Yang et al., 2019). In addition, this review delves into future developments and prospects for NF technology. Figure 1 provides the schematic of advanced nanofiltration technology.

2 Fundamentals of NF

Among the prevailing pressure-driven membrane-based techniques, NF and RO are widely used in the treatment of industrial wastewater because of the low energy requirement and sophisticated process involved (Wang et al., 2018a). The characteristics of MF, UF, NF and RO are presented in Table 1. The pore sizes of NF membranes are in the range of 1–10 nm, which is slightly less than pore sizes of UF membranes and slightly higher than that of RO membranes. NF membranes can

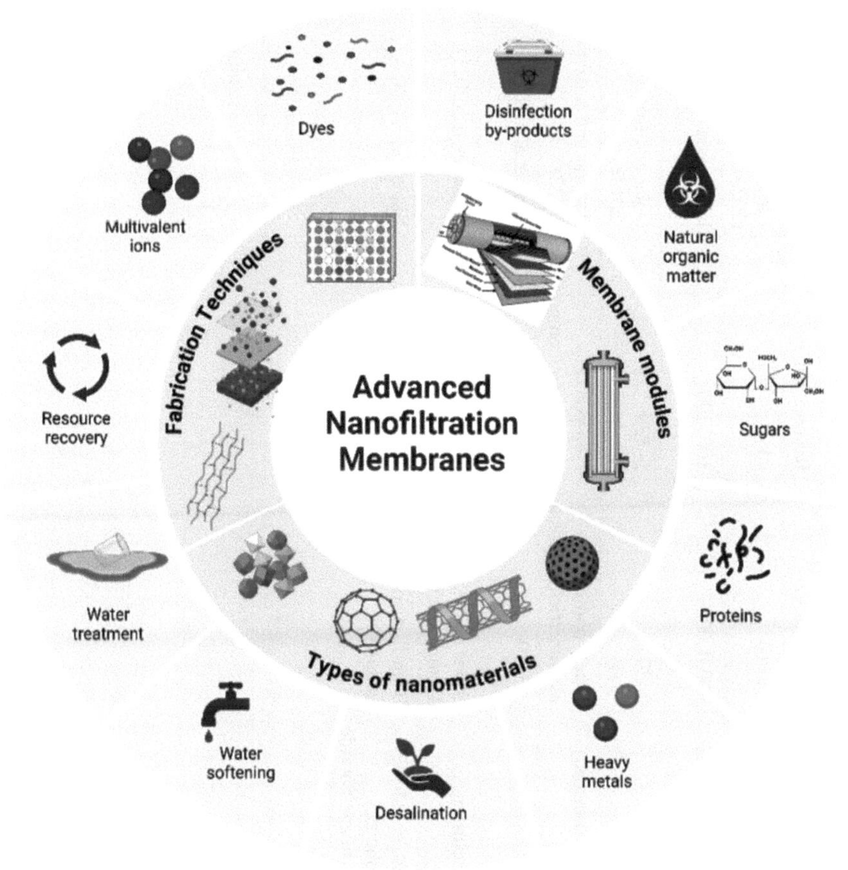

Fig. 1 Schematic of advanced nanofiltration technology: fabrication, nanomaterial integration and applications.

reject multivalent ions and specifically separate small organic components of diverse molecular weights, which make them suitable for lower energy and higher throughput uses where excessive removal of monovalent salts is not essential. NF membranes are widely employed in food, pharmaceutical, dairy and biotechnological industries; desalination; and the treatment of wastewater (Zhang et al., 2020).

2.1 NF Separation Mechanisms

Separation by NF membranes occurs via convective and diffusive passage mechanisms. Convective passage of ions with the solvent flux occurs because of the pressure differential between feed and effluent chambers, whereas diffusive transfer occurs

Table 1 Characteristics of various membrane processes

Characteristics	Microfiltration	Ultrafiltration	Nanofiltration	Reverse osmosis	References
Pore size (nm)	100–10,000	10–100	1 to 10	<1	Farhad and Ozan (2023)
Molecular weight cutoff (Da)	1,00,000	2,000–1,00,000	300–1,000	100–300	Imbrogno et al., (2024)
Pressure (bar)	0.1–2	0.1–5	3–20	5–120	Dolar and Košutić (2013)
Membrane type	Porous, isotropic	Porous, asymmetric	Finely porous, asymmetric/composite	Non-porous, asymmetric/composite	Bhaskar and Kaleekkal (2022)
Permeability (L/h m^2 bar)	>1,000	10–1,000	1.5–30	0.05–1.5	Yang et al., (2019)
Mechanism	Steric exclusion	Steric exclusion	Steric exclusion, Donnan exclusion	Solution diffusion	Sarkar et al., (2021)
Membrane material	Ceramics, polypropylene, polysulphone, polyvinylidene, fluoride, polytetrafluoroethylene, polyacrylonitrile	Ceramics, cellulosics, polysulphone, polyvinylidene fluoride	Thin-film composites, celluloses	Thin-film composites, cellulosics, polysulphonates, polysulphone	Xia et al., (2018)

(continued)

Table 1 (continued)

Characteristics	Microfiltration	Ultrafiltration	Nanofiltration	Reverse osmosis	References
Module configuration	Tubular, hollow fibre	Tubular, hollow fibre, spiral wound, plate and frame	Tubular, spiral wound, plate and frame	Tubular, spiral wound, plate and frame	Abdullah et al., (2018)
Application (can reject)	Particulate matter, bacteria, algae	Particulate matter, bacteria, algae, virus	Particulate matter, bacteria, algae, multivalent ions Natural organic matter, dyes, disinfection by-products, sugars, proteins, ions removal	Salts (desalination), particulate matter, bacteria, algae, multivalent ions monovalent ions, >99% of TDS, trace pollutants	Zhang et al., (2020)

because of the concentration gradient created by the rejection of solutes (Sigurdardottir et al., 2020). In NF membranes, extremely intricate processes like hydrodynamic and interfacial actions occur on the surface and pores of membranes. They reject the pollutants based on the principles of steric hindrance, Gibbs–Donnan effect, adsorption and electrostatic and hydrophobic repulsion (Huang et al., 2019; Sarkar et al., 2021).

Neutral ions sized greater than or equal to the pore sizes of NF membranes get rejected via steric hindrance (size exclusion) (Li et al., 2024a). The equilibrium and interactions among charged molecules and interface of the charged membranes are described by the Gibbs–Donnan effect where the membrane rejects ions of the same charge (Bonner et al., 2019). The charge on membranes occurs due to the separation of acidic/basic ion groups on the surface and inside the pore structures. NF membranes also have inherent weak ion exchange capacities and in certain cases ions get adsorbed from feed onto the surface and lead to modified membrane charge. Electrostatic repulsion/attraction occurs due to the local concentration of charged molecules and charge of the NF membrane (Du et al., 2016; Guo et al., 2020). Further, hydrophobic molecules get adsorbed onto the surface of hydrophobic membranes and result in effective rejection till saturation (Rehman et al., 2020). Dielectric exclusion is the result of the interaction of ionisable components with the bound charges induced by ions at interfaces among media of diverse dielectric constants. Molecules passing in solvent experience drag forces (expressed in terms of both convection and diffusion) that arise because of the flow of solvents in the restrained pore structures (Zhu et al., 2019).

2.2 *Conventional NF Membranes*

Rejection of pollutants by NF membranes is mainly reliant on the properties of the nanomembrane. Other less significant properties include physicochemical properties of solute, operating factors and feed wastewater characteristics (Cristóvão et al., 2019). Hence, the membrane materials play an important role in enhancing the NF process. Potent membrane materials should be (i) resistant to chemical and thermal attacks, (ii) robust, (iii) thin, (iv) free from defects and (v) economical (Fane et al., 2011; Joseph et al., 2023). Traditional nanomembrane materials comprise polymers, ceramics and thin-film composites (Xia et al., 2018). Polymeric membranes are the most widely used because of their high tensile strength and the feasibility of arranging into flat sheets or hollow fibre structures. Polymeric nanomembranes are synthesised mostly by the phase inversion technique. The polymers that are generally used in nanomembrane synthesis include cellulose acetate, polyvinylidene fluoride, polyacrylonitrile, polypropylene, polyethersulphone and polysulphone (Koyuncu et al., 2015). Based on the properties, polymeric materials can be chosen for specific applications as they have certain advantages and disadvantages. For example, hydrophilic materials evade microbial adsorption but are less efficient compared to hydrophobic polymeric materials. Cellulose acetate shows efficient permeation and rejection but is prone to hydrolysis and has inadequate resistance to pH, Cl and fouling. Instead, polyethersulphone, polyvinylidene fluoride, polysulphone and polyacrylonitrile can be adapted and blended to make them more permeable and flexible. Polypropylene cannot blend well and is susceptible to oxidation (Fane et al., 2011).

Ceramic materials are superior to polymers because of their narrow pore size distribution, high porosity, improved mechanical and chemical stability and less membrane fouling. Also, it is easy to recover ceramic membranes from fouling as ceramics are resistant to harsh cleaning practices. However, compared to polymeric membranes, the rejection of dissolved organics and ions is not satisfactory. Also, the ceramic membranes are costly when applied at an industrial scale (Sawunyama et al., 2023). Characteristic ceramic membranes are asymmetric in nature and are synthesised by the sol–gel method in which monomers get converted into a colloidal solution which acts as the precursor to obtain an integrated network (gel) of discrete/particulate materials (Shang et al., 2017). Common materials used in ceramic membranes are Zr, Si, Al, porcellanite, oxides and sintered metals (Dommati et al., 2019).

Thin-film composite membranes are characterised by their asymmetric structure, consisting of a support layer that provides mechanical strength, while allowing low resistance to permeate flow and an active thin-film layer that offers high selectivity for separation. These layers are synthesised by interfacial polymerisation or coating/cross-linking and each layer can be individually manipulated to enhance the stability, selectivity and permeability (Zhang et al., 2017b). Commonly used thin-film composite membrane materials include polyurea, polyamide and polyuria-amide (Nazia et al., 2020). The support layer can be fabricated using polymeric materials. Efficiency of thin-film membranes can be improved by amending monomers, surfactants and/or additives which gives rise to new structures (Ang et al., 2020). Thin-film membranes are widely employed in desalination, treatment of water and brackish water, removal of dissolved organics and softening of water.

2.3 Methodologies for the Preparation of the NF Membrane

This section comprehensively explores various methodologies for the fabrication of NF membranes, encompassing phase inversion, the layer-by-layer (LBL) method, interfacial polymerisation and additional techniques.

2.3.1 Phase Inversion Method

The phase inversion method is a chemical phenomenon exploited in the creation of thin-film composite NF membranes. This method typically involves extracting the solvent from a solution of liquid polymer, resulting in the formation of a solid, porous membrane (Algieri et al., 2021). Non-solvent-induced phase separation (NIPS) or thermally induced phase separation (TIPS) technologies are majorly employed (Du et al., 2022). The transformation of liquid polymer into a solid state is achieved through processes like TIPS, controlled evaporation, precipitation from the vapour phase and immersion precipitation (Tang et al., 2021). Post-treatment, such as chemical cross-linking, is common for optimising membrane performance by preventing polymer chain compaction or matrix swelling under organic solvents (Shi et al., 2021).

2.3.2 The LBL Method

The LBL method involves the alternating electrostatic adsorption of cationic and anionic polyelectrolytes on the surface of a support membrane, resulting in the self-assembly of a layered composite (Guzman et al., 2020). However, the deposition time crucially influences the separation performance of the LBL membranes, with increased deposition time leading to decreased water permeability because of the increase in the thickness of the layered structure. To address the time-consuming nature of the LBL method, researchers have explored several alternatives such as dopamine materials, tannic acid and Jeffamine to enhance membrane hydrophilicity (Wang et al., 2022). The LBL assembly technique is crucial in crafting NF membranes, providing adjustable charge and pore size, excellent water permeability and effective anti-fouling properties. These features make these membranes well-suited for applications such as resource recovery, seawater desalination and various other fields.

2.3.3 Interfacial Polymerisation

Interfacial polymerisation stands out as another valuable technique for the preparation of NF membrane, forming an ultra-thin selective polyamide layer on the surface of a microporous support membrane, creating a thin-film composite membrane with favourable water flux and solute retention characteristics (Bhaskar & Kaleekkal, 2022). Successful interfacial polymerisation depends on controlling the polymerisation process to form a polymer network at the interface of two immiscible phases (Li et al., 2021).

2.3.4 Other Techniques for the Preparation of NF Membrane

Beyond the mentioned methodologies, diverse approaches contribute to the preparation of the NF membrane. Electron beam irradiation is utilised to induce changes in membrane structure and properties, impacting separation efficiency (Gu et al., 2022). UV/photo-grafting involves using ultraviolet light to initiate graft polymerisation on the membrane surface, enhancing hydrophilicity, reducing roughness and introducing antibacterial layers (Suresh et al., 2021). Plasma treatment is employed to modify the surface properties of NF membranes, enhancing hydrophilicity, introducing functional groups and improving the overall performance (Mohammed et al., 2021). Incorporation of organic and inorganic particles is a versatile approach allowing for blending of materials to achieve desired membrane properties (Wang et al., 2021).

In summary, the combination of phase inversion, LBL, interfacial polymerisation and additional techniques offers a diverse array of strategies for tailoring NF membranes to specific applications, enhancing their performance and ensuring stability under various conditions.

2.4 *Characterisation of NF Membranes*

Characterisation of the membrane surface and nanopores is important to understand the mechanism of separation. This characterisation can be achieved using a variety of methodologies. In the Brunauer–Emmett–Teller method, adsorption and desorption of gaseous nitrogen onto membrane surface helps to analyse and measure surface area and pore size distribution of porous materials (Nosheen, 2019). In atomic force microscopy (AFM), the cantilever/tip assembly interacts with the membrane surface and gives information about pore sizes and distribution, surface roughness, topography and force interactions of membrane and solutes (Zhang et al., 2017c). In transmission electron microscopy (TEM), a beam of electrons is transmitted through the membrane which gives the direct measurement of pore size and distribution. Similarly, in scanning electron microscopy (SEM), a beam of focussed electrons scans the surface of a membrane for images of the membrane surface, cross section and fouled layers (Wang et al., 2024). On the other hand, indirect measurement of pore sizes can be obtained by neutral solute rejection studies. Other simple and cost-effective technologies like contact angle to understand hydrophilicity or hydrophobicity and cutting-edge spectroscopic techniques for morphology and structural analysis are also available (Mohammad et al., 2015). Since there are several methods available to characterise nanomembranes, the desired results can be obtained by selecting the correct technique/combinations and suitable resolution.

2.5 *Membrane Modules*

Large-scale application of NF membranes requires several thousands of square feet of membranes which is not practically possible to accommodate as a straight forward conventional geometry. Hence, to decrease the cost and footprint, membranes are packed into compact modules. Membranes are present in diverse configurations and are mainly categorised into hollow fibre tubular, capillary, flat sheet, plate-and-frame and spiral wound type, based on their construction The most widely employed NF membranes are tubular and spiral wound modules. The ideal module should also have a porous support channel to decrease the saturation, minimise pressure losses and deliver suitable flow conditions to hinder the phenomenon of concentration polarisation. In the spiral wound module, flat sheets are wrapped around a centre tube. The membranes are fastened alongside the boundaries of a permeating spacer, which provides support to the membrane and diverts permeate to the centre tube. Mesh-like feed spacers are also used to offer a hydrodynamic environment on the surface of the membrane which in turn hinders the concentration polarisation. The entire module is wrapped in a casing layer to prevent telescoping (Chen et al., 2020). Tubular modules contain bundles of tubes with an active membrane layer facing the interior. To provide extra support, this set-up is then inserted into perforated metal tubes. This design gives a turbulent flow, which ensures less concentration polarisation. However, these modules are energy-intensive and have limited burst pressure, which limits their application to the filtration of feed with particulates (Xue et al., 2021b). In the plate-and-frame module, the plates have grooved channels that direct the flow of the permeate. When the feed solution enters the module from one end, water molecules pass through the membrane and are gathered as permeate solution in a central permeate collection manifold (Sepehri, 2023).

Despite being widely employed in field-scale processes and carrying out substantial research to improve the performance of the membrane, conventional NF membranes still have certain drawbacks such as low permeance at higher solute concentrations, membrane fouling, low flux, reject disposal and costly membranes (Imbrogno et al., 2019). Several strategies like the usage of feed channel spacers, membrane vibration/rotation, pulse feeding, gas bubbling, etc., can be used to decrease the extent of concentration polarisation and membrane fouling. However, these are not viable on an industrial scale. To overcome these disadvantages, researchers have come up with innovative nanomembranes with a progression in hi-tech membrane fabrication techniques.

3 Novel Nanomaterial-Based NF Membranes

Recently, with the advancement in nanotechnology, the traditional polymer-based membranes were modified by incorporating diverse nanoparticles (metal/metal oxide nanoparticles, carbon-based nanoparticles, metal organic frameworks (MOFs), water channel proteins, organic nanoparticles, etc.) on the surface of the membrane (Ji et al., 2017; Khdary et al., 2023). These nanoparticles are blended or entrapped membranes which are called mixed-matrix membranes (MMMs). They are prepared by mixing desired nanoparticles in the monomer or polymer sol during synthesis. Another type of nanomaterial-based membranes called thin-film nanocomposite membranes are fabricated through dip coating or pressure-driven deposition techniques where nanofillers get self-assembled/embedded into polyamide thin film layers (Jain et al., 2021). These nanomaterial-based NF membranes offer several advantages such as (i) effective solubility of solvents, (ii) effective membrane selectivity, (iii) increased pore connectivity, (iv) hinders macro-voids formation, (v) increased surface-to-volume ratio, (vi) excellent mechanical and thermal stabilities and (vii) improved anti-fouling properties. Besides, the nanoparticles can disinfect, adsorb and degrade organics (Koseoglu-Imer et al., 2013). Later studies concentrated on more advanced nanomaterial-based NF membranes fabricated through state-of-the-art technologies. These new fabrication techniques allow to synthesise advanced NF membranes with the following characteristics: (i) excellent dispersion of nanoparticles on membrane, (ii) use of nanoparticles that are compatible with polymers, (iii) stable interactions among nanoparticles and polymer matrix.

3.1 Metal/Metal Oxide-Based NF Membranes

Metal/metal oxide nanoparticles that are most widely used in the fabrication of NF membranes include zeolites, SiO_2, TiO_2, Ag, ZnO, ZrO_2, CdS, Fe_3O_4, etc. These nanoparticles can be synthesised by various methods such as ball milling, sputtering, hydrothermal, sol–gel, laser-ablation, pyrolysis and biosynthesis. Adding these nanoparticles directly into the system is not economic as a considerable number of materials are needed. Hence, the metal/metal oxide nanoparticle powders are coated or embedded in the membranes. During synthesis, these externally added nanoparticles interact with membrane surfaces through Van der Waals interactions, bridge forces, depletion, electric double layer overlap, hydration forces and adsorbed polymer steric interactions (Bashambu et al., 2021).

TiO_2 nanoparticles are commonly employed fillers in MMMs because of their exceptional qualities such as (i) capable of increasing flux, (ii) removing salts, (iii) degrading organic compounds and dyes because of photocatalytic properties, (iv) creating new hydrophilic structures and (v) averting membrane fouling, as it can decompose microbial cells (Sakarkar et al., 2021). Batool et al. (2021) fabricated novel mixed-matrix nanocomposite membrane for water desalination by combining TiO_2 nanoparticle fillers (0–2% wt), polyvinylpyrrolidone pore formers and polymeric membranes (polyethersulphone and cellulose acetate). Effective dispersal of TiO_2 particles on membranes was attained by the chemical precipitation process, and an asymmetric design was attained through the phase inversion technique. The resultant NF membrane has high salt rejection, polymer chain rigidity, flux, hydrophilicity, thermal stability and increased mean pore size. This membrane attained an effective salt rejection of 80% and water flux of 90 $L/m^2/h$. NF membranes modified with oxides of Ti and Zr have displayed potential for usage in high-performance filtration under tough environmental conditions because of their excellent stability in broad pH range (pH 1–13) and resistance to corrosion. Hence, TiO_2 and ZrO_2 embedded NF membranes are preferred in corrosive conditions (Anisah et al., 2021). Worou et al. (2021) synthesised a faultless, enormously loose NF membrane by coating the Zr nanoparticles on the polyethersulphone membrane via the in situ formation method and applied in the removal of multivalent ions. This membrane has a mean pore size of 0.9 nm and permeates water flux of 62 $L/m^2/h$. It has shown a rejection rate of 93% and 38%, respectively, for Ca^+ and Na^+. Xue et al. (2021a) developed a novel, multifunctional Zr-based nanocomposite NF membrane using the in situ deposition technique. This NF membrane has exceptional properties such as rough exterior, super wettability, super hydrophilicity and underwater super oleophobicity, and it can efficiently retain 99% of emulsified oils that are insoluble in water. It has a high flux of 829–1542 L/m^2/h.

Fouling of NF membranes reduces its permeability and selectivity, drastically affects its performance and increases operational and membrane cleaning costs. Nanocomposites comprising Ag nanoparticles have displayed excellent anti-fouling properties because of their critical biocidal properties against bacteria, fungi and virus. Ag nanoparticles are also used to selectively transport elements and act as a barrier to certain pollutants. Polyamide and cellulose acetate membranes are mostly used to incorporate Ag nanoparticles. Abadikhah et al. (2019) synthesised an NF membrane by filling the polyamide layer with reduced graphene oxide (GO), TiO_2 and Ag nanoparticles, which showed efficient desalination, dye removal and antimicrobial properties. Similarly, Mansourpanah et al. (2021) used a thin-film NF membrane embedded with Ag, GO and polyamidoamine for water desalination. The fabricated NF membrane has superior hydrophilicity, permeability, pore size and surface roughness, and shows nearly 90%, 77% and 99% rejection of Na_2SO_4, $Cr(NO_3)_3$ and crystal-violet dye, respectively.

Al_2O_3 nanoparticles are widely employed as fillers/additives in polymeric membranes to enhance the anti-fouling nature, fracture toughness, tensile strength and weariness force of epoxy polymers (Opelt et al., 2015). Kotp (2021) fabricated a low-cost, high-flux NF membrane by incorporating Al_2O_3 nanoparticles into polyamide layers via the interfacial polymerisation technique, which is capable of water softening and removal of multivalent cations. Addition of Al improved membrane negativity and hydrophilicity and results showed >90% rejection of NaCl. Sherugar et al. (2021) synthesised a mixed-matrix NF membrane with a very high surface area of 262 m^2/g using the solution combustion technique by incorporating Zn doped Al_2O_3 nanoparticles into the polysulphone membrane for the rejection of heavy metals from water.

Fe/Fe-oxide nanoparticles make ideal fillers for NF membranes because of non-toxicity, biocompatibility, easy availability, hydrophilicity and exceptional magnetic, catalytic, mechanical and thermal properties. Kamari and Shahbazi (2021) fabricated a chitosan functionalised Fe_3O_4 SiO_2 nanocomposite and used it as a filler in a polyethersulphone NF membrane. The modified membrane can remove salts, heavy metals and dyes efficiently without facing any membrane fouling. By increasing the dose of the Fe nanocomposite filler by 0.5% wt, the water flux considerably increased to about 1.5 fold from 47 to 70 L/m^2/h. Ghaemi et al. (2015) revealed that >90% of Cu can be removed from wastewater using the Fe-oxide (0.1 % wt) nanocomposite embedded on a polyethersulphone membrane.

Silica is broadly used in nanotechnological applications because of its surface modifiability via functionalisation. Silica nanoparticles are generally embedded on polymer membranes via the sol–gel synthesis method. Si nanoparticles are also employed as carriers and anchors for anti-fouling ligands in the membrane matrix. The addition of Si nanocomposites into the membrane reduces the leaching and/or migration of ligands and enhances the wettability of modified NF membranes (Samavati et al., 2023). Filling of Si nanoparticles into dope solutions considerably enhance the solution viscosity, which permits the easy manipulation of membrane structures during phase inversion (Mavukkandy et al., 2017). The most notable feature of a Si nanoparticle is its ability to enhance the pore size and porosity. Fernandes et al. (2021) used hydrophobic Si as an additive, and results indicate that Si in low concentrations could enhance the formation of a large number of pores which gives effective hydraulic performance. Huang et al. (2012) used mesoporous Si to prepare a nanocomposite membrane with increased pore sizes and good pore interconnectivity and achieved 96% retention of target contaminants.

Cu and Zn nanocomposites possess excellent anti-fouling properties and can be used to modify NF membranes. Zhang et al. (2017a) developed a Cu-modified NF membrane with higher salt rejection and increased resistance to microbes. On the other hand, Cu can also improve the mechanical properties of membranes (Bashambu et al., 2021). Zareei and Hosseini (2019) developed a composite polyethersulphone membrane incorporating copper (Cu) and iron (Fe). The addition of CuO enhanced the formation of smooth surface and hydrophilicity. The rejection of salts also increased drastically in the modified NF membrane. ZnO is a semiconductor with a high band gap. It is very reactive, chemically stable, non-toxic and has excellent photocatalytic properties. These exceptional physicochemical properties make ZnO an ideal filler when used in NF membranes. ZnO nanoparticle-embedded membranes exhibit improved membrane topography and stability (Agboola et al., 2021). Khan et al. (2015) successfully developed a ZnO-based NF membrane which not only improved the selectivity and water flux recovery but also increased the anti-fouling properties. ZnO was also reported to increase the hydrophilicity, water flux recovery, permeability and membrane stability (da Silva Biron et al., 2021).

3.2 *Carbon Nanomaterial-Based NF Membranes*

Carbon is a highly malleable element, and innovative nanomaterials with diverse structures can be easily synthesised from Carbon. Carbon-based nanomaterials like carbon nanotubes (CNTs), graphene, GO, reduced graphene oxide (rGO) and fullerenes can be used as alternatives to fabricate state-of-the-art NF membranes (Karimi-Maleh et al., 2020). These nanomaterials have exceptional intrinsic electrical, thermal, mechanical and separation features owing to their size, high surface area and porosity (Kumari et al., 2019). CNTs are one-dimensional nanomaterials formed by rolling up graphite sheets into a cylindrical structure. The CNTs made of single graphene sheets are called single-walled CNTs (SWCNTs), while CNTs with several graphene sheets are called multiwalled CNTs (MWCNTs) (Machado & Lima, 2018). SWCNTs are microporous in nature and MWCNTs are mesoporous in nature. The hydrophobic and smooth interior walls of CNTs enable easy passage of water (Li et al., 2023). Further, functionalised (coatings of surfactants, synthetic polymers or biopolymers) and engineered CNTs have enhanced pore sizes, pore volumes, high selectivity, hydrophilicity and chemical stability. Nanocomposite NF membranes have CNTs coated on the membrane matrix or bonded to metal/metal oxide nanoparticles. These MMMs possess high surface functionality, exhibit enhanced rejection and have anti-fouling properties. The well-structured and robust pores of CNTs act as energy barriers and provide passage channels which aid in the swift transfer of solvents and rejection of salts at the same time. Similarly, GO has distinctive characteristics like ultra-thin structure, hydrophilicity, two-dimensional nanoconstruction, flexibility like polymers and chemical, mechanical and thermal stabilities, which make them ideal candidates to synthesise advanced NF membranes (Wu & Ting, 2013). Mainly, GO nanosheets comprise adequate oxygen-rich functional groups on the basal planes and lateral ends, which act as potential reactive sites for alteration. The thickness of GO films can also be easily modified by varying the doses of GO nanosheets. The two-dimensional nanostructures prepared by stacking continuous membranes lead to channels between GO sheets, which allows for faster permeation of water (Dai et al., 2016). In a recent study, low-pressure and loose NF membranes were synthesised by the intercalation of CNTs on GO laminates reinforced by cellulose acetate matrix. This arrangement enlarges the space between GO layers and decreases the resistance of water flux and salt permeation. Cross-linked bonds among GO layers were produced to generate stable size sieving channel necks of the tortuous path for selective penetration. GO composite membranes with 80% wt of CNT and 0.5 mg of GO dosage with a flux of 26 L/m^2/h/bar achieved >94% rejection of textile dyes (Huang et al., 2021).

3.3 *Metal Organic Framework (MOF)-Based NF Membranes*

MOFs are a group of porous organic and inorganic hybrid polymers synthesised by connecting metal ionic nodes and organic polydentate linkers, formed into multidimensional structures with excellent characteristics like relatively flexible construction, high surface area, porosity and easy modifiability (Joseph et al., 2019; Yang & Gates, 2019). These features can be exploited to prepare novel composite NF membranes for high-precision molecular separation (Rasheed et al., 2020). Incorporation of MOFs into thin-film NF membranes provide few passage channels including (i) polyamide layer pores, (ii) MOF pores and (iii) around MOF and polyamide interface (Dai et al., 2022). Xu et al. (2021) fabricated high-performance, self-cleaning membranes by the interfacial polymerisation technique. The melamine membrane and a woven-like polyamide layer amended with Zn-TCPP MOF laminate displayed high permeation of 63 L/m^2/h/bar with efficient dye rejection of >97%.

3.4 Catalytic NF Membranes

In advanced oxidation processes, the generated hydroxyl radicals can degrade a wide range of organic pollutants. Alternately, catalysts can be incorporated into NF membranes to generate catalytic membranes, which act as heterogeneous Fenton-like reaction by combined action of inorganic nanoparticles and physicochemical features of organic polymeric membranes. The catalytic membranes have improved flux and selectivity, anti-fouling properties and increased thermal/mechanical properties. These catalytic membranes can also be recycled. The catalytic materials like TiO_2, ZnO, goethite, pyrite, haematite, ferrihydrite, lepidocrocite and magnetite have been extensively used in catalytic membranes. Ma et al. (2014) successfully synthesised a nanocomposite membrane that exhibits efficient separation, catalytic and anti-fouling properties. Other types of reactive catalytic membranes like semiconductor-based membranes get activated by sunlight/UV to degrade organic pollutants. In semiconductor materials, valence bands act as oxidising agents, while conduction bands act as reducing agents. With semiconductor-doped NF membranes, several pollutants including heavy metals and emerging contaminants can be removed. In a recent study, Fe_3O_4-PVDF three-channel hollow fibre catalytic membrane was synthesised by non-solvent-induced phase inversion process. This membrane has high hydrophilicity, enhanced separation, anti-fouling and catalytic activity with a high water flux of 176 L/m^2/h and efficient rejection of methylene blue (98%) through Fenton catalytic reaction (Huang et al., 2020).

3.5 Block Copolymer NF Membranes

Block copolymers consist of a sequence of monomer unions of similar composition positioned adjacently in polymeric molecules, and they are again attached to another sequence of chemically distinctive monomeric unities though covalent bonds (Nunes, 2020). These macromolecules can be self-assembled into well-ordered structures when mixed in appropriate solvents. The compactly arranged tubular pores are perfect for water separation. Porous NF membranes with diverse morphologies can be obtained using block copolymers. Main synthesis techniques are (i) etching of dense films or limited solubilisation and (ii) self-assembled and non-solvent-induced phase separation. Luo et al. (2020) developed novel nanoporous membrane using oriented 100 nm thick model cylindrical block copolymer films (polystyrene *block* methyl methacrylate) that can be applied as the top active layer in oil–water filtration. Firstly, the membrane was treated using a roll-to-roll technique of cold-zone annealing to achieve vertical orientation. Then, selective etching of PMMA cylinders was carried out via UV-radiation to obtain vertically aligned nanopores. The developed nanochannels have high hydrophilicity, uniform pore sizes of 23 nm, narrow pore size distribution and improved nanopore density of 420/μ^2. Polyethersulphone is used as the bottom support layer.

3.6 MXene-Based NF Membrane

Transition metal carbide, also called MXene (Formula = $M_{n+1}X_nT_x$), made up of transition metals and carbides or nitrides, is an attractive two-dimensional material with exceptional characteristics and varied uses. MXenes have relatively rigid lamellar structures and hydrophilic and easily functionalised functional groups. The intrinsically different bonds like covalent, metal and ionic bonds have terrific potential in fabricating advanced NF membranes for effective molecular separation (Xu et al., 2020). Shao et al. (2021) fabricated MXene membrane by pillaring MXene nanosheets with CNTs. Robust CNTs are used for the intercalation of neighbouring MXene nanosheets. The synthesised MXene/CNT-CTAB membrane has high confined mass transfer effect and improved d-spacing with fivefold higher permeation compared to pristine membranes. The membrane was able to reject 95% of dyes with diverse charges and sizes in numerous green solvents.

3.7 Tannic Acid-Based NF Membranes

Polyamide synthetic coatings used in thin-film composite membranes have few disadvantages like (i) being toxic, (i) poor interaction between polymer and substrate leads to polymer peeling and (iii) meagre resistance to chloride. Though the non-polyamide membranes overcome these hurdles, the efficiency is less. In this context, bio-polyphenols obtained from plants are being used as coatings for NF membranes because of their wide chemical flexibility and metal-chelating and antimicrobial characteristics (Chakrabarty et al., 2017). Tannic acid, which is copiously present in natural substances like tea leaves, fruits, wood, etc., is a widely used plant-derived polyphenol (Mallakpour & Azadi, 2021). It is characterised by functional groups like gallol/catechol, which bind with the membrane surface via covalent/non-covalent bonds and rapidly form thin polymerised layers. Chakrabarty et al. (2017) reported an easy and economical technique to coat porous membranes through complexation of tannic acid and cupric acetate by co-deposition. The synthesised membrane showed excellent hydrophilicity and water permeance.

3.8 Aquaporin-Based NF Membranes

Cellular membranes show ultra-high perm-selectivity by allowing the transfer of water, ions or small organic molecules across the lipid bilayer controlled by the water channel proteins called aquaporins (Sengur-Tasdemir et al., 2020). The lipid bilayers contain aquaporins and display very high permeability and selectivity compared to other commercial membranes and hence these are used in the synthesis of bio mimic NF membranes. Wagh et al. (2019) developed a membrane by adding cysteine functional groups onto aquaporin Z to aid in the immobilisation of proteins to the polymeric membrane and align in the flow direction. This membrane showed a high salt rejection compared to conventional membranes.

4 Conclusions and Future Prospects

To battle with the ever-increasing pollution of water, NF has emerged as a potential technique. It is widely employed in the treatment of saline and brackish water, water and wastewater and product separation in industries. Significant progress was seen in the fabrication of novel nanoparticle-based NF membranes with the advancement in the techniques of nanomembrane synthesis. Nanomaterial-based NF membranes are considered to overcome the problem of 'trade-off' between permeability and selectivity (Ji et al., 2017). Numerous nanomaterials, such as metal/metal-oxide nanoparticles, CNTs, graphene, GO, rGO, hybrid nanocomposites, bioinspired nanomaterials, MXene and their combinations have tremendous potential to act as ideal materials for NF membranes. Besides being highly efficient, these novel NF membranes offer several other advantages such as (i) effective solubility of solvents; (ii) effective membrane selectivity; (iii) increased pore connectivity; (iv) effective hindering of the formation of macro-voids; (v) increased surface-to-volume ratio; (vi) excellent mechanical, chemical, thermal and optical stabilities and (vii) improved anti-fouling properties. Thus, nanotechnology has an immense role to play in the removal of pollutants through NF. Despite the advantages, there is still a need to further reduce the energy requirement and operational costs and also address the challenges faced in the fabrication of nanomaterial-based NF membranes such as poor dispersion and agglomeration. Also, most advanced NF membranes have only been synthesised at the lab/pilot scale. For large-scale application, advanced instrumentation and simulation technologies have to be developed to characterise and analyse the variations of membrane microstructures in real-field applications. It is also imperative to clearly understand the molecule-level interactions between the matrix and nanoparticles, the transport phenomenon and anti-fouling mechanisms of the NF membranes. Because of the increasing research interest in NF, it is expected that a new generation of commercial nanomaterial-based NF membranes will be available soon.

References

Abadikhah, H., Naderi Kalali, E., Khodi, S., Xu, X., & Agathopoulos, S. (2019). Multifunctional thin-film nanofiltration membrane incorporated with reduced graphene oxide@TiO_2@ Ag nanocomposites for high desalination performance, dye retention, and antibacterial properties. *ACS Applied Materials & Interfaces, 11*, 23535–23545. https://doi.org/10.1021/acsami.9b03557

Abdullah, N., Rahman, M. A., Dzarfan Othman, M. H., Jaafar, J., & Ismail, A. F. (2018). Chapter 2—Membranes and membrane processes: Fundamentals. In A. Basile, S. Mozia, & R. Molinari (Eds.), *Current trends and future developments on (bio-)membranes* (pp. 45–70). Elsevier.

Agboola, O., Fayomi, O., Oni, B., Ayeni, A., Sadiku, E., Popoola, A., Moropeng, L., Adegbola, S., & Daramola, D. (2021). Performance of mixed matrix nanofiltration membranes in wastewater treatment: A review. In *IOP conference series: materials science and engineering* (p. 012007). IOP Publishing.

Algieri, C., Chakraborty, S., Pal, U. (2021). Efficacy of phase inversion technique for polymeric membrane fabrication. *Journal of Phase Change Materials, 1*.

Ang, M. B. M. Y., Tang, C. L., De Guzman, M. R., Maganto, H. L. C., Caparanga, A. R., Huang, S. H., Tsai, H. A., Hu, C. C., Lee, K. R., & Lai, J. Y. (2020). Improved performance of thin-film nanofiltration membranes fabricated with the intervention of surfactants having different structures for water treatment. *Desalination, 481*, 114352. https://doi.org/10.1016/j.desal.2020.114352

Anisah, S., Kanezashi, M., Nagasawa, H., & Tsuru, T. (2021). Effect of the Ti/Zr ratio on the hydrothermal and chemical stability of TiO_2-ZrO_2 nanofiltration membranes. *Separation and Purification Technology, 274*, 119060. https://doi.org/10.1016/j.seppur.2021.119060

Arola, K., Van der Bruggen, B., Mänttäri, M., & Kallioinen, M. (2019). Treatment options for nanofiltration and reverse osmosis concentrates from municipal wastewater treatment: A review. *Critical Reviews in Environmental Science and Technology, 49*, 2049–2116. https://doi.org/10.1080/10643389.2019.1594519

Bashambu, L., Singh, R., & Verma, J. (2021). Metal/metal oxide nanocomposite membranes for water purification. *Materials Today: Proceedings, 44*, 538–545. https://doi.org/10.1016/j.matpr.2020.10.213

Batool, M., Shafeeq, A., Haider, B., & Ahmad, N. M. (2021). TiO_2 Nanoparticle filler-based mixed-matrix PES/CA nanofiltration membranes for enhanced desalination. *Membranes, 11*, 433. https://doi.org/10.3390/membranes11060433

Bhaskar, V. V., & Kaleekkal, N. J. (2022). Next-generation thin-film composite nanofiltration membranes for water remediation: A review. *Emergent Materials, 5*, 1373–1390.

Bonner, R., Germishuizen, C., & Franzsen, S. (2019). Prediction of nanofiltration rejection performance in brackish water reverse osmosis brine treatment processes. *Journal of Water Process Engineering, 32*, 100900. https://doi.org/10.1016/j.jwpe.2019.100900

Chakrabarty, T., Pérez-Manríquez, L., Neelakanda, P., & Peinemann, K. V. (2017). Bioinspired tannic acid-copper complexes as selective coating for nanofiltration membranes. *Separation and Purification Technology, 184*, 188–194. https://doi.org/10.1016/j.seppur.2017.04.043

Chen, B. Z., Ju, X., Liu, N., Chu, C. H., Lu, J. P., Wang, C., & Sun, S. P. (2020). Pilot-scale fabrication of nanofiltration membranes and spiral-wound modules. *Chemical Engineering Research and Design, 160*, 395–404. https://doi.org/10.1016/j.cherd.2020.06.011

Cristóvão, M. B., Torrejais, J., Janssens, R., Luis, P., Van der Bruggen, B., Dubey, K. K., Mandal, M. K., Bronze, M. R., Crespo, J. G., & Pereira, V. J. (2019). Treatment of anticancer drugs in hospital and wastewater effluents using nanofiltration. *Separation and Purification Technology, 224*, 273–280. https://doi.org/10.1016/j.seppur.2019.05.016

da Silva Biron, D., Santos, V. D., & Bergmann, C. P. (2021). Tubular ceramic membranes coated with ZnO and applied in the disinfection of water contaminated with Staphylococcus aureus. *Ceramics International*, *47*(19), 27082–27090. https://doi.org/10.1016/j.ceramint.2021.06.121

Dai, H., Xu, Z., & Yang, X. (2016). Water permeation and ion rejection in layer-by-layer stacked graphene oxide nanochannels: A molecular dynamics simulation. *The Journal of Physical Chemistry C, 120*, 22585–22596. https://doi.org/10.1021/acs.jpcc.6b05337

Dai, R., Han, H., Zhu, Y., Wang, X., & Wang, Z. (2022). Tuning the primary selective nanochannels of MOF thin-film nanocomposite nanofiltration membranes for efficient removal of hydrophobic endocrine disrupting compounds. *Frontiers of Environmental Science & Engineering, 16*, 1–13. https://doi.org/10.1007/s11783-021-1474-7

de Almeida, F. R., de Mello Ferreira, I. L., & dos Reis, R. A. (2024). Recent progress on the development and application of polymeric nanofiltration membranes: A mini-review. *Mini-Reviews in Organic Chemistry, 21*, 3–21.

Dolar, D., & Košutić, K. (2013). Removal of pharmaceuticals by ultrafiltration (UF), nanofiltration (NF), and reverse osmosis (RO). *Comprehensive Analytical Chemistry, 62*, 319–344. https://doi.org/10.1016/B978-0-444-62657-8.00010-0

Dommati, H., Ray, S. S., Wang, J. C., & Chen, S. S. (2019). A comprehensive review of recent developments in 3D printing technique for ceramic membrane fabrication for water purification. *RSC Advances, 9*, 16869–16883. https://doi.org/10.1039/C9RA00872A

Du, Y., Lv, Y., Qiu, W. Z., Wu, J., & Xu, Z. K. (2016). Nanofiltration membranes with narrowed pore size distribution via pore wall modification. *Chemical Communications, 52*, 8589–8592. https://doi.org/10.1039/C6CC03842E

Du, Y., Pramanik, B. K., Zhang, Y., Dumee, L., & Jegatheesan, V. (2022). Recent advances in the theory and application of nanofiltration: A review. *Current Pollution Reports, 8*, 51–80.

Fane, A., Tang, C., & Wang, R. (2011). Membrane technology for water: Microfiltration, ultrafiltration, nanofiltration, and reverse osmosis. *Treatise on Water Science, 4*, 301–335. https://doi.org/10.1016/B978-0-444-53199-5.00091-9

Farhad, A., & Ozan, N. C. (2023). Purification and fractionation of bioactive peptides through membrane filtration: A critical and application review. *Trends in Food Science & Technology, 131*, 118–128. https://doi.org/10.1016/j.tifs.2022.11.024

Feng, Y., Yang, S., Xia, L., Wang, Z., Suo, N., Chen, H., Long, Y., Zhou, B., & Yu, Y. (2019). In-situ ion exchange electrocatalysis biological coupling (i-IEEBC) for simultaneously enhanced degradation of organic pollutants and heavy metals in electroplating wastewater. *Journal of Hazardous Materials, 364*, 562–570. https://doi.org/10.1016/j.jhazmat.2018.10.068

Fernandes, C. S., Md Nordin, N. A. H., Bilad, M. R., Matsuura, T., Putra, Z. A., Wirzal, M. D. H., & Jaafar, J. (2021). Explication of hydrophobic silica as effective pore former for membrane fabrication. *Applied Surface Science Advances, 3*, 100051. https://doi.org/10.1016/j.apsadv.2020.100051

García Doménech, N., Purcell-Milton, F., & Gun'ko, Y. K. (2020). Recent progress and future prospects in development of advanced materials for nanofiltration. *Materials Today Communications, 23*, 100888. https://doi.org/10.1016/j.mtcomm.2019.100888

Ghaemi, N., Madaeni, S. S., Daraei, P., Rajabi, H., Zinadini, S., Alizadeh, A., Heydari, R., Beygzadeh, M., & Ghouzivand, S. (2015). Polyethersulfone membrane enhanced with iron oxide nanoparticles for copper removal from water: Application of new functionalized Fe_3O_4 nanoparticles. *Chemical Engineering Journal, 263*, 101–112. https://doi.org/10.1016/j.cej.2014.10.103

Gu, Y., Zhao, J., Zhou, H., Jiang, H., Li, J., & Zhang, B. (2022). Engineering robust RGO/PVA composite membrane for acid recovery via electron beam irradiation. *Carbon, 191*, 243–254.

Guo, Y. S., Ji, Y. L., Wu, B., Wang, N. X., Yin, M. J., An, Q. F., & Gao, C. J. (2020). High-flux zwitterionic nanofiltration membrane constructed by in-situ introduction method for monovalent salt/antibiotics separation. *Journal of Membrane Science, 593*, 117441. https://doi.org/10.1016/j.memsci.2019.117441

Gupta, S., Singh, A., Sharma, T., Kaur, R., Khandelwal, V., Rawat, K. D., Pathak, S., Sharma, M. K., Singh, J., & Shah, M. P. (2024). Applications of ultrafiltration, nanofiltration, and reverse osmosis in pharmaceutical wastewater treatment. *Development in wastewater treatment research and processes* (pp. 33–49). Elsevier.

Guzman, E., Rubio, R. G., & Ortega, F. (2020). A closer physico-chemical look to the layer-by-layer electrostatic self-assembly of polyelectrolyte multilayers. *Advances in Colloid and Interface Science, 282*, 102197.

Huang, J., Zhang, K., Wang, K., Xie, Z., Ladewig, B., & Wang, H. (2012). Fabrication of polyethersulfone-mesoporous silica nanocomposite ultrafiltration membranes with antifouling properties. *Journal of Membrane Science, 423–424*, 362–370. https://doi.org/10.1016/j.memsci.2012.08.029

Huang, L., Li, Z., Luo, Y., Zhang, N., Qi, W., Jiang, E., Bao, J., Zhang, X., Zheng, W., An, B., & He, G. (2021). Low-pressure loose GO composite membrane intercalated by CNT for effective dye/salt separation. *Separation and Purification Technology, 256*, 117839. https://doi.org/10.1016/j.seppur.2020.117839

Huang, Z. H., Zhang, X., Wang, Y. X., Sun, J. Y., Zhang, H., Liu, W. L., Li, M. P., Ma, X. H., & Xu, Z. L. (2020). Fe_3O_4/PVDF catalytic membrane treatment organic wastewater with simultaneously improved permeability, catalytic property and anti-fouling. *Environmental Research, 187*, 109617. https://doi.org/10.1016/j.envres.2020.109617

Huang, Z., Gong, B., Huang, C. P., Pan, S. Y., Wu, P., Dang, Z., & Chiang, P. C. (2019). Performance evaluation of integrated adsorption-nanofiltration system for emerging compounds removal: Exemplified by caffeine, diclofenac and octylphenol. *Journal of Environmental Management, 231*, 121–128. https://doi.org/10.1016/j.jenvman.2018.09.092

Imbrogno, A., Samanta, P., & Schäfer, A. I. (2019). Fate of steroid hormone micropollutant estradiol in a hybrid magnetic ion exchange resin-nanofiltration process. *Environmental Chemistry, 16*, 630–640. https://doi.org/10.1071/EN19126

Imbrogno, A., Calvo, J. I., Breida, M., Schwaiger, R., & Schäfer, A. I. (2024). Molecular weight cut off (MWCO) determination in ultra- and nanofiltration: Review of methods and implications on organic matter removal. *Separation and Purification Technology, 354*(P2), 128612. https://doi.org/10.1016/j.seppur.2024.128612

Jain, V., Patel, S., Patel, P., Patel, K., & Shah, D. (2021). Study of molybdenum trioxide thin film deposited using dip coating method. *Materials Today: Proceedings, 42*, 1700–1703. https://doi.org/10.1016/j.matpr.2020.09.209

Ji, Y., Qian, W., Yu, Y., An, Q., Liu, L., Zhou, Y., & Gao, C. (2017). Recent developments in nanofiltration membranes based on nanomaterials. *Chinese Journal of Chemical Engineering, 25*, 1639–1652. https://doi.org/10.1016/j.cjche.2017.04.014

Joseph, L., Jun, B. M., Jang, M., Park, C. M., Muñoz-Senmache, J. C., Hernández-Maldonado, A. J., Heyden, A., Yu, M., & Yoon, Y. (2019). Removal of contaminants of emerging concern by metal-organic framework nanoadsorbents: A review. *Chemical Engineering Journal, 369*, 928–946. https://doi.org/10.1016/j.cej.2019.03.173

Joseph, T. M., Al-Hazmi, H. E., Śniatała, B., Esmaeili, A., & Habibzadeh, S. (2023). Nanoparticles and nanofiltration for wastewater treatment: From polluted to fresh water. *Environmental Research*, 117114.

Kamari, S., & Shahbazi, A. (2021). High-performance nanofiltration membrane blended by Fe_3O_4@SiO_2–CS bionanocomposite for efficient simultaneous rejection of salts/heavy metals ions/dyes with high permeability, retention increase and fouling decline. *Chemical Engineering Journal, 417*, 127930. https://doi.org/10.1016/j.cej.2020.127930

Karimi-Maleh, H., Shafieizadeh, M., Taher, M. A., Opoku, F., Kiarii, E. M., Govender, P. P., Ranjbari, S., Rezapour, M., & Orooji, Y. (2020). The role of magnetite/graphene oxide nano-composite as a high-efficiency adsorbent for removal of phenazopyridine residues from water samples, an experimental/theoretical investigation. *Journal of Molecular Liquids, 298*, 112040. https://doi.org/10.1016/j.molliq.2019.112040

Khan, S. B., Alamry, K. A., Bifari, E. N., Asiri, A. M., Yasir, M., Gzara, L., & Ahmad, R. Z. (2015). Assessment of antibacterial cellulose nanocomposites for water permeability and salt rejection. *Journal of Industrial and Engineering Chemistry, 24*, 266–275. https://doi.org/10.1016/j.jiec.2014.09.040

Khdary, N. H., Almuarqab, B. T., & El Enany, G. (2023). Nanoparticle-embedded polymers and their applications: A review. *Membranes, 13*, 537.

Koseoglu-Imer, D. Y., Kose, B., Altinbas, M., & Koyuncu, I. (2013). The production of polysulfone (PS) membrane with silver nanoparticles (AgNP): Physical properties, filtration performances, and biofouling resistances of membranes. *Journal of Membrane Science, 428*, 620–628. https://doi.org/10.1016/j.memsci.2012.10.046

Kotp, Y. H. (2021). High-flux TFN nanofiltration membranes incorporated with camphor-Al_2O_3 nanoparticles for brackish water desalination. *Chemosphere, 265*, 128999. https://doi.org/10.1016/j.chemosphere.2020.128999

Koyuncu, I., Sengur, R., Turken, T., Guclu, S., & Pasaoglu, M. E. (2015). 3—Advances in water treatment by microfiltration, ultrafiltration, and nanofiltration. In A. Basile, A. Cassano, & N. K. Rastogi (Eds.), *Advances in membrane technologies for water treatment* (pp. 83–128). Woodhead Publishing.

Kumari, P., Alam, M., & Siddiqi, W. A. (2019). Usage of nanoparticles as adsorbents for waste water treatment: An emerging trend. *Sustainable Materials and Technologies, 22*, e00128. https://doi.org/10.1016/j.susmat.2019.e00128

Li, Q., Wang, X., Liu, Y., Ma, Z., Wang, Q., Xu, D., Gao, J., Wu, R., Sun, H., & Gao, X. (2023). Enhanced separation of tetrafluoropropanol from water via carbon nanotubes membranes: insights from molecular dynamics simulations. *Frontiers of Environmental Science & Engineering, 17*, 140.

Li, X., Chen, X., Loh, C.H., Fu, K., & Yang, L. (2024a). A novel in-situ hydrolysis approach to prepare ultra-selective and ultrasmooth nanofiltration membrane for efficient and sustainable ion separation. *Journal of Membrane Science*, 122433.

Li, X., Lin, L., Zhang, X., Dai, R., Wu, Z., & Wang, Z. (2024b). Performance and fouling characteristics in a pilot-scale reverse osmosis membrane system for microelectronic wastewater treatment. *Separation and Purification Technology*, 126333.

Li, X., Wang, Z., Han, X., Liu, Y., Wang, C., Yan, F., & Wang, J. (2021). Regulating the interfacial polymerization process toward high-performance polyamide thin-film composite reverse osmosis and nanofiltration membranes: A review. *Journal of Membrane Science, 640*, 119765.

Luo, Y., Wang, X., Zhang, R., Singh, M., Ammar, A., Cousins, D., Hassan, M. K., Ponnamma, D., Adham, S., & Al-Maadeed, M. A. A. (2020). Vertically oriented nanoporous block copolymer membranes for oil/water separation and filtration. *Soft Matter, 16*, 9648–9654. https://doi.org/10.1039/D0SM00526F

Ma, X. H., Zhang, H. X., Gu, S. W., Cao, Y., Wen, X., & Xu, Z. L. (2014). Process optimization and modeling of membrane reactor using self-sufficient catalysis and separation of difunctional ceramic composite membrane to produce methyl laurate. *Separation and Purification Technology, 132*, 370–377. https://doi.org/10.1016/j.seppur.2014.04.038

Machado, F. M., & Lima, É. C. (2018). Carbon nanoadsorbents for removal of organic contaminants from water. In *Bio- and nanosorbents from natural resources* (pp. 21–53). Springer.

Mallakpour, S., & Azadi, E. (2021). Novel methodologies and materials for facile fabrication of nanofiltration membranes. *Emergent Materials*, 1–26. https://doi.org/10.1007/s42247-021-00278-3

Mansourpanah, Y., Ghanbari, A., Yazdani, H., Mohammadi, A. G., & Rahimpour, A. (2021). Silver-polyamidoamine/graphene oxide thin film nanofiltration membrane with improved antifouling and antibacterial properties for water purification and desalination. *Desalination, 511*, 115109. https://doi.org/10.1016/j.desal.2021.115109

Mavukkandy, M. O., Bilad, M. R., Kujawa, J., Al-Gharabli, S., & Arafat, H. A. (2017). On the effect of fumed silica particles on the structure, properties and application of PVDF membranes. *Separation and Purification Technology, 187*, 365–373. https://doi.org/10.1016/j.seppur.2017.06.077

Mohammad, A. W., Teow, Y. H., Ang, W. L., Chung, Y. T., Oatley-Radcliffe, D. L., & Hilal, N. (2015). Nanofiltration membranes review: Recent advances and future prospects. *Desalination, 356*, 226–254. https://doi.org/10.1016/j.desal.2014.10.043

Mohammed, S., Hegab, H. M., Ou, R., Liu, S., Ma, H., Chen, X., Sridhar, T., & Wang, H. (2021). Effect of oxygen plasma treatment on the nanofiltration performance of reduced graphene oxide/cellulose nanofiber composite membranes. *Green Chemical Engineering, 2*, 122–131.

Nazia, S., Sekhar, S. C., Jegatheesan, V., Bhargava, S. K., & Sridhar, S. (2020). Performance of chemically resistant polyurea reverse osmosis membrane in the treatment of highly alkaline

industrial wastewater containing sodium aluminate. *Water Science and Technology, 82*, 2259–2270. https://doi.org/10.2166/wst.2020.495
Nosheen, S. (2019). Nanomembrane applications in environmental engineering. In *Nanotechnology applications in environmental engineering* (pp. 103–120). IGI Global.
Nunes, S. P. (2020). Chapter 11—Block copolymer membranes. In G. Szekely & A. Livingston (Eds.), *Sustainable nanoscale engineering* (pp. 297–316). Elsevier.
Opelt, C. V., Becker, D., Lepienski, C. M., & Coelho, L. A. F. (2015). Reinforcement and toughening mechanisms in polymer nanocomposites—Carbon nanotubes and aluminum oxide. *Composites Part b: Engineering, 75*, 119–126. https://doi.org/10.1016/j.compositesb.2015.01.019
Park, M., & Snyder, S. A. (2020). Chapter 6—Attenuation of contaminants of emerging concerns by nanofiltration membrane: Rejection mechanism and application in water reuse. In A. J. Hernández-Maldonado & L. Blaney (Eds.), *Contaminants of emerging concern in water and wastewater* (pp. 177–206). Butterworth-Heinemann.
Purkait, M. K., Sinha, M. K., Mondal, P., & Singh, R. (2018). Chapter 1—Introduction to membranes. In M. K. Purkait, M. K. Sinha, P. Mondal, & R. Singh (Eds.), *Interface science and technology* (Vol. 25, pp. 1–37). Elsevier.
Rasheed, T., Bilal, M., Hassan, A. A., Nabeel, F., Bharagava, R. N., Romanholo Ferreira, L. F., Tran, H. N., & Iqbal, H. M. N. (2020). Environmental threatening concern and efficient removal of pharmaceutically active compounds using metal-organic frameworks as adsorbents. *Environmental Research, 185*, 109436. https://doi.org/10.1016/j.envres.2020.109436
Rehman, F., Thebo, K. H., Aamir, M., & Akhtar, J. (2020). Chapter 8—Nanomembranes for water treatment. In A. Amrane, S. Rajendran, T. A. Nguyen, A. A. Assadi, & A. M. Sharoba (Eds.), *Nanotechnology in the beverage industry* (pp. 207–240). Elsevier.
Sakarkar, S., Muthukumaran, S., & Jegatheesan, V. (2021). Tailoring the effects of titanium dioxide (TiO_2) and polyvinyl alcohol (PVA) in the separation and antifouling performance of thin-film composite polyvinylidene fluoride (PVDF) membrane. *Membranes, 11*, 241.
Samavati, Z., Samavati, A., Goh, P. S., Ismail, A. F., & Abdullah, M. S. (2023). A comprehensive review of recent advances in nanofiltration membranes for heavy metal removal from wastewater. *Chemical Engineering Research and Design, 189*, 530–571.
Santos, A. F., Lopes, D. V., Alvarenga, P., Gando-Ferreira, L. M., & Quina, M. J. (2024). Phosphorus removal from urban wastewater through adsorption using biogenic calcium carbonate. *Journal of Environmental Management, 351*, 119875.
Sarkar, P., Modak, S., & Karan, S. (2021). Ultraselective and highly permeable polyamide nanofilms for ionic and molecular nanofiltration. *Advanced Functional Materials, 31*, 2007054. https://doi.org/10.1002/adfm.202007054
Sawunyama, L., Oyewo, O. A., Seheri, N., Onjefu, S. A., & Onwudiwe, D. C. (2023). Metal oxide functionalized ceramic membranes for the removal of pharmaceuticals in wastewater. *Surfaces and Interfaces*, 102787.
Sengur-Tasdemir, R., Pekgenc, E., Urper-Bayram, G. M., Ergon-Can, T., Tutuncu, H. E., Zeytuncu, B., Gul-Karaguler, N., Ates-Genceli, E., & Koyuncu, I. (2020). Determination of the effect of proteoliposome concentration on Aquaporin Z incorporated nanofiltration membranes. *Environmental Technology, 41*, 2229–2239.
Sepehri, A. (2023). Separation processes: Advancements in membrane module design and performance optimization (Doctoral dissertation, University of Toledo).
Shahkaramipour, N., Tran, T. N., Ramanan, S., & Lin, H. (2017). Membranes with surface-enhanced antifouling properties for water purification. *Membranes, 7*, 13. https://doi.org/10.3390/membranes7010013
Shang, R., Goulas, A., Tang, C. Y., de Frias Serra, X., Rietveld, L. C., & Heijman, S. G. J. (2017). Atmospheric pressure atomic layer deposition for tight ceramic nanofiltration membranes: Synthesis and application in water purification. *Journal of Membrane Science, 528*, 163–170. https://doi.org/10.1016/j.memsci.2017.01.023
Shao, D. D., Zhang, Q., Wang, L., Wang, Z. Y., Jing, Y. X., Cao, X. L., Zhang, F., & Sun, S. P. (2021). Enhancing interfacial adhesion of MXene nanofiltration membranes via pillaring carbon

nanotubes for pressure and solvent stable molecular sieving. *Journal of Membrane Science, 623*, 119033. https://doi.org/10.1016/j.memsci.2020.119033
Sherugar, P., Naik, N. S., Padaki, M., Nayak, V., Gangadharan, A., Nadig, A. R., & Déon, S. (2021). Fabrication of zinc doped aluminium oxide/polysulfone mixed matrix membranes for enhanced antifouling property and heavy metal removal. *Chemosphere, 275*, 130024. https://doi.org/10.1016/j.chemosphere.2021.130024
Shi, G. M., Feng, Y., Li, B., Tham, H. M., Lai, J. Y., & Chung, T. S. (2021). Recent progress of organic solvent nanofiltration membranes. *Progress in Polymer Science, 123*, 101470.
Siddique, T., Dutta, N. K., & Roy Choudhury, N. (2020). Nanofiltration for arsenic removal: Challenges, recent developments, and perspectives. *Nanomaterials, 10*, 1323. https://doi.org/10.3390/nano10071323
Sigurdardottir, S. B., DuChanois, R. M., Epsztein, R., Pinelo, M., & Elimelech, M. (2020). Energy barriers to anion transport in polyelectrolyte multilayer nanofiltration membranes: Role of intra-pore diffusion. *Journal of Membrane Science, 603*, 117921. https://doi.org/10.1016/j.memsci.2020.117921
Suresh, D., Goh, P. S., Ismail, A. F., & Hilal, N. (2021). Surface design of liquid separation membrane through graft polymerization: A state of the art review. *Membranes, 11*, 832.
Tang, Y. J., Wang, L. J., Xu, Z. L., & Zhang, H. Z. (2018). Novel chitosan-piperazine composite nanofiltration membranes for the desalination of brackish water and seawater. *Journal of Polymer Research, 25*, 1–12. https://doi.org/10.1007/s10965-018-1514-6
Tang, Y., Lin, Y., Ford, D. M., Qian, X., Cervellere, M. R., Millett, P. C., & Wang, X. (2021). A review on models and simulations of membrane formation via phase inversion processes. *Journal of Membrane Science, 640*, 119810.
Wagh, P., Zhang, X., Blood, R., Kekenes-Huskey, P. M., Rajapaksha, P., Wei, Y., & Escobar, I. C. (2019). Increasing salt rejection of polybenzimidazole nanofiltration membranes via the addition of immobilized and aligned aquaporins. *Processes, 7*, 76. https://doi.org/10.3390/pr7020076
Wang, C., Park, M. J., Seo, D. H., Drioli, E., Matsuyama, H., & Shon, H. (2021). Recent advances in nanomaterial-incorporated nanocomposite membranes for organic solvent nanofiltration. *Separation and Purification Technology, 268*, 118657.
Wang, C., Park, M. J., Yu, H., Matsuyama, H., Drioli, E., & Shon, H. K. (2022). Recent advances of nanocomposite membranes using layer-by-layer assembly. *Journal of Membrane Science*, 120926.
Wang, J., Nie, L., Zhang, C., & Wang, B. (2024). Free radical graft polymerization of hydroxyethyl methacrylate and acrylic acid on polyetherimide membrane surface via redox system to improve anti-fouling and oil resistance performance. *Materials Today Communications*, 108197.
Wang, Z., Wang, Z., Lin, S., Jin, H., Gao, S., Zhu, Y., & Jin, J. (2018a). Nanoparticle-templated nanofiltration membranes for ultrahigh performance desalination. *Nature Communications, 9*, 2004. https://doi.org/10.1038/s41467-018-04467-3
Wang, Z., Wu, A., Colombi Ciacchi, L., & Wei, G. (2018b). Recent advances in nanoporous membranes for water purification. *Nanomaterials, 8*, 65. https://doi.org/10.3390/nano8020065
Worou, C. N., Kang, J., Alamou, E. A., Degan, A., Yan, P., Gong, Y., Guene, R. L., & Chen, Z. (2021). High flux membrane based on in-situ formation of zirconia layer coated the polyethersulfone substrate for ions separation. *Water Supply, 21*(7), 3344–3355. https://doi.org/10.2166/ws.2021.092
Wu, B. (2019). Membrane-based technology in greywater reclamation: A review. *Science of the Total Environment, 656*, 184–200. https://doi.org/10.1016/j.scitotenv.2018.11.347
Wu, T. T., & Ting, J. M. (2013). Preparation and characteristics of graphene oxide and its thin films. *Surface and Coatings Technology, 231*, 487–491. https://doi.org/10.1016/j.surfcoat.2012.05.066
Xia, L., Ren, J., Weyd, M., & McCutcheon, J. R. (2018). Ceramic-supported thin film composite membrane for organic solvent nanofiltration. *Journal of Membrane Science, 563*, 857–863. https://doi.org/10.1016/j.memsci.2018.05.069

Xu, D., Zhu, X., Luo, X., Guo, Y., Liu, Y., Yang, L., Tang, X., Li, G., & Liang, H. (2020). MXene nanosheet templated nanofiltration membranes toward ultrahigh water transport. *Environmental Science & Technology, 55*, 1270–1278.

Xu, M., Feng, X., Liu, Z., Han, X., Zhu, J., Wang, J., Bruggen, B. V. D., & Zhang, Y. (2021). MOF laminates functionalized polyamide self-cleaning membrane for advanced loose nanofiltration. *Separation and Purification Technology, 275*, 119150. https://doi.org/10.1016/j.seppur.2021.119150

Xue, J., Xu, M., Gao, J., Zong, Y., Wang, M., & Ma, S. (2021a). Multifunctional porphyrinic Zr-MOF composite membrane for high-performance oil-in-water separation and organic dye adsorption/photocatalysis. *Colloids and Surfaces a: Physicochemical and Engineering Aspects, 628*, 127288. https://doi.org/10.1016/j.colsurfa.2021.127288

Xue, Y. L., Zhang, R., Cao, B., & Li, P. (2021b). Chapter 20—Tubular membranes and modules. In T. S. Chung & Y. Feng (Eds.), *Hollow fiber membranes* (pp. 431–448). Elsevier.

Yang, D., & Gates, B. C. (2019). Catalysis by metal organic frameworks: Perspective and suggestions for future research. *ACS Catalysis, 9*, 1779–1798.

Yang, W. S., Qiu, Z. L., Yu, W. H., Fang, L. F., & Zhu, B. K. (2024). Integrating binary organic phosphonic acid into nanofiltration membrane for high precision separation of mono-/divalent anions. *Separation and Purification Technology, 336*, 126287.

Yang, Z., Zhou, Y., Feng, Z., Rui, X., Zhang, T., & Zhang, Z. (2019). A review on reverse osmosis and nanofiltration membranes for water purification. *Polymers, 11*, 1252. https://doi.org/10.3390/polym11081252

Zareei, F., & Hosseini, S. M. (2019). A new type of polyethersulfone based composite nanofiltration membrane decorated by cobalt ferrite-copper oxide nanoparticles with enhanced performance and antifouling property. *Separation and Purification Technology, 226*, 48–58. https://doi.org/10.1016/j.seppur.2019.05.077

Zhang, A., Zhang, Y., Pan, G., Xu, J., Yan, H., & Liu, Y. (2017a). In situ formation of copper nanoparticles in carboxylated chitosan layer: Preparation and characterization of surface modified TFC membrane with protein fouling resistance and long-lasting antibacterial properties. *Separation and Purification Technology, 176*, 164–172. https://doi.org/10.1016/j.seppur.2016.12.006

Zhang, H., He, Q., Luo, J., Wan, Y., & Darling, S. B. (2020). Sharpening nanofiltration: Strategies for enhanced membrane selectivity. *ACS Applied Materials & Interfaces, 12*, 39948–39966. https://doi.org/10.1021/acsami.0c11136

Zhang, Q., Zhang, Z., Dai, L., Wang, H., Li, S., & Zhang, S. (2017b). Novel insights into the interplay between support and active layer in the thin film composite polyamide membranes. *Journal of Membrane Science, 537*, 372–383. https://doi.org/10.1016/j.memsci.2017.05.033

Zhang, X., Marschewski, E., Penner, P., Beyer, A., & Gölzhäuser, A. (2017c). Investigation of electronic transport through ultrathin carbon nanomembrane junctions by conductive probe atomic force microscopy and eutectic Ga–In top contacts. *Journal of Applied Physics, 122*, 055103. https://doi.org/10.1063/1.4995533

Zhu, Y., Zhu, H., Li, G., Mai, Z., & Gu, Y. (2019). The effect of dielectric exclusion on the rejection performance of inhomogeneously charged polyamide nanofiltration membranes. *Journal of Nanoparticle Research, 21*, 1–13. https://doi.org/10.1007/s11051-019-4665-4

Advanced Oxidation Processes Using Nanomaterials

Maurício Dalla Costa Rodrigues da Silva, Pâmela Cristine Ladwig Muraro, Gabriela Carvalho Collazzo, Daniel Moro Druzian, Giovani Pavoski, Raisi Natalia Lenz Baldez, and William Leonardo da Silva

1 Introduction

Water is an essential resource for a variety of functions. The use of water for different purposes, including anthropological, agricultural and industrial activities, requires certain standards of quality. However, because of population explosion and extensive agro-industrial development, the availability of clean and safe water is becoming a rarity. Pollution of water bodies because of the unabated discharge of hazardous substances and the limitations of the conventional wastewater treatment system to remove such substances are the main reasons for this environmental problem that has been bothering authorities worldwide (Cardoso et al., 2023).

There are two types of conventional treatment systems: water treatment and wastewater treatment. These systems often use chemical, physical and biological operations to remove effluents so that the water can be used for various purposes (Kurian, 2021). However, some pollutants cannot be effectively removed by these

M. D. C. R. da Silva · D. M. Druzian · W. L. da Silva (✉)
Applied Nanomaterials Research Group (GPNAp), Franciscan University (UFN), Santa Maria, RS, Brazil
e-mail: w.silva@ufn.edu.br

P. C. L. Muraro · R. N. L. Baldez
Northern Regional Technological Institute, Technological University of Uruguay (UTEC), Rivera, UY, Uruguay

G. C. Collazzo
Research Group on Adsorptive and Catalytic Process Engineering (ENGEPAC), Federal University of Santa Maria (UFSM), Santa Maria, RS, Brazil

G. Pavoski
Polytechnical School of Chemical Engineering, University of the Sao Paulo (USP), São Paulo, SP, Brazil

Department of Materials Engineering, The University of British Columbia (UBC), Vancouver, BC, Canada

© Capital Publishing Company, New Dehli, India. 2025
R. Neelancherry and S. M. Kumar, *Nanotechnology in Wastewater Treatment*,
https://doi.org/10.1007/978-3-031-94246-4_8

operations. These pollutants called emerging organic pollutants (EOPs) consist of persistent organic pollutants (POPs), pharmaceutical products, personal care products, pesticides and hormones. Thus, EOPs can cause health and environmental problems because of their accumulation and toxicity (Bracamontes-Ruelas et al., 2022). Therefore, new technologies have been developed to tackle them in the last few years.

One of the most prominent technologies is the advanced oxidation processes (AOPs). The AOPs involve the production and use of reactive radical species, such as hydroxyl radical (OH), superoxide (O_2^-) and hydroperoxyl (HO_2), to oxidate the target organic pollutants into less toxic by-products. The complete oxidation of these pollutants leads to the mineralisation of the organic molecules and the formation of water, carbon dioxide and inorganic salts that could be further removed by traditional processes (Deng & Zhao, 2015). The main AOP process includes photocatalysis, sonocatalysis, ozonation, Fenton systems (photo-Fenton and electro-Fenton) and a combination of AOPs in hybrid systems.

A new horizon in the field of AOPs has been flourishing with the advances in nanotechnology. Nanomaterials present a high specific surface area in a small particle volume, allowing for a boost in active sites for the adsorption and generation of reactive radicals for application in the AOPs (Zhang et al., 2019a). They are often used as (nano)-catalysts in AOPs, boosting their efficiency to degrade EOPs and reducing the quantity of toxic by-products.

Research with organic, inorganic, carbon-based and composite nanomaterials presented significant results for the degradation of these harmful compounds combined with AOPs, evidencing their great potential (Bethi et al., 2016). In addition, treatments applying nanomaterials can achieve efficient removal of EOPs, meeting the quality standards required for water consumption, water recovery and wastewater regulations (Domingues et al., 2021). In this context, this chapter aims to explore the fundamentals and applications of nanomaterials in AOPs. Moreover, it evaluates the current scenario, opportunities and challenges for the development of AOPs along with the advances in nanotechnology.

2 Advanced Oxidation Processes

AOPs represent a class of highly effective techniques for removing POPs and toxic compounds from wastewater. This section aims to explore the integration of nanomaterials into AOPs, highlighting their role in promoting the degradation of organic contaminants and expanding the applications of these processes in water treatment systems (Ameta et al., 2018; Preethi et al., 2024).

The fundamentals of AOPs are essential for understanding how these promising techniques can eliminate POPs from water. Oxygen reactive radicals (also known as reactive oxygen species, ROS) presented at least one unpaired electron, playing a pivotal role in oxidation processes. The main free radicals that deserve a mention are the hydroxyl radical (·OH), singlet oxygen, hydroxide ion (OH^-), peroxide ion

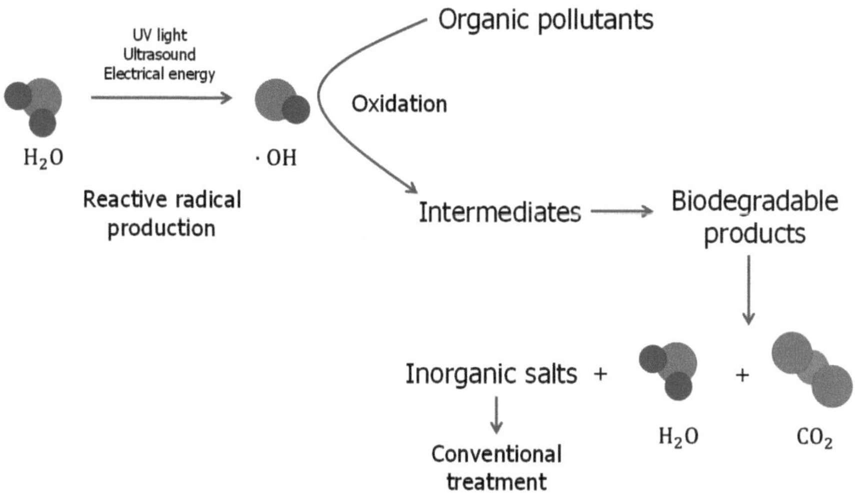

Fig. 1 Mechanistic overview illustrates the intricate workings of the advanced oxidation process, highlighting the multi-step pathway involved in pollutant degradation and transformation

(O_2^{2-}), superoxide anion radical ($O_2^{\cdot -}$), peroxyl radical ($RO_2^{\cdot}$), carbonate radical (CO_3^{-}) and sulphate radical ($SO_4^{\cdot -}$) (Kumari & Kumar, 2023; Vinayagam et al., 2024).

AOPs operate through a sequence of stages. In the initial step, the ROS are produced, possessing elevated reduction potentials, in adequate concentrations because of oxidation reactions. Oxidants might be produced through the utilisation of catalysts, application of energy or chemical reactions. The following are some examples of energies involved in AOPs: UV light (Pizzichetti et al., 2023), ultrasound (Lei et al., 2023) and electrical energy (Vucic et al., 2021). The various free radicals generated react with pollutants to form intermediates ($O_2^{\cdot -}$, ·OH, $SO_4^{\cdot -}$, $RO_2^{\cdot}$, etc.). The chemical reactions for each radical differ; for instance, ·OH reactions with organic molecules may involve (i) hydrogen removal from NH, OH or CH groups, (ii) interactions among radicals, such as the incorporation of molecular oxygen resulting in the generation of the $RO_2^{\cdot}$ or (iii) direct electron transfer producing oxidised intermediates. Conversely, the mechanism for the sulphate radical involves the transfer of an electrophilic electron. In addition, the oxidised intermediates produce biodegradable compounds that undergo mineralisation, converting them into stable substances such as water, carbon dioxide and inorganic salts. These salts can be further removed using conventional treatments, such as physical (e.g., sedimentation), chemical (e.g., precipitation) and biologic operations (e.g., anaerobic digestion). Figure 1 illustrates the fundamental mechanism underlying the AOP (Kanakaraju et al., 2018; Zhou et al., 2019).

AOPs can be categorised based on the mode of oxidation. Oxidation can occur through chemical processes (utilising chemical reagents and catalysts), electrochemical methods (employing electrical sources), photochemical reactions (utilising solar energy/UV sources) and sonochemical processes (utilising ultrasound) (Oturan & Aaron, 2014).

2.1 Direct or Indirect Photolysis of Hydrogen Peroxide

The direct photolysis of hydrogen peroxide (H_2O_2) occurs when it is directly irradiated with UV or visible light, resulting in the breaking of the O–H bond and the formation of two ·OH. These ·OH are remarkably reactive and react with organic pollutants, oxidising them into less toxic products.

On the other hand, in indirect photolysis, semiconductors such as titanium dioxide (TiO_2) are used as catalysts. When TiO_2 is irradiated with UV light, it absorbs enough energy to generate electron/hole pairs. The holes (gaps in the valence band) react with water or H_2O_2 adsorbed on the surface of TiO_2, generating ·OH and singlet oxygen ($1O_2$), which can also oxidise organic pollutants (Dong et al., 2022; Masjoudi & Mohseni, 2023).

2.2 Photocatalysis Using Semiconductors

Photocatalysis involves the utilisation of semiconductors as catalysts to facilitate oxidation reactions of organic pollutants under the influence of light. TiO_2 is the most widely studied and utilised semiconductor because of its chemical stability, low cost and high photocatalytic efficiency. When TiO_2 is exposed to UV light, electron excitation to the conduction band occurs, creating holes in the valence band. These holes can react with water or (H_2O_2) adsorbed on the surface of TiO_2, generating ·OH, which are highly reactive in the oxidation of organic pollutants (Jabbar et al., 2024; Yang et al., 2024).

2.3 Electrochemistry

In electrochemistry, anodic oxidation is employed to generate reactive species, including ·OH. Typically, electrodes made of platinum, graphite or metal oxides are polarised with an electric current, thereby inducing the oxidation of water or compounds present in the solution. This leads to the formation of ·OH and other oxidising species with potential for degradation of organic pollutants.

These three routes of ·OH generation demonstrate the versatility and efficacy of AOPs in removing organic pollutants from water, paving the way for the development of more efficient and sustainable technologies for treating contaminated wastewater (Nie et al., 2016).

3 Nanomaterials for AOPs

AOPs harness the power of oxidants generated in situ to break down organic compounds. Central to this process is the formation of potent transient species such as the OH, which can effectively degrade a broad spectrum of pollutants (Ganiyu et al., 2022). The efficacy of AOPs depends on the generation of very reactive and non-selective free radicals (Ratchnashree et al., 2023). AOPs, especially when enhanced with nanomaterials as photocatalysts, have shown exceptional efficacy in breaking down numerous hazardous pollutants. The integration of nanomaterials in AOPs is revolutionising research, offering promising avenues for more effective and sustainable water purification solutions.

Nanomaterials, defined by their dimensions, exhibit unique properties that distinguish them from their bulk counterparts. With characteristic dimensions ranging from 1 to 100 nm, these materials have better surface and interface characteristics over bulk properties. This results in a significantly increased specific surface area and surface energy, leading to an enhancement in catalytic activities. Small particle sizes ensure maximised exposure of surface area to reactants, facilitating a greater number of reactions (Hodges et al., 2018). The interesting properties of nanomaterials, such as heightened reactivity, rapid dissolution rates, extensive specific surface area and robust sorption capabilities, enable the development of cost-effective, safe and energy-efficient water treatment systems (Nishu & Kumar, 2023). These systems contribute to a more sustainable urban water cycle by minimising energy consumption. In recent years, nanotechnology has proven to be a promising avenue for innovation in the treatment of wastewater, with a variety of nanomaterials being engineered for this purpose. There has been a concerted effort to harness advanced nanotechnology for the enhancement of water purification processes.

Nanotechnology holds the promise of creating cutting-edge materials and devices by leveraging unique properties at the nanoscale (Ugwuany et al., 2024). Copper-based nanocatalysts, for instance, have shown exceptional efficacy in AOPs for the treatment of water. Researchers are exploring treatment of wastewater through various nanocomposites, such as $CuFe_2O_4$, CoCuO magnetic nanoparticles and $BiAg_xO_y$ (bismuth silver oxide, BSO) (Vaiano et al., 2020). For these nanomaterials that utilise AOPs to effectively eliminate degradable substances, a preliminary step often involves the adsorption of the substance onto the nanoparticle surface. This proximity or a synergistic effect between the adsorbent and the catalyst enhances the catalytic reaction. Utilising nanocomposites made of both an adsorbent and a catalyst can be particularly effective for degrading organic compounds like dyes (Mahadevan et al., 2023). The process of dye removal from water can involve both the adsorption and degradation of these organic molecules. Semiconductor nanomaterials, such as TiO_2, ZnO, $BiFeO_3$, $BiVO_4$, SnO_2 and CdO, are also frequently employed in photocatalytic applications due to their efficiency and effectiveness in this domain (Kaus et al., 2021).

Thus, nanomaterials, encompassing nanoparticles, nanocatalysts, nanofilms and other nanostructures, have revolutionised heterogeneous photocatalysis with their superior surface area-to-volume ratios and heightened surface reactivity (Gaur et al., 2022). These properties significantly enhance the efficiency of the degradation processes of AOPs compared to conventional AOPs.

4 Application of Nanomaterials in AOPs

Nanotechnology is regarded as one of the most important and fascinating fields of science mainly because of its enormous potential for developing novel nanostructured materials with new properties (Charitidis et al., 2014). It is worth noting that nanostructured materials are used alone or doped with a second less active material as catalytic support increasing their photocatalytic activity because of the reduction in band gap energy and consequently the process efficiency (Praxedes et al., 2021). Therefore, in the next sections, the applications of nanomaterials in AOPs are explored and discussed.

4.1 Photocatalysis

Photocatalysis involves exploiting the energy displaced from light for initiating chemical reactions, which are divided into four steps: (i) absorption of photons through the incident light from electron–hole pairs; (ii) charge generation; (iii) transfer of electrons and holes on the catalyst surface; and (iv) generation of the hydroxyl radical (OH) from redox reactions (Li & Wu, 2015). In this sense, the recombination of electrons from the catalyst's conduction band dispersed forming energy in the form of heat, where the charges generated on the catalyst surface endorse redox reactions depending on the organic pollutant on the adsorbed surface (Gershon et al., 2014). Figure 2 presents the basic principles of photocatalysis and possible reactions using a TiO_2 nanocatalyst as an example.

Photocatalysis is divided into two types: (i) homogeneous, based on the catalyst dissolved in the solution containing the pollutant forming a single phase, for example, transition metal oxides and photo-Fenton systems of Fe^{2+} and ZnO containing H_2O_2 (Liang et al., 2020a), and (ii) heterogeneous, characterised for a catalyst not dissolved in the solution containing the pollutant (e.g., TiO_2, CeO_2 and CuO) in particular materials with high surface area and thermal and chemical stability (Bellotti et al., 2023; Druzian et al., 2023; Yin et al., 2022). Thus, both types of photocatalysis are characterised by economical and scalable processes.

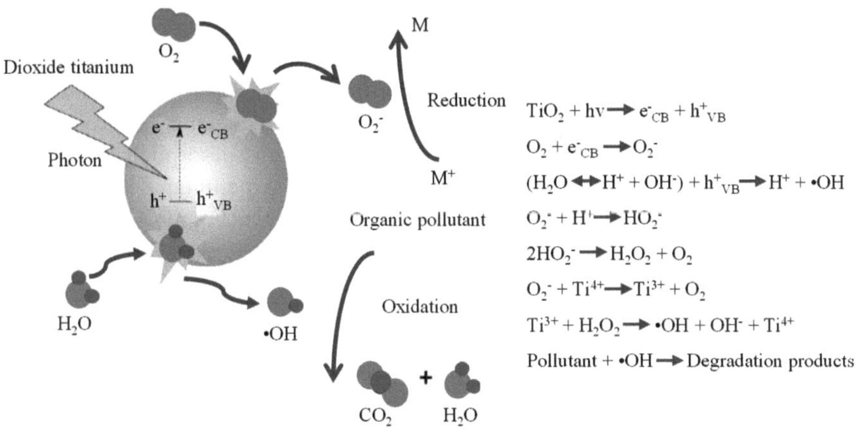

Fig. 2 Schematic of the photocatalysis using titanium dioxide as an example for the photodegradation of organic pollutants

Photocatalysis can be termed as an interfacial reaction, where organic pollutants are effectively mineralised when they diffuse to the surface of nanocatalysts. However, for photodegradation to occur, the pollutants must be adsorbed. In this sense, both processes take place at the interface of nanocatalysts, where studies demonstrate that the synergy of adsorption and photocatalysis increases the efficiency of the removal of pollutants. It is noteworthy that it is important to discover the synergistic adsorption-photocatalysis mechanism and the medium and low adsorption capabilities to enhance the overall removal effect (Sheng et al., 2019; Yu et al., 2020). The main variables influencing photocatalysis are given in Table 1.

Table 1 Main variables influencing the degradation result of organic pollutants by photocatalysis

Variables	Comments	References
Irradiation time	Increasing the irradiation time increments the degradation rate	Prasert et al. (2020)
Light source and intensity	The increase in intensity provides a high energy of photons producing an increase in ·OH denoting a high rate of degradation	Din et al. (2021)
pH	The changing of the pH value modifies the surface of the nanocatalyst varying the adsorption process and generally helping oxidation directly for h^+_{VB}	Azeez et al. (2018)
Nanocatalyst doping	The degradation efficiency is increased due to the decline in band gap energy depending on the doping of the nanocatalyst	Chen et al. (2022)
Surface area	High surface areas increase photocatalytic activity in virtue of providing more active sites by adsorption	Wei et al. (2018)

In the study of photodegradation and photocatalytic reactions in aqueous media, two types of kinetic models are generally used: Langmuir–Hinshelwood (L–H) and pseudo-first order (Swenson & Stadie, 2019). In this way, the L–H model demonstrated in Eqs. (1)–(3) is applied to solid catalytic reactions showing stages of adsorption, dissociation and reactions involved in the process and desorption of products, respectively, where it is generally more compatible with experimental kinetic values (Cho et al., 2024). However, the first-order model presented in Eq. (4) is used to investigate the rate of photodegradation when there is a difference in organic production and pollutant concentration versus time (Abu-Dalo et al., 2021).

$$-r_i = -\frac{dC_i}{dt} = \frac{k_s * K * C_i}{1 + K * C_i}. \tag{1}$$

$$-r_i = - = k_s * k * C_i \tag{2}$$

$$C_i = C_{io} * \mathrm{e}^{-k*t} \tag{3}$$

$$C_i = C_{io} * \mathrm{e}^{-k_1*t} \tag{4}$$

where $-r_i$ is the reaction rate (mol time^{-1} volume^{-1}); C_i is the sample concentration (mol volume^{-1}); is the reaction time (min); k_s is the apparent reaction constant; K is the adsorption constant; C_{io} is the initial concentration (mol volume^{-1}); k is the specific reaction rate (min^{-1}); and k_i is the first-order rate constant (min^{-1}).

4.2 Sonocatalysis

Sonocatalysis demonstrates a promising treatment related to the removal of pollutants contained in water, which is based on employing physical treatment wherein mechanical energy facilitates the transfer of electrons to wastewater. In this sense, the subjacent mechanisms of the process of sonocatalysis change the physical–chemical properties of the treated environment, agglomerating pollutants and mineralising because of the high-frequency vibrations of the process generating sound waves, with small ruptures of the bubbles effectively mixing the pollutants and chemical reactions in the water, thereby achieving the goal of treating water (Sviridov et al., 2015; Xu et al., 2023a). It is worth noting that the sonoactivation process is shown in two phases: (i) acoustic phase, which constitutes the formation and rupture of bubbles, and (ii) chemical phase, which consists of reactions containing radicals in the process (Bao et al., 2023). Figure 3 demonstrates a schematic of sonocatalysis and possible types of reactions.

The cavitation mechanism exerted by sonocatalysis stands out, which generates modified pressures and temperatures, with denominated ultrasonic waves causing the dissociation of molecules forming oxidative radicals, generally the ·OH. Thus, at

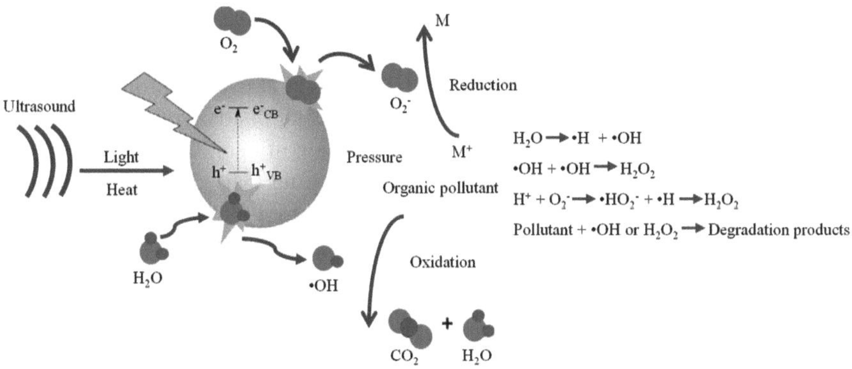

Fig. 3 Schematic of the sonocatalysis process using nanomaterials and possible types of reactions related to oxidative radicals degrading the organic pollutants

low pressures, the vapour bubbles are formed collapsing at high pressures inducing different chemical reactions. In addition, at lower temperatures, the ·OH groups can generate H_2O_2 thereby increasing the degradation efficiency of the nanocatalyst (Theerthagiri et al., 2020). Optimising sonocatalysis is fundamental involving different parameters such as the frequency and potential of the sound waves. For example, low wave frequencies are associated with larger cavities inducing a violent collapse of the bubbles and intensifying chemical reactions (Yan et al., 2015). However, excessive potential waves result in the fusion of cavities reducing the effect of degradation and may lead to the decomposition of water molecules into smaller compounds or ions (Wang et al., 2022a).

Sonocatalysis has numerous advantages such as strong oxidation capacity, low energy consumption, absence of added products and secondary pollution. However, it is a relatively expensive process requiring the use of equipment (Vasseghian et al., 2022). It is worth noting that studies with nanomaterials for sonocatalysis have demonstrated high degradation (>80%) of organic pollutants, for example, the Ag_2O/$CuWO_4$ nanocomposite for the degradation of acid orange 7 reaching 99.01% after 15 min (Liu et al., 2023) and the BiOBr/$FeWO_4$ nanomaterials in the degradation of tetracycline showing 86.51% after 120 min (Xu et al., 2023b).

4.3 *Ozonation*

Ozonation is a chemical oxidation technique that promotes the complete or partial decomposition of organic pollutants that are difficult to degrade. This strong ozone oxidant potential (2.07 V) (Parson, 2005) is an advantage of this process. Ozone is relatively unstable in an aqueous solution and its half-life is short; it should be produced in situ.

When ozone is dissolved in water, it begins to react as an oxidising agent because of its high redox potential (Ikehata & Li, 2018); the reactivity of ozone in the aqueous medium depends on conditions such as pH and temperature. According to Fig. 4, ozonation can directly or indirectly oxidise organic compounds. The direct reaction, favoured by acid-neutral pH, is the result of the action of ozone as a molecule (O_3) with organic compounds (M), generating oxidised matter (M_{oxid}) and releasing ions OH^-. The direct reaction, also called ozonolysis, is more selective and reactions happen slowly. The indirect reaction of ozone in a basic medium is the result of the action of free radical species resulting from the decomposition of ozone to produce the hydroxyl radical (·OH), which is the main species responsible for indirect reactions with organic pollutants (Ikehata & El-Din, 2004). In this type of reaction, ozone rapidly decomposes in a chain process divided into different steps: initiation, chain reactions and termination.

The reactions of ozone with hydroxyl and hydroperoxide ions are considered the fundamental initiation reactions for the ozone decomposition mechanism in aqueous media, where the O_3 molecule reacts with the hydroxyl ion forming an anion $O_2^{-\cdot}$ and the radical $OH_2^{\cdot}$.

The second step is the chain reaction, in which the $O_3^{-\cdot}$ anion is formed by the reaction of O_2^- and the O_3 molecule. At the same time, several chemical reactions produce radicals $HO_2^{\cdot}$ and/or O_2^-, which leads to new reactions. Substances with

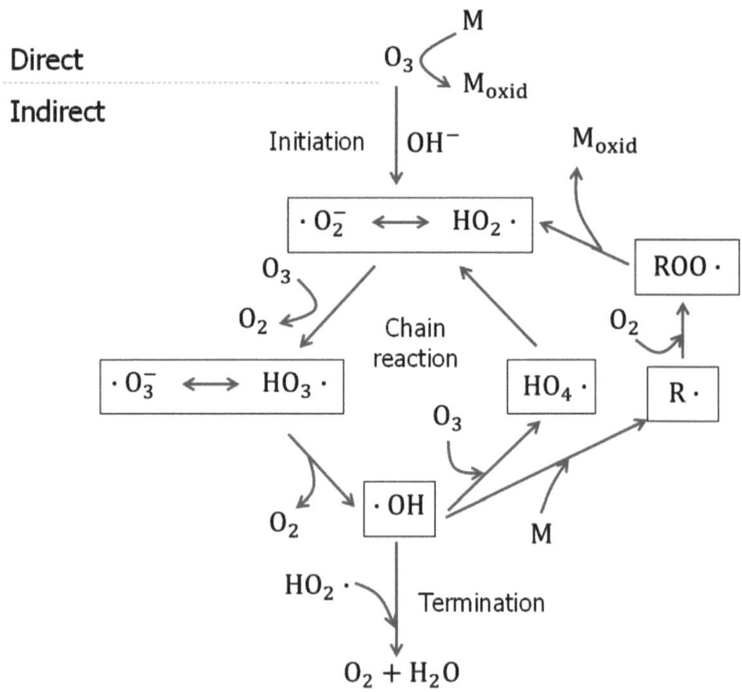

Fig. 4 Reactions involved in direct and indirect ozonation

different natures can also contribute to the formation or inhibition of free radicals. These substances are named initiators and inhibitors. Initiators are those, such as the hydroperoxide ion, which react directly with ozone to produce the radical and inhibitors are those components that generally terminate chain reactions (Gottschalk et al., 2009).

Furthermore, ozone does not oxidise some refractory organic compounds and has a low reaction rate. Thus, ozone can be used in combination with H_2O_2, UV light, catalyst, photocatalyst and ultrasound to increase the production of $\cdot OH$.

The catalytic ozonation systems have shown satisfactory performance, and it has several advantages compared to ozonation alone. Heterogeneous catalytic ozonation has increasingly received interest in the academic community in recent years because of the operation of mild conditions and the application on the degradation of several toxic and refractory organic pollutants (Hu et al., 2008). Furthermore, the application of solid catalysts has become the key for heterogeneous catalytic ozonation, as it accelerates the ozone decomposition into highly reactive oxygen species, leading to the decomposition of the structure of organic pollutants and intermediates, favouring further mineralisation. Catalysts commonly used in catalytic ozonation are transition metal oxides, molecular sieves, activated carbon, minerals (e.g., zeolites) and supported metal/metal oxides (Álvarez et al., 2007).

An increasing number of researchers focused on the use of nanocatalysts, with potential applications in heterogeneous catalytic ozonation. In comparison to conventional catalysis, nanocatalysis presents high activity, greater selectivity and excellent stability. Therefore, the synthesis of the nanocatalysts is pivotal for tailoring the size, structure, morphology, composition, electronic properties and chemical and thermal stability of the nanomaterial for further applications (Polshettiwar et al., 2011).

4.4 *Fenton*

The Fenton reaction was reported for the first time in 1894 by Fenton, characterised as the oxidative potential with increasing H_2O_2, using iron (Fe) as a catalyst under acidic conditions. The reactions involved in the process are (Haber & Willstätter, 1931) (Eqs. (5)–(10)):

$$Fe^{2+} + H_2O_2 \rightarrow Fe^{3+} + \cdot OH + OH^- \quad (5)$$

$$\cdot OH + H_2O_2 \rightarrow HO_2^{\cdot} + H_2O \quad (6)$$

$$Fe^{2+} + \cdot OH \rightarrow Fe^{3+} + OH^- \quad (7)$$

$$Fe^{3+} + HO_2^{\cdot} \rightarrow Fe^{2+} + O_2 + H^+ \quad (8)$$

$$\cdot OH + \cdot OH \rightarrow H_2O_2 \quad (9)$$

$$\text{Organic pollutant} + \cdot OH \rightarrow \text{Degraded products} \quad (10)$$

The Fenton reaction is separated into three main reaction stages: (i) chain initiation, (ii) chain propagation and (iii) chain termination (Oj et al., 2023). In general, ROS, such as hydroxyl and HO_2 radicals, are formed in the chain propagation phase. ROS begin to react with organic compounds generating alkyl and alkyl $RO_2^{\cdot}$, responsible for promoting the degradation of chemicals and other structural transformations. Finally, the reactive intermediates and the reaction process stop at the chain termination phase (Lai et al., 2021; Tian et al., 2021). Despite promising applications, the efficiency of the Fenton process depends on several parameters such as pH, low formation of Fe^{2+}, very rapid consumption of H_2O_2 in reactions and generation of secondary pollution (Zhang et al., 2019b).

To minimise the deficiencies occurring in the application of the Fenton process, modification and/or integration methods are developed to modulate the Fenton reactions, such as the application of nanomaterials that have specific optical, physical and chemical properties because of the nanoscale (Etaiw & El-Bendary, 2012; Ganiyu et al., 2018; Wang et al., 2022b). Studies have demonstrated that the use of nanomaterials presented significant advantages with the Fenton reaction, such as the cessation of pH limitation, thus expanding the use of the process (Bolobajeri et al., 2015; Boruah et al., 2017; Cheng et al., 2018). Another advantage of the Fenton reaction is that nanoscale materials can be modified according to the needs of the environment in which the process is being applied, thereby improving the effectiveness and specific conditions of Fenton reactions (Li et al., 2019). A series of nanomaterials were applied in the Fenton process, as shown in Table 2.

According to Table 2, using nanomaterials associated with the Fenton process is a promising approach because of the specific properties of nanomaterials such as high surface area, stability and reuse of catalysts and mainly chemical reactivity, which increases the efficiency of the Fenton process. Thus, the combination of nanomaterials and the Fenton process represents a promising strategy for the treatment of contaminated water, offering the potential for more efficient and sustainable removal of organic pollutants, contributing to the improvement of water quality and preservation of the environment.

4.5 *Photo-Fenton*

Photo-Fenton processes were verified in the 1950s, by the oxidation of organic compounds under UV irradiation, which resulted in the production of ·OH, responsible for the oxidation process (Nogueira et al., 2007), according to the following reactions (Eqs. (11)–(14)):

Table 2 Relationship of nanomaterials and target molecules in the Fenton process

Nanomaterial	Target molecule	Degradation efficiency (%)	Optimum operating parameters	References
Titanium/carbon nanotube	Bisphenol A	100	[Bisphenol A] = 50 mg L^{-1} pH = 6.30 [H_2O_2] = 2.5 g L^{-1} T = 70 °C [Catalyst] = 100 mg L^{-1}	Mirzaee et al. (2019)
Zero-valent iron	Rhodamine B	100	[Rhodamine B] = 0.1 mmol L^{-1} pH = 4 t = 30 min [H_2O_2] = 8 mmol L^{-1} [Catalyst] = 9 mmol L^{-1}	Liang et al. (2020b)
Iron oxide	Textile dye	92.2	Colour = 280 Pt–Co pH = 6 [H_2O_2] = 8 mmol L^{-1} [Catalyst] = 0.57 g L^{-1} t = 60 min	Nguyen et al. (2021)
Porous iron oxide	Levofloxacin	~90	[Levofloxacin] = 20 μmol L^{-1} [H_2O_2] = 10 mmol L^{-1} t = 90 min [Catalyst] = 1% wt	Paiva et al. (2023)
$FeWO_4$	Methylene blue	99.5	[Methylene blue] = 20 mg L^{-1} t = 14 min [H_2O_2] = 0.02% vol [Catalyst] = 1 g L^{-1}	Sun et al. (2022)

(continued)

Table 2 (continued)

Nanomaterial	Target molecule	Degradation efficiency (%)	Optimum operating parameters	References
Iron oxide	Methyl orange	>90	[Methyl orange] = $4.7 \cdot 10^{-5}$ mol L^{-1} $[H_2O_2]$ = 0.69 mol L^{-1} pH = 6.9 t = 5 h [Catalyst] = 0.5 g L^{-1}	Franco et al. (2023)
Fe_2O_3-Ni	Florfenicol	78	[Florfenicol] = 15 mg L^{-1} pH = 7 $[H_2O_2]$ = $6.5 \cdot 10^{-3}$ mol L^{-1} t = 180 min [Catalyst] = 300 mg L^{-1}	Sepuheda et al. (2023)
Titanium nanotube	Methyl orange	99.7	[Methyl orange] = 20 mg L^{-1} pH = 3 $[H_2O_2]$ = 1 mL of 6% H_2O_2 t = 60 min 2 × 1 cm^2 catalyst film	Zavahir et al. (2023)

$$Fe^{3+} + H_2O \rightarrow Fe(OH)^{2+} + H^+ \tag{11}$$

$$Fe(OH)^{2+} + hv \rightarrow Fe^{2+} + \cdot OH \tag{12}$$

$$H_2O_2 + hv \rightarrow 2 \cdot OH \tag{13}$$

$$\cdot OH + H^+ + e^- \rightarrow H_2O \tag{14}$$

As it is one of the most studied processes among AOPs, the photo-Fenton process appears as an attractive alternative because of the possibility of using solar energy during the reaction (Bacardit et al., 2007). However, some parameters limit the widening of its application, such as the pH value, as it is responsible for the formation of ·OH. Also, the light source emission spectrum, concentration and absorbance of photoactive species and quantum efficiencies influence the contribution of a specific reaction of the complex photo-Fenton mechanism (Clarizia et al., 2017; Kim &

Vogelpohl, 1999; Umar et al., 2010). Thus, the reactivity, reaction rate and path of Fe in the presence of some organic substance need to be evaluated (Giannakis, 2019).

To improve the efficiency and application of photo-Fenton processes, nanotechnology offers significant advantages and improvements, contributing mainly to the efficiency and effectiveness of the degradation of organic compounds (Wang et al., 2023). Nanomaterials, such as TiO_2 (Zheng et al., 2023) and iron oxides (Fe_2O_3, Fe_3O_4) (Zhu et al., 2019), exposed to UV-visible irradiation generate electron pairs that can react with H_2O_2 to generate highly reactive ·OH, resulting in a greater production of radicals and consequently, a faster and more efficient degradation rate of organic compounds (Feng et al., 2021).

Furthermore, nanomaterials can improve the energy efficiency of the photo-Fenton process by reducing the need for external energy sources and making the process more sustainable (Ganiyu et al., 2018). Fe_2NiO_4 nanoparticles achieved increased stability and reusability, with the advantage of reducing costs and sustainability of the process (Ansari et al., 2015). Thus, in graphene oxide-based nanomaterials, the photo-Fenton process improved the selectivity of the process, directing ·OH to specific organic compounds, being valid for the selective removal of specific pollutants (Liu et al., 2018).

It is concluded that the application of nanomaterials plays a fundamental role in the photo-Fenton process, offering significant advantages, such as improving selectivity, reuse and energy efficiency and contributing to the advancement of the technique and process of degradation of organic pollutants, making it a promising and beneficial choice to improve the effectiveness and sustainability of the process.

4.6 *Electro-Fenton*

The electro-Fenton is an advanced electrochemical oxidation method used for the degradation of organic compounds in water, in which an electrolyte containing Fe is electrolysed to form Fe ions (Fe^{2+}) that act as a catalyst in the decomposition of H_2O_2 into ·OH (Moreira et al., 2017). The electro-Fenton process is demonstrated in Eqs. (15)–(18) (Khataee et al., 2017):

$$Fe^{2+} + H_2O_2 + H^+ \rightarrow Fe^{3+} + \cdot OH + H_2O \tag{15}$$

$$O_{2(g)} + 2H^+ + 2e^- \rightarrow H_2O_2 \tag{16}$$

$$O_{2(g)} + 4H^+ + 4e^- \rightarrow 2H_2O \tag{17}$$

$$Fe^{3+} + e^- \rightarrow Fe^{2+} \tag{18}$$

The electro-Fenton method has some disadvantages such as the production of H_2O_2 being slow because oxygen has low solubility in water and the efficiency of the current under acidic pH (pH = 2) (Li et al., 2021). Therefore, the application of nanomaterials in the electro-Fenton process represents an effective approach because the introduction of nanomaterials, such as Fe nanoparticles (Seror et al., 2022), metal oxides (Zadeh-Kalantar & Lei, 2022) or carbon nanotubes (Lu et al., 2014), which have specific properties increase the generation of ·OH and improve process efficiency (Gormez et al., 2019; Hayer et al., 2022; Su et al., 2019; Zhang et al., 2018). Furthermore, the chemical reactivity of nanomaterials and the high surface area favour the adsorption of organic pollutants, facilitating their subsequent degradation by ·OH (Niu et al., 2017; Park et al., 2019; Syn et al., 2019).

Another advantage is the possibility of the precise control of the properties of nanomaterials, such as size, morphology and composition, to optimise their catalytic activity and selectivity in the degradation of different types of pollutants (Cui et al., 2022; Ghauch et al., 2011; Pelaez et al., 2012; Yu et al., 2011; Zou et al., 2023). Furthermore, the stability and regeneration capacity of nanomaterials can contribute to the reuse of the catalyst, reducing operational costs and minimising waste generated during treatment (Akhavan, 2010; Luo et al., 2024; Wang et al., 2024). Thus, Table 3 presents published studies on the application of nanomaterials in the electro-Fenton process.

The use of various nanomaterials in the electro-Fenton process has demonstrated promising results in terms of efficiently degrading organic pollutants. These nanomaterials, such as metal oxides and nanostructured carbon, can increase the oxidation rate of organic compounds, resulting in greater efficiency of the electro-Fenton process. Furthermore, the use of these nanomaterials can help reduce treatment time and the generation of toxic by-products.

4.7 Hybrid Systems

The use of combined AOPs has improved the treatment of wastewater due to the high degree of energy efficiency and production of higher amounts of ·OH as compared to the individual oxidation process. Many organic pollutants can be degraded by using combined AOPs. However, despite their excellent performance, the implementation and optimisation of these methods are often challenging and costly.

Although AOPs surpass conventional treatment methods, increasingly strict regulations for the discharge of wastewater have made the application of a single AOP insufficient to reduce pollutant concentrations to meet regulation limits. Moreover, even if a single AOP achieves the limits proposed by the regulations, it is not economically favourable to upgrade the entire treatment plant. Therefore, the use of hybrid or integrated AOPs is an alternative to achieve effluent quality standards in a cost-effective manner (Ahmed et al., 2015; Shestakova & Sillanpää, 2013).

For instance, the application of an O_3/UV/TiO_2 oxidation system at pH 7 was effective for the removal of diclofenac, a popular non-steroidal anti-inflammatory

Table 3 Published articles that apply several types of nanomaterials in the electro-Fenton process

Nanomaterials	Target molecule	Degradation efficiency (%)	Optimum operating parameters	References
Fe_3O_4	Mesotrione	93	[Mesotrione] = 20 mg L^{-1} pH = 3.6 T – 34 °C t = 9 h Air flow rate = 140 mL min^{-1} [Catalyst] = 1 g L^{-1} Voltage = 50 mV	Zhao et al. (2023)
Nitrogen-doped carbon	Chlorophenoxy acid herbicides	92–98.6	[Herbicides] = 1 mmol L^{-1} 8 × 4 cm^2 electrode Air flow rate = 30 mL min^{-1} t = 300 min Current density = 40 mA cm^{-2}	Kamaraj et al. (2020)
FeOCl	Sulphamethoxazole	100	[Sulphamethoxazole] = 20 mmol L^{-1} pH = 3.0 t = 30 min 2 × 2 cm^2 electrode O_2 flow rate = 300 mL min^{-1} Applied potential = –0.6 V	Zhong et al. (2022)
Fe-graphene	Pesticide	~91	[Pesticide] = 10 mg L^{-1} pH = 6.86 t = 360 min Applied voltage = 15 V [Catalyst] = 1 g L^{-1}	Ghanbarlou et al. (2020)
Graphene	Ciprofloxacin	>99.9	[Ciprofloxacin] = 10 mg L^{-1} pH = 3 t = 15 min 5 × 2 cm^2 electrode Applied voltage = –2.5 V	Divyapriya et al. (2018)

(continued)

Table 3 (continued)

Nanomaterials	Target molecule	Degradation efficiency (%)	Optimum operating parameters	References
Tri-metal-doped carbon	Enrofloxacin	99.8	[Enrofloxacin] = 100 mg L^{-1} pH = 3 Current = 40 mA t = 30 min 2 × 2 cm^2 electrode	Li et al. (2022)
Gold	Tetracycline	83.3	[Tetracycline] = 0.04 mmol L^{-1} pH = 6.70 t = 120 min Applied voltage = − 2.5 V [Catalyst] = 1 mmol L^{-1}	Liu et al. (2020)

pharmaceutical drug, from water (García-Araya et al., 2010). Therefore, the combined process was more effective than the application of the ozonation process alone for the removal of pharmaceuticals from wastewater with effluents.

The combination of AOPs can be applied to increase the mineralisation efficiency of the organic compounds as well as the AOPs' conventional treatment combination. Integrated and hybrid AOPs systems help to reduce the overall cost and use chemicals in the treatment of wastewater along with the reduction of storage facilities of chemicals in operations plants (Sun et al., 2007). However, the application of these systems requires a comprehensive assessment in terms of economic viability, ecological impact and sustainability supported by environmental public policies (Asheghmoalla & Mehrvar, 2024).

5 Challenges and Future Perspectives

Globally, researchers are addressing the critical issue of wastewater management by developing various treatment techniques that efficiently eliminate contaminants. AOPs are particularly noted for their efficiency, making them suitable for large-scale wastewater treatment, and common oxidants employed include chlorine, air oxygen, ozone and H_2O_2 (Buthiyappan et al., 2016). Often, AOPs serve as either a preliminary step to convert stubborn pollutants into biodegradable forms or as a final treatment before releasing water into receiving bodies, considering the cost of the process. Their flexibility allows them to be used either as standalone water treatment technologies or in tandem with other methods. AOPs are favoured for treating toxic, intricate and bio-resistant contaminants, offering benefits such as high efficiency, non-selectivity, excellent reproducibility and operational simplicity.

Despite the progress in nanotechnology facilitating advancements in AOPs, challenges persist, including large-scale controlled synthesis, prolonged reaction times, suboptimal efficiency, limited photocatalyst recyclability, high recombination rates, inadequate adsorption on the nanoparticles' active surface and lack of knowledge of reaction mechanism and by-products.

Because of their size, nanomaterials are often supported to prevent their leaching during operations. Therefore, the evaluation of ecotoxicity, phytotoxicity, genotoxicity and cytotoxicity are important to avoid toxicological effects on animals, plants and humans due to the application of nanomaterials in AOPs.

Nevertheless, the integration of nanotechnology in the treatment of water has transformed purification methods, enabling the effective elimination of pathogens such as bacteria and viruses (Shayo et al., 2024). The contribution of nanotechnology extends to cost-effective system production, making clean water accessible to economically disadvantaged communities (Kazemi et al., 2023). The effective implementation of AOPs that use nanomaterials for water remediation is a task that integrates different domains of knowledge. This integration ranges from materials science, focussed on the development of innovative catalysts, to water treatment engineering, responsible for designing reactors that are both efficient and practical. The challenges present in this field are as significant as the opportunities for advancement, offering a fertile ground for research and application of emerging technologies in water purification, for example.

6 Final Remarks

Conventional strategies for the treatment of water and wastewater have been implemented as biological, physical and chemical methods to solve the problems of water contaminated with pollutants. However, the interrelated practices are often expensive, unproductive and non-destructive with regard to carbon-based contaminants, necessitating the use of advanced oxidative processes combined with nanotechnology.

The application of nanomaterials in AOPs gave a new perspective on the degradation of hazardous compounds that often pollute the water resources, due to their intrinsic properties, such as the higher amount of active site for adsorption and generation of reactive radicals that boost the efficiency of AOPs. The incorporation of nanomaterials among different AOPs has generated great advances and new perspectives for the degradation of a wide variety of organic pollutants.

While nanomaterials offer significant advantages in AOPs, challenges persist, including the controlled and scalable synthesis of nanomaterials, management of solid wastes generated during the process and assessment of the potential environmental and health impacts associated with the use of these materials. Furthermore, a deeper understanding of the degradation mechanisms and kinetics of reactions involved in AOPs with nanomaterials is needed to optimise the efficiency and applicability of these processes in various environmental and industrial contexts.

Another point is that the evaluation of the ecotoxicity, phytotoxicity, cytotoxicity and genotoxicity need further studies to ensure that the nanomaterials applied in AOPs did not cause side effects. However, despite these challenges, the use of AOPs combined with the advances in nanotechnology has the potential to push boundaries for the development of new and more efficient systems for the degradation of organic pollutants.

Acknowledgements The authors would like to thank the Applied Nanomaterials Research Group (GPNAp) at the Franciscan University (UFN); Recycling, Waste Treatment and Extraction Laboratory (LAREX) at the University of São Paulo (USP); Ciencias e Ingeniería de Materiales de la Unidad Tecnológica de Automatismos y Sistemas Inteligentes at Technological University of Uruguay (UTEC) and Research Group On Adsorptive and Catalytic Process Engineering (ENGEPAC) at Federal University of Santa Maria (UFSM) for their support in carrying out this work. This study was financed by the Conselho Nacional de Desenvolvimento Científico e Tecnológico—Brazil (CNPq/MCTI/FNDCT no. 23/2022)—InovaNióbio (no. 408422/2022-0).

References

Abu-Dalo, M. A., Al-Rosan, S. A., & Albiss, B. A. (2021). Photocatalytic degradation of methylene blue using polymeric membranes based on cellulose acetate impregnated with ZnO nanostructures. *Polymers, 13*, 3451–3467. https://doi.org/10.3390/polym13193451

Ahmed, M. B., Zhou, J. L., Ngo, H. H., & Guo, W. (2015). Adsorptive removal of antibiotics from water and wastewater: Progress and challenges. *Science of the Total Environment, 532*, 112–126. https://doi.org/10.1016/j.scitotenv.2015.05.130

Akhavan, O. (2010). Thickness dependent activity of nanostructured $TiO_2/\alpha\text{-}Fe_2O_3$ photocatalyst thin films. *Applied Surface Science, 257*, 1724–1728. https://doi.org/10.1016/j.apsusc.2010.09.005

Álvarez, P. M., Beltrán, F. J., Pocostales, J. P., et al. (2007). Preparation and structural characterization of Co/Al_2O_3 catalysts for the ozonation of pyruvic acid. *Applied Catalysis b: Environment and Energy, 72*, 322–330. https://doi.org/10.1016/j.apcatb.2006.11.009

Ameta, R., Solanki, M. S., Benjamin, S. et al. (2018). Photocatalysis. In: S. C. Ameta, & R. Ameta (Eds.), *Advanced oxidation processes for waste water treatment* (pp. 135–175). Academic Press. https://doi.org/10.1016/B978-0-12-810499-6.00006-1

Ansari, S., Moghadassi, A., & Hosseini, S. (2015). Fabrication of novel poly(phenylene ether ether sulfone) based nanocomposite membrane modified by Fe_2NiO_4 nanoparticles and ethanol as organic modifier. *Desalination, 357*, 189–196. https://doi.org/10.1016/j.desal.2014.11.022

Asheghmoalla, M., & Mehrvar, M. (2024). Integrated and hybrid processes for the treatment of actual wastewaters containing micropollutants: A review on recent advances. *Processes, 12*(2), 339–376. https://doi.org/10.3390/pr12020339

Azeez, F., Al-Hetlani, E., Arafa, M., et al. (2018). The effect of surface charge on photocatalytic degradation of methylene blue dye using chargeable titania nanoparticles. *Scientific Reports, 8*, 7104–7113. https://doi.org/10.1038/s41598-018-25673-5

Bacardit, J., Stotzner, J., Chamarro, E., et al. (2007). Effect of salinity on the Photo-Fenton process. *Industrial & Engineering Chemistry Research, 46*, 7615–7619. https://doi.org/10.1021/ie070154o

Bao, J., Guo, S., Fan, D., et al. (2023). Sonoactivated nanomaterials: A potent armament for wastewater treatment. *Ultrasonics Sonochemistry, 99*, 106569–106588. https://doi.org/10.1016/j.ultsonch.2023.106569

Bellotti, V., Daldossi, C., Perilli, D., et al. (2023). Mechanism of sustainable photocatalysis based on doped-titanium dioxide nanoparticles for UV to visible light induced PET-RAFT photopolymerization. *Journal of Catalysis, 428*, 115074–115088. https://doi.org/10.1016/j.jcat.2023.07.015

Bethi, B., Sonowane, S. H., Bhanvase, B. A., et al. (2016). Nanomaterials-based advanced oxidation processes for wastewater treatment: A review. *Chemical Engineering and Processing: Process Intensification, 109*, 178–189. https://doi.org/10.1016/j.cep.2016.08.016

Bolobajeri, J., Trapido, M., Goi, A., et al. (2015). Improvement in iron activation ability of alachlor Fenton-like oxidation by ascorbic acid. *Chemical Engineering Journal, 281*, 566–574. https://doi.org/10.1016/j.cej.2015.06.115

Boruah, P., Sharma, B., Karbhal, I., et al. (2017). Ammonia-modified graphene sheets decorated with magnetic Fe_3O_4 nanoparticles for the photocatalytic and photo-Fenton degradation of phenolic compounds under sunlight irradiation. *Journal of Hazardous Materials, 325*, 90–100. https://doi.org/10.1016/j.jhazmat.2016.11.023

Bracamontes-Ruelas, A. R., Ordaz-Díaz, L. A., Bailón-Salas, A. M., et al. (2022). Emerging pollutants in wastewater, advanced oxidation processes as an alternative treatment and perspectives. *Processes, 10*, 1041–1064. https://doi.org/10.3390/pr10051041

Buthiyappan, A., Abdul Aziz, A. R., & Wan Daud, W. M. A. (2016). Recent advances and prospects of catalytic advanced oxidation process in treating textile effluents. *Reviews in Chemical Engineering, 32*(1), 1–47. https://doi.org/10.1515/revce-2015-0034

Cardoso, I. M. F., Da Silva, L. P., & Da Silva, J. C. G. E. (2023). Nanomaterial-based advanced oxidation/reduction processes for the degradation of PFAS. *Nanomaterials, 13*(10), 1668–1690. https://doi.org/10.3390/nano13101668

Charitidis, C., Pantelitsa, G., Malamatenia, A., et al. (2014). Manufacturing nanomaterials: From research to industry. *Manufacturing Review, 1*, 1–19. https://doi.org/10.1051/mfreview/2014009

Chen, W. J., Hsu, K. C., Fang, T. H., et al. (2022). Characteristics and heterostructure of metal-doped TiO_2/ZnO nanocatalysts. *Current Applied Physics, 38*, 1–6. https://doi.org/10.1016/j.cap.2022.03.001

Cheng, M., Lai, C., Liu, Y., et al. (2018). Metal-organic frameworks for highly efficient heterogeneous Fenton-like catalysis. *Coordination Chemistry Reviews, 368*, 80–92. https://doi.org/10.1016/j.ccr.2018.04.012

Cho, Y. S., Nguyen, T. T. H., Nguyen, H., et al. (2024). Modeling and experimental study of hollow cylindrical catalysts for prediction of transient behaviors of suspension photoreaction systems by Langmuir-Hinshelwood kinetics. *Catalysis Today, 425*, 114345–114362. https://doi.org/10.1016/j.cattod.2023.114345

Clarizia, L., Russo, D., & Samma, I. (2017). Homogeneous photo-Fenton processes at near neutral pH: A review. *Applied Catalysis b: Environment and Energy, 209*, 358–371. https://doi.org/10.1016/j.apcatb.2017.03.011

Cui, L., Zhao, X., Xi, H., et al. (2022). Overcoming the activity–stability trade-off in heterogeneous electro-Fenton catalysis: Encapsulating carbon cloth-supported iron oxychloride within graphitic layers. *ACS Catalysis, 12*, 13334–133348. https://doi.org/10.1021/acscatal.2c03571

Deng, Y., & Zhao, R. (2015). Advanced oxidation processes (AOPs) in wastewater treatment. *Current Pollution Reports, 1*, 167–176. https://doi.org/10.1007/s40726-015-0015-z

Din, M. I., Khalid, R., Najeeb, J., et al. (2021). Fundamentals and photocatalysis of methylene blue dye using various nanocatalytic assemblies—A critical review. *Journal of Cleaner Production, 298*, 126567–126585. https://doi.org/10.1016/j.jclepro.2021.126567

Divyapriya, G., Nambi, I., & Senthilnathan, J. (2018). Ferrocene functionalized graphene based electrode for the electro-Fenton oxidation of ciprofloxacin. *Chemophere, 209*, 113–123. https://doi.org/10.1016/j.chemosphere.2018.05.148

Domingues, E., Fernandes, E., Gomes, J., et al. (2021). Advanced oxidation processes perspective regarding swine wastewater treatment. *Science of the Total Environment, 776*, 145958–145975. https://doi.org/10.1016/j.scitotenv.2021.145958

Dong, S., Yan, X., Yue, Y., et al. (2022). H_2O_2 concentration influenced the photoaging mechanism and kinetics of polystyrene microplastic under UV irradiation: Direct and indirect photolysis. *Journal of Cleaner Production, 380*(2), 135046. https://doi.org/10.1016/j.jclepro.2022.135046

Druzian, D. M., Oviedo, L. R., Loureiro, S. N., et al. (2023). Cerium oxide nanoparticles: Biosynthesis, characterization, antimicrobial, ecotoxicity and photocatalytic activity. *Journal of Photochemistry and Photobiology, 443*, 114773–114785. https://doi.org/10.1016/j.jphotochem.2023.114773

Etaiw, S., & El-Bendary, M. (2012). Degradation of methylene blue by catalytic and photo-catalytic processes catalyzed by the organotin-polymer $3\infty[(Me_3Sn)_4Fe(CN)_6]$. *Applied Catalysis b: Environment and Energy, 126*, 326–333. https://doi.org/10.1016/j.apcatb.2012.07.032

Feng, J., Chu, C., Ma, Z., et al. (2021). Fenton and Fenton-like catalysts for electrochemical immunoassay: A mini review. *Electrochemistry Communications, 125*, 106970–106976. https://doi.org/10.1016/j.elecom.2021.106970

Franco, R., Silva, A., Licea, Y. et al. (2023). Effect of Camellia sinensis origin and heat treatment in the iron oxides nanomaterials composition and Fenton degradation of methyl orange. *Journal of the Brazilian Chemical Society, 34*, 694–704. https://doi.org/10.21577/0103-5053.20220140

Ganiyu, S., Zhou, M., Martínez-Huitle, C., et al. (2018). Heterogeneous electro-Fenton and photoelectro-Fenton processes: A critical review of fundamental principles and application for water/wastewater treatment. *Applied Catalysis b: Environment and Energy, 235*, 103–129. https://doi.org/10.1016/j.apcatb.2018.04.044

Ganiyu, S. O., Sable, S., & El-Din, M. G. (2022). Advanced oxidation processes for the degradation of dissolved organics in produced water: A review of process performance, degradation kinetics and pathway. *Chemical Engineering Journal, 429*, 132492. https://doi.org/10.1016/j.cej.2021.132492

García-Araya, J. F., Beltrán, F. J., & Aguinaco, A. (2010). Diclofenac removal from water by ozone and photolytic TiO_2 catalyzed processes. *Journal of Chemical Technology & Biotechnology, 85*, 798–804. https://doi.org/10.1002/jctb.2363

Gaur, N., Dutta, D., Singh, A., et al. (2022). Recent advances in the elimination of persistent organic pollutants by photocatalysis. *Frontiers in Environmental Science, 10*, 872514. https://doi.org/10.3389/fenvs.2022.872514

Gershon, T., Shin, B., Bojarczuk, N., et al. (2014). The role of sodium as a surfactant and suppressor of non-radiative recombination at internal surfaces in Cu_2ZnSnS_4. *Advanced Energy Materials, 5*, 1400849–1400857. https://doi.org/10.1002/aenm.201400849

Ghanbarlou, H., Nasernejad, B., Fini, M., et al. (2020). Synthesis of an iron-graphene based particle electrode for pesticide removal in three-dimensional heterogeneous electro-Fenton water treatment system. *Journal of Chemical Engineering, 395*, 125025–125040. https://doi.org/10.1016/j.cej.2020.125025

Ghauch, A., Baydoun, H., & Dermesropian, P. (2011). Degradation of aqueous carbamazepine in ultrasonic/Fe^0/H_2O_2 systems. *Journal of Chemical Engineering, 172*, 18–27. https://doi.org/10.1016/j.cej.2011.04.002

Giannakis, D. (2019). A review of the concepts, recent advances and niche applications of the (photo) Fenton process, beyond water/wastewater treatment: Surface functionalization, biomass treatment, combatting cancer and other medical uses. *Applied Catalysis b: Environment and Energy, 248*, 309–319. https://doi.org/10.1016/j.apcatb.2019.02.025

Gormez, F., Gormez, O., Gormez, B., et al. (2019). Degradation of chloramphenicol and metronidazole by electro-Fenton process using graphene oxide-Fe_3O_4 as heterogeneous catalyst. *Journal of Environmental Chemical Engineering, 7*, 102990–103002. https://doi.org/10.1016/j.jece.2019.102990

Gottschalk, C., Libra, J. A., & Saupe, A. (2009). *Ozonation of water and waste water: A practical guide to understanding ozone and its applications*. John Wiley & Sons. https://doi.org/10.1002/9783527628926

Haber, F., & Willstätter, R. (1931). Unpaarigheit und radikalketten im Reaktion-Mechanismus organischer und enzymatischer Vorgänge. *Ber Deu Química Gés, 64*, 2844–2856. https://doi.org/10.1002/cber.19310641118

Hayer, A., Hasan, J., Bright, R., et al. (2022). Spiked titanium nanostructures that inhibit anaerobic dental pathogens. *ACS Applied Nano Materials, 9*, 12051–12062. https://doi.org/10.1021/acsanm.1c04073

Hodges, B. C., Cates, E. L., & Kim, J. H. (2018). Challenges and prospects of advanced oxidation water treatment processes using catalytic nanomaterials. *Nature Nanotechnology, 13*, 642–650. https://doi.org/10.1038/s41565-018-0216-x

Hu, C., Xing, S., & Qu, J. (2008). Catalytic ozonation of herbicide 2,4-D over cobalt oxide supported on mesoporous zirconia. *Journal of Physical Chemistry C, 112*(15), 5978–5983. https://doi.org/10.1021/jp711463e

Ikehata, K., & El-Din, M. G. (2004). Degradation of recalcitrant surfactants in wastewater by ozonation and advanced oxidation processes: A review. *Ozone: Science & Engineering, 26*(4), 327–343. https://doi.org/10.1080/01919510490482160

Ikehata, K., & Li, Y. (2018). Ozone-based processes. In S. C. Ameta, & R. Ameta (Eds.), *Advanced oxidation processes for waste water treatment* (pp. 115–134). Academic Press. https://doi.org/10.1016/B978-0-12-810499-6.00005-X

Jabbar, Z. H., Graimed, B. H., Ammar, S. H., et al. (2024). The latest progress in the design and application of semiconductor photocatalysis systems for degradation of environmental pollutants in wastewater: Mechanism insight and theoretical calculations. *Materials Science in Semiconductor Processing, 173*, 108153. https://doi.org/10.1016/j.mssp.2024.108153

Kamaraj, R., Nesakumar, N., & Vasuderan, S. (2020). Nitrogen doped carbon nanomaterial as electrocatalyst for oxygen reduction reaction in acidic media: To use in electro-Fenton. *ChemistrySelect, 5*, 10034–10040. https://doi.org/10.1002/slct.202002413

Kanakaraju, D., Glass, B. D., & Oelgemöller, M. (2018). Advanced oxidation process-mediated removal of pharmaceuticals from water: A review. *Journal of Environmental Management, 219*, 189–207. https://doi.org/10.1016/j.jenvman.2018.04.103

Kaus, N. H. M., Rithwan, A. F., Adnan, R., et al. (2021). Effective strategies, mechanisms, and photocatalytic efficiency of semiconductor nanomaterials incorporating rGO for environmental contaminant degradation. *Catalysts, 11*(3), 302–327. https://doi.org/10.3390/catal11030302

Kazemi, S., Hosseingholian, A., Gohari, S. D., et al. (2023). Recent advances in green synthesized nanoparticles: From production to application. *Materials Today Sustainability, 24*, 100500–100522. https://doi.org/10.1016/j.mtsust.2023.100500

Khataee, A., Hasanzadeh, A., et al. (2017). Modified cathodes with carbon-based nanomaterials for Electro-Fenton process. *Electro-Fenton Process, 61*, 111–143. https://doi.org/10.1007/698_2017_74

Kim, S. M., & Vogelpohl, A. (1999). Degradation of organic pollutants by the photo-Fenton-process. *Chemical Engineering & Technology, 2*, 187–192. https://doi.org/10.1002/(SICI)1521-4125(199802)21:2%3C187::AID-CEAT187%3E3.0.CO;2-H

Kumari, P., & Kumar, A. (2023). Advanced oxidation process: A remediation technique for organic and non-biodegradable pollutant. *Results in Surfaces and Interfaces, 11*, 100122–100142. https://doi.org/10.1016/j.rsurfi.2023.100122

Kurian, M. (2021). Advanced oxidation processes and nanomaterials—A review. *Cleaner Engineering and Technology, 2*, 100090–100103. https://doi.org/10.1016/j.clet.2021.100090

Lai, C., Shi, X., Li, L., et al. (2021). Enhancing iron redox cycling for promoting heterogeneous Fenton performance: A review. *Science of the Total Environment, 776*, 145850–145867. https://doi.org/10.1016/j.scitotenv.2021.145850

Lei, Y., Hou, J., Fang, C., et al. (2023). Ultrasound-based advanced oxidation processes for landfill leachate treatment: Energy consumption, influences, mechanisms and perspectives. *Ecotoxicology and Environmental Safety, 263*, 155366–155384. https://doi.org/10.1016/j.ecoenv.2023.115366

Li, C., Wu, J., Peng, W., et al. (2019). Peroxymonosulfate activation for efficient sulfamethoxazole degradation by Fe_3O_4/β-FeOOH nanocomposites: Coexistence of radical and non-radical reactions. *Journal of Chemical Engineering, 356*, 904–914. https://doi.org/10.1016/j.cej.2018.09.064

Li, J., & Wu, N. (2015). Semiconductor-based photocatalysts and photoelectrochemical cells for solar fuel generation: A review. *Catalysis Science & Technology, 5*, 1360–1384. https://doi.org/10.1039/C4CY00974F

Li, X., Xiao, X., Ruan, X., et al. (2022). Enrofloxacin degradation in a heterogeneous electro-Fenton system using a tri-metal-carbon nanofibers composite cathode. *Journal of Chemical Engineering, 427*, 130927–130941. https://doi.org/10.1016/j.cej.2021.130927

Li, Y., Liu, L., Zhang, O., et al. (2021). Hybrid electro-Fenton and peroxi-coagulation process for high removal of 2,4-dichlorophenoxiacetic acid with low iron sludge generation. *Electrochimica Acta, 382*, 138304–138312. https://doi.org/10.1016/j.electacta.2021.138304

Liang, H., Li, T., Zhang, J., et al. (2020a). 3-D hierarchical Ag/ZnO@CF for synergistically removing phenol and Cr(VI): Heterogeneous vs. homogeneous photocatalysis. *Journal of Colloid and Interface Science, 558*, 85–94. https://doi.org/10.1016/j.jcis.2019.09.105

Liang, L., Cheng, L., Zhang, Y., et al. (2020b). Efficiency and mechanisms of rhodamine B degradation in Fenton-like systems based on zerovalent iron. *RSC Advances, 10*, 28509–28515. https://doi.org/10.1039/d0ra03125a

Liu, F., Liu, Y., Yao, Q., et al. (2020). Supported atomically-precise gold nanoclusters for enhanced flow-through electro-Fenton. *Environmental Science & Technology, 54*, 5913–5921. https://doi.org/10.1021/acs.est.0c00427

Liu, X., Zhai, Y., Zhang, I., et al. (2018). Insight into electro-Fenton and photo-Fenton for the degradation of antibiotics: Mechanism study and research gaps. *Journal of Chemical Engineering, 347*, 379–397. https://doi.org/10.1016/j.cej.2018.04.142

Liu, Y. C., Liu, X., Wang, X., et al. (2023). Hybrid persulfate/sonocatalysis for degradation of acid orange 7 in the presence of Ag_2O/$CuWO_4$ composite: Operating parameters and sonocatalytic mechanism. *Journal of Cleaner Production, 394*, 136287–136301. https://doi.org/10.1016/j.jclepro.2023.136287

Lu, Z., Chen, F., He, M., et al. (2014). Microwave synthesis of a novel magnetic imprinted TiO_2 photocatalyst with excellent transparency for selective photodegradation of enrofloxacin hydrochloride residues solution. *Journal of Chemical Engineering, 249*, 15–26. https://doi.org/10.1016/j.cej.2014.03.077

Luo, X., Zhu, R., Zhao, L., et al. (2024). Defective nitrogen doped carbon material derived from nano-ZIF-8 for enhanced in-situ H_2O_2 generation and tetracycline hydrochloride degradation in electro-Fenton system. *Environmental Research, 251*, 118644–118652. https://doi.org/10.1016/j.envres.2024.118644

Mahadevan, R., Palanisamy, S., & Sakthivel, P. (2023). Role of nanoparticles as oxidation catalyst in the treatment of textile wastewater: Fundamentals and recent advances. *Sustainable Chemistry for the Environment, 4*, 1000044. https://doi.org/10.1016/j.scenv.2023.100044

Masjoudi, M., & Mohseni, M. (2023). Photolysis of chloramines in vacuum-UV and vacuum-UV/chlorine advanced oxidation processes for removal of 1,4-dioxane: Effect of water matrix, kinetic modeling, and implications for potable reuse. *Journal of Hazardous Materials, 454*, 131454. https://doi.org/10.1016/j.jhazmat.2023.131454

Mirzaee, S., Jaafarzadeh, N., Gomes, H., et al. (2019). Magnetic titanium/carbon nanotube nanocomposite catalyst for oxidative degradation of Bisphenol A from high saline polycarbonate plant effluent using catalytic wet peroxide oxidation. *Journal of Chemical Engineering, 370*, 372–389. https://doi.org/10.1016/j.cej.2019.03.202

Moreira, F., Boaventura, R., Brillais, E., et al. (2017). Electrochemical advanced oxidation processes: A review on their application to synthetic and real wastewaters. *Applied Catalysis b: Environment and Energy, 202*, 217–261. https://doi.org/10.1016/j.apcatb.2016.08.037

Nguyen, T., Huynh, K., Padungthon, S., et al. (2021). Synthesis of natural flowerlike iron-alum oxide with special interaction of Fe-Si-Al oxides as an effective catalyst for heterogeneous

Fenton process. *Journal of Environmental Chemical Engineering, 9*, 105732–105741. https://doi.org/10.1016/j.jece.2021.105732

Nie, C., Shao, N., Wang, B., et al. (2016). Fully solar-driven thermo- and electrochemistry for advanced oxidation processes (STEP-AOPs) of 2-nitrophenol wastewater. *Chemosphere, 154*, 604–612. https://doi.org/10.1016/j.chemosphere.2016.04.020

Nishu, & Kumar, S. (2023). Smart and innovative nanotechnology applications for water purification. *Hybrid Advances, 3*, 1000044–100065. https://doi.org/10.1016/j.hybadv.2023.100044

Niu, Q., Guo, J., Chen, B., et al. (2017). Bimetal-organic frameworks/polymer core-shell nanofibers derived heteroatom-doped carbon materials as electrocatalysts for oxygen reduction reaction. *Carbon, 114*, 250–260. https://doi.org/10.1016/j.carbon.2016.12.016

Nogueira, R., Trovó, A., Silva, M., et al. (2007). Fundaments and environmental applications of Fenton and photo-Fenton processes. *Química Nova, 30*, 400–408. https://doi.org/10.1590/S0100-40422007000200030

Oj, J., Jiang, G., Wan, Y., et al. (2023). Nanomaterials-modulated Fenton reactions: Strategies, chemodynamic therapy and future trends. *Journal of Chemical Engineering, 466*, 142960–142970. https://doi.org/10.1016/j.cej.2023.142960

Oturan, M. A., & Aaron, J. J. (2014). Advanced oxidation processes in water/wastewater treatment: Principles and applications. A review. *Critical Reviews in Environmental Science and Technology, 44*(23), 2577–2641. https://doi.org/10.1080/10643389.2013.829765

Paiva, A. E., Guin, J. P., Vasquez, J. F. B., et al. (2023). Degradation of an antibiotic via a photo-Fenton process on block copolymer-nanopatterned iron oxide coated substrates. *Chemical Engineering Journal, 466*, 142925–142936. https://doi.org/10.1016/j.cej.2023.142925

Park, H., Oh, S., Lee, S., et al. (2019). Cobalt- and nitrogen-codoped porous carbon catalyst made from core–shell type hybrid metal–organic framework (ZIF-L@ZIF-67) and its efficient oxygen reduction reaction (ORR) activity. *Applied Catalysis b: Environment and Energy, 246*, 322–329. https://doi.org/10.1016/j.apcatb.2019.01.083

Parson, S. (2005). *Advanced oxidation processes for water and wastewater treatment*. IWA Publishing. https://doi.org/10.2166/9781780403076

Pelaez, M., Falaras, P., Kontos, A., et al. (2012). A comparative study on the removal of cylindrospermopsin and microcystins from water with NF-TiO_2-P25 composite films with visible and UV-vis light photocatalytic activity. *Applied Catalysis b: Environment and Energy, 121*, 30–39. https://doi.org/10.1016/j.apcatb.2012.03.010

Pizzichetti, R., Reynolds, K., Pablos, C., et al. (2023). Removal of diclofenac by UV-B and UV-C light-emitting diodes (LEDs) driven advanced oxidation processes (AOPs): Wavelength dependence, kinetic modelling and energy consumption. *Chemical Engineering Journal, 471*, 144520–144531. https://doi.org/10.1016/j.cej.2023.144520

Polshettiwar, V., Luque, R., Fihri, A., et al. (2011). Magnetically recoverable nanocatalysts. *Chemical Reviews, 111*, 3036–3075. https://doi.org/10.1021/cr100230z

Prasert, A., Sontikaew, S., Sriprapai, D., et al. (2020). Polypropylene/ZnO nanocomposites: Mechanical properties, photocatalytic dye degradation, and antibacterial property. *Materials, 13*, 914–930. https://doi.org/10.3390/ma13040914

Praxedes, F. R., Nobre, M. A. L., Poon, P. S., et al. (2021). Nanostructured $K_xNa_{1-x}NbO_3$ hollow spheres as potential materials for the photocatalytic treatment of polluted water. *Applied Catalysis b: Environment and Energy, 298*, 120502–120502. https://doi.org/10.1016/j.apcatb.2021.120502

Preethi, Shanmugavel, S. P., Kumar, G. et al. (2024). Recent progress in mineralization of emerging contaminants by advanced oxidation process: A review. *Environmental Pollution, 314*, 122842–122858. https://doi.org/10.1016/j.envpol.2023.122842

Ratchnashree, S. R., Karmegam, N., Selvam, M., et al. (2023). Advanced technologies for the determination of quantitative structure-activity relationships and degradation efficiency of micropollutants and their removal in water—A review. *Science of the Total Environment, 904*, 166563. https://doi.org/10.1016/j.scitotenv.2023.166563

Sepuheda, P., Campos, S., Calzadilla, W., et al. (2023). Pd-Based nanoparticles as catalysts for improved removal of florfenicol via heterogeneous Fenton and photo-Fenton(-like) processes. *ACS Applied Nano Materials, 6*, 12177–12189. https://doi.org/10.1021/acsanm.3c01915

Seror, S., Shami, D., Alboy, Y., et al. (2022). Elucidation of a mechanism for the heterogeneous electro-Fenton process and its application in the green treatment of azo dyes. *Chemosphere, 286*, 131832–131841. https://doi.org/10.1016/j.chemosphere.2021.131832

Shayo, G. M., Elimbinzi, E., & Shao, G. N. (2024). Water-based technologies for improving water quality at the point of use: A review. *Wires Nanomedicine and Nanobiotechnology, 16*(1), e1940. https://doi.org/10.1002/wnan.1940

Sheng, Y., Wei, Z., Miao, H. et al. (2019). Enhanced organic pollutant photodegradation via adsorption/photocatalysis synergy using a 3D g-C_3N_4/TiO_2 free-separation photocatalyst. *Journal of Chemical Engineering, 370*, 287–294. https://doi.org/10.1016/j.cej.2019.03.197

Shestakova, M., & Sillanpää, M. (2013). Removal of dichloromethane from ground and wastewater: A review. *Chemosphere, 93*, 1258–1267. https://doi.org/10.1016/j.chemosphere.2013.07.022

Su, P., Zhou, M., Lu, X., et al. (2019). Electrochemical catalytic mechanism of N-doped graphene for enhanced H_2O_2 yield and in-situ degradation of organic pollutant. *Applied Catalysis b: Environment and Energy, 245*, 583–593. https://doi.org/10.1016/j.apcatb.2018.12.075

Sun, D., Igbal, N., & Liao, W. (2022). Efficient degradation of MB dye by 1D $FeWO_4$ nanomaterials through the synergistic effect of piezo-Fenton catalysis. *Ceramics International, 48*, 25465–25473. https://doi.org/10.1016/j.scitotenv.2020.141020

Sun, J. H., Sun, S. P., Sun, J. Y., et al. (2007). Degradation of azo dye Acid black 1 using low concentration iron of Fenton process facilitated by ultrasonic irradiation. *Ultrasonics Sonochemistry, 14*(6), 761–766. https://doi.org/10.1016/j.ultsonch.2006.12.010

Sviridov, A. P., Osminkina, L. A., Nikolaev, A. L., et al. (2015). Lowering of the cavitation threshold in aqueous suspensions of porous silicon nanoparticles for sonodynamic therapy applications. *Applied Physics Letters, 107*, 123107–123112. https://doi.org/10.1063/1.4931728

Swenson, H., & Stadie, N. P. (2019). Langmuir's theory of adsorption: A centennial review. *Langmuir, 35*, 5409–5426. https://doi.org/10.1021/acs.langmuir.9b00154

Syn, W., Sun, Y., Shah, K., et al. (2019). Electrocatalytic oxidation of tetracycline by Bi-Sn-Sb/γ-Al_2O_3 three-dimensional particle electrode. *Journal of Hazardous Materials, 370*, 24–32. https://doi.org/10.1016/j.jhazmat.2018.09.085

Theerthagiri, J., Madhavan, J., Lee, S. J., et al. (2020). Sonoelectrochemistry for energy and environmental applications. *Ultrasonics Sonochemistry, 63*, 104960–104978. https://doi.org/10.1016/j.ultsonch.2020.104960

Tian, O., Xue, F., Wanf, Y., et al. (2021). Recent advances in enhanced chemodynamic therapy strategies. *Nano Today, 39*, 101162–101180. https://doi.org/10.1016/j.nantod.2021.101162

Ugwuany, E. D., Nwokediegwu, Z. Q. S., Dada, M. A. et al. (2024). The impact of nanomaterials in enhancing wastewater treatment processes: A review. *Magna Scientia Advanced Research and Reviews, 10*(01), 273–285. https://doi.org/10.30574/msarr.2024.10.1.0030

Umar, M., Aziz, H., & Yusoff, M. (2010). Trends in the use of Fenton, electro-Fenton and photo-Fenton for the treatment of landfill leachate. *Waste Management, 30*, 2113–2121. https://doi.org/10.1016/j.wasman.2010.07.003

Vaiano, V., Sannino, D., & Sacco, O., et al. (2020). The use of nanocatalysts (and nanoparticles) for water and wastewater treatment by means of advanced oxidation processes. In A. Amrane, S. Rajendran, & T. A. Nguyen (Eds.), *Micro and nano technologies, Nanotechnology in the beverage industry* (pp. 241–264). Elsevier.

Vasseghian, Y., Dragoi, E. N., Almomani, F., et al. (2022). A comprehensive review on MXenes as new nanomaterials for degradation of hazardous pollutants: Deployment as heterogeneous sonocatalysis. *Chemosphere, 287*, 132387–132397. https://doi.org/10.1016/j.chemosphere.2021.132387

Vinayagam, V., Palani, K. N., Ganesh, S., et al. (2024). Recent developments on advanced oxidation processes for degradation of pollutants from wastewater with focus on antibiotics and organic dyes. *Environmental Research, 240*(2), 117500–117543. https://doi.org/10.1016/j.envres.2023.117500

Vucic, M. R., Baosic, R., Mitrovic, J., et al. (2021). Comparison of the advanced oxidation processes in the degradation of pharmaceuticals and pesticides in simulated urban wastewater: Principal component analysis and energy requirements. *Process Safety and Environmental Protection, 149*, 786–793. https://doi.org/10.1016/j.psep.2021.03.039

Wang, F. X., Wang, C. C., Du, X., et al. (2022a). Efficient removal of emerging organic contaminants via photo-Fenton process over micron-sized Fe-MOF sheet. *Journal of Chemical Engineering, 49*, 132495–132503. https://doi.org/10.1016/j.cej.2021.132495

Wang, J., Li, N., Jiang, Y., et al. (2024). High-throughput screening of N-doped curved graphene-loaded transition metal single-atom catalysts for nitrogen reduction reaction. *Applied Surface Science, 658*, 159858–159865. https://doi.org/10.1016/j.apsusc.2024.159858

Wang, J., Zhang, Z., Wu, F., et al. (2023). Facile fabrication of Fe_3O_4-biochar hybrid nanomaterials as catalysts for photo-Fenton degradation of tetracycline. *Optical Materials, 143*, 114156–114166. https://doi.org/10.1016/j.optmat.2023.114156

Wang, Y., Li, B., Guo, Y., et al. (2022b). Effects of ultrasound on the structural and emulsifying properties and interfacial properties of oxidized soybean protein aggregates. *Ultrasonics Sonochemistry, 87*, 106046–106060. https://doi.org/10.1016/j.ultsonch.2022.106046

Wei, J. Q., Chen, X. J., Wang, P. F., et al. (2018). High surface area TiO_2/SBA-15 nanocomposites: Synthesis, microstructure and adsorption-enhanced photocatalysis. *Chemical Physics, 510*, 47–53. https://doi.org/10.1016/j.chemphys.2018.05.012

Xu, L., An, H., Liu, N., et al. (2023a). Construction and characterization of ternary BiOBr/Bi/$CoWO_4$ composites and study on sonocatalysis degradation of tetracycline. *Composites Part b: Engineering, 263*, 110870–110884. https://doi.org/10.1016/j.compositesb.2023.110870

Xu, L., Wu, X. Q., Li, C. Y., et al. (2023b). Sonocatalytic degradation of tetracycline by BiOBr/$FeWO_4$ nanomaterials and enhancement of sonocatalytic effect. *Journal of Cleaner Production, 394*, 136275–136293. https://doi.org/10.1016/j.jclepro.2023.136275

Yan, J. K., Pei, J., Ma, H. L., et al. (2015). Effects of ultrasound on molecular properties, structure, chain conformation and degradation kinetics of carboxylic curdlan. *Carbohydrate Polymers, 121*, 64–70. https://doi.org/10.1016/j.carbpol.2014.11.066

Yang, Y., Guo, L., Wang, X., et al. (2024). Halogen modified organic porous semiconductors in photocatalysis: Mechanism, synthesis, and application. *Advanced Powder Materials, 3*(2), 100178. https://doi.org/10.1016/j.apmate.2024.100178

Yin, T., Meng, X., Wang, S., et al. (2022). Study on the adsorption of low-concentration VOCs on zeolite composites based on chemisorption of metal-oxides under dry and wet conditions. *Separation and Purification Technology, 280*, 119634–119648. https://doi.org/10.1016/j.seppur.2021.119634

Yu, X., Lui, S., Yu, J., et al. (2011). Superparamagnetic γ-Fe_2O_3@SiO_2@TiO_2 composite microspheres with superior photocatalytic properties. *Applied Catalysis b: Environment and Energy, 104*, 12–20. https://doi.org/10.1016/j.apcatb.2011.03.008

Yu, Y., Xu, W., Fang, J., et al. (2020). Soft-template assisted construction of superstructure TiO_2/SiO_2/g-C_3N_4 hybrid as efficient visible-light photocatalysts to degrade berberine in seawater via an adsorption-photocatalysis synergy and mechanism insight. *Applied Catalysis b: Environment and Energy, 268*, 118751–118764. https://doi.org/10.1016/j.apcatb.2020.118751

Zadeh-Kalantar, K., & Lei, T. (2022). CS applied nano materials Australia forum. *ACS Applied Nano Materials, 5*, 12049–12050. https://doi.org/10.1021/acsanm.2c03752

Zavahir, S., Elmakki, T., Ismail, N., et al. (2023). Degradation of organic methyl orange (MO) dye using a photocatalyzed non-ferrous Fenton reaction. *Nanomaterials, 13*, 1–18. https://doi.org/10.3390/nano13040639

Zhang, M. H., Dong, H., Zhao, L., et al. (2019a). A review on Fenton process for organic wastewater treatment based on optimization perspective. *Science of the Total Environment, 670*, 110–121. https://doi.org/10.1016/j.scitotenv.2019.03.180

Zhang, W., Zhang, D., & Liang, Y. (2019b). Nanotechnology in remediation of water contaminated by poly- and perfluoroalkyl substances: A review. *Environmental Pollution, 247*, 266–276. https://doi.org/10.1016/j.envpol.2019.01.045

Zhang, Z., Meng, H., Wang, Y., et al. (2018). Fabrication of graphene@graphite-based gas diffusion electrode for improving H_2O_2 generation in electro-Fenton process. *Electrochimica Acta, 9*, 12051–12062. https://doi.org/10.1016/j.electacta.2017.11.048

Zhao, H., Zhang, L., Zhou, Y., et al. (2023). Recycling of Fe_3O_4 nanomaterial from coal fly ash as catalyst to develop green and sustainable bio-electro Fenton: Characterization, optimization, and performance. *Journal of Environmental Chemical Engineering, 11*, 110678–110688. https://doi.org/10.1016/j.jece.2023.110678

Zheng, H., Li, M., Jiang, S., et al. (2023). MIL-88A/carbon quantum dots nanomaterials promote the photo-Fenton reaction to enhance the fouling resistance of PVDF membrane. *Journal of Membrane Science, 684*, 121855–121873. https://doi.org/10.1016/j.memsci.2023.121855

Zhong, S., Zhu, Z. S., Zhou, P., et al. (2022). FeOCl nanoparticles loaded onto oxygen-enriched carbon nanotubes and nickel-foam-based cathodes for the electro-Fenton degradation of pollutants. *ACS Applied Nano Materials, 5*, 12095–12106. https://doi.org/10.1021/acsanm.2c01413

Zhou, Z., Liu, X., Sun, K., et al. (2019). Persulfate-based advanced oxidation processes (AOPs) for organic-contaminated soil remediation: A review. *Chemical Engineering Journal, 372*, 836–851. https://doi.org/10.1016/j.cej.2019.04.213

Zhu, Y., Zhu, R., Xi, Y., et al. (2019). Strategies for enhancing the heterogeneous Fenton catalytic reactivity: A review. *Applied Catalysis b: Environment and Energy, 255*, 117739–117755. https://doi.org/10.1016/j.apcatb.2019.05.041

Zou, W., Dong, J., Ji, M., et al. (2023). Synthesis of $Bi_{25}FeO_{40}$ nanoparticles with oxygen vacancies via ball milling for Fenton oxidation of tetracycline hydrochloride and reduction of Cr(VI). *ACS Applied Nano Materials, 6*, 4309–4318. https://doi.org/10.1021/acsanm.2c05448

Application of Nano Magnesium Oxide in Advanced Oxidation Processes for the Degradation of Organic Pollutants

Onkar Sudhir Ekande, Gopal Achari, and S. Mathava Kumar

1 Introduction

In the past few decades, there has been extensive research on advanced oxidation processes (AOPs) and their application in the degradation of organic pollutants in water (Hollman et al., 2020; Izadifard et al., 2013, 2018; Malkhasian et al., 2014; Radwan et al., 2016). Besides oxidation (Mehrabani-Zeinabad et al., 2016), research has also been done on reductive processes with promising results (Izadifard et al., 2008). Various catalysts have been documented and used to increase the performance of AOPs and advanced reduction processes (ARPs) (Jourshabani et al., 2019). Metal oxides, such as TiO_2, ZnO, Bi_2O_3, SnO_2 and WO_3, their composites, doping and modifications (Langford et al., 2014) have been studied as effective catalysts for environmental remediation (Medhi et al., 2020). Nano magnesium oxide (nMgO) is an underrated cost-effective material that is widely available and effective in the adsorption of anions such as phosphate, fluoride, arsenic (Li et al., 2016; Wan et al., 2019; Yao et al., 2011) and others. nMgO combined with nitrogen-doped biochar is reported to be an effective adsorbent for heavy metals such as Pb^{2+} and Cd^{2+} (Ling et al., 2017). Mesoporous nanosheets of MgOs are used as electrocatalysts for the detection of Pb^{2+} ion in aqueous solutions (Liu et al., 2019b). nMgO has also been effective in CO_2 sequestration, methane oxidation and nitric oxide reduction (Hao et al., 2020; Liu et al., 2013; Olanders & Stroemberg, 1995). Recently, graphitic

O. S. Ekande
Department of Civil and Environmental Engineering, Texas A&M University, College Station, Texas, USA

G. Achari
Department of Civil Engineering, University of Calgary, University Drive NW, Calgary, Canada

S. M. Kumar (✉)
Environmental Engineering Division, Department of Civil Engineering, Indian Institute of Technology Madras, Chennai, Tamil Nadu, India
e-mail: mathav@iitm.ac.in

© Capital Publishing Company, New Dehli, India. 2025

R. Neelancherry and S. M. Kumar, *Nanotechnology in Wastewater Treatment*,
https://doi.org/10.1007/978-3-031-94246-4_9

carbon nitride (g-C_3N_4) has been gaining popularity because of its activity in the visible light range (Akintunde et al., 2022). Mg atom coordinated with nitrogen atom in the graphitic structure and acted as an electrocatalyst in the reduction of oxygen (Liu et al., 2020b). nMgO combined with g-C_3N_4 and ZnO are effective in photocatalytic water splitting (Mao & Jiang, 2019; Wang et al., 2017b)

The sol–gel process, electron spray pyrolysis, electrospun nanofibre, hydrothermal process, solvothermal process, chemical vapour deposition (CVD) and solution combustion synthesis (SCS) (Demirci et al., 2015; Devaraja et al., 2014; Li et al., 2015; Pilarska et al., 2017; Possato et al., 2020) are used to synthesise nMgO. nMgO has enhanced catalytic properties because of accelerated electron transfer caused by exposed defects and low coordinated sites (Hao et al., 2020; Ruiz et al., 2017). nMgO performs quite effectively in various AOPs such as photocatalysis, ozonation and Fenton-like processes with different morphologies such as nanofibre, nanorod, nanoflakes and flower-like structure (Mantilaka et al., 2018; Najafi, 2017; Zheng et al., 2019). Recently, a number of studies have reported the activation of persulphate (PS) to generate sulphate radicals (SO_4^{-}) and singlet oxygen ($^1O_2$) by nMgO in the absence of light (Hao et al., 2017; Jawad et al., 2020).

This review provides a summary of nMgO applications as catalyst in the degradation of persistent organic pollutants (POPs) in water through AOPs. The catalytic activity is enhanced by nMgO because of high specific surface area, exposed oxygen and magnesium defects, low coordinating oxygen sites and exposed facet (111). Recent advances in the application of nMgO in photocatalysis, ozonation, Fenton-like process and PS activation are discussed. Finally, some perspectives on the application of nMgO in AOPs are included.

2 Properties of nMgO

2.1 *Physicochemical Properties of nMgO*

nMgO is a rock salt with a cubic structure in which Mg^{2+} and O^{2-} form an ionic bond with a lattice parameter of 4.2 Å (Hinuma et al., 2018; Pacchioni & Freund, 2013). nMgO has high electric resistance and is used in the production of insulators (Pilarska et al., 2017). It is insoluble in water but is soluble in acid. The point of zero charge (pHpzc) of nMgO is ~12, indicating that its surface is positively charged in aqueous solution. nMgO has a bandgap energy in the range of 3.5–6 eV because of its structural defects (Demirci et al., 2015; Fujita et al., 2020). The terrace sites of nMgO have a low tendency to donate electrons, which increases three times at low coordinated sites such as at the corners and edges (Di Valentin et al., 2006; Sterrer et al., 2003). The oxygen vacancy (OV), formation energy (E_{ov}) of nMgO is ~6 eV, which is higher than the widely used oxidative metal oxide catalysts such as TiO_2, Bi_2O_3 and ZnO (Hinuma et al., 2018; Ruiz Puigdollers et al., 2017).

High Madelung potential of bulk MgO has led to its high stability and therefore a low propensity to accept and donate electrons (Pacchioni & Freund, 2013). Corners and edges (low coordinated sites) of MgO have low Madelung potential, which leads to low stability. At low coordinated sites, oxygen acts as Lewis base (electron donor) and magnesium as Lewis acid (electron acceptor) (Table 1) (Pacchioni & Freund, 2013). This leads to the formation of surface bound electron–hole pairs which facilitate the redox reactions (Fig. 1) (Sterrer et al., 2003). In nMgO, valence band (VB) consists of O 2p orbital marginally mixed with Mg 3s and 3p orbital while conduction band (CB) is made of Mg 3s–3p levels.

OVs, also known as F centres, present as F (two electrons at the vacancy site), F^+ (one electron at the vacancy site) and F^{++} (no electron at the vacancy site). They are formed when oxygen escapes from the lattice of nMgO (Fig. 2) (Choudhury et al., 2014; Di Valentin et al., 2006; Kumar et al., 2011). Adjacent F centres combine to form a pair of F centres known as dimmers. These are denoted as F_2 (four electrons), F_2^+ (three electrons) and F_2^{2+} (two electrons) (Choudhury et al., 2014).

These F centres and dimmers contribute to the catalytic activity of nMgO because of the formation of intermediate energy level in the forbidden region of nMgO (Fig. 2) (Choudhury et al., 2014; Kumar et al., 2011). Interstitial oxygen defects (O_i) form the energy level near VB, as shown in Fig. 2. The filled electronic energy levels of the O_i can transfer electrons to empty F centres, which can lead to the formation of holes in the O_i with high redox potential (Fig. 2) (Pathak et al., 2016).

Table 1 Madelung constant at different coordinated sites of MgO

Surface	Madelung constant
Bulk MgO	1.747
MgO (001)	1.681
Four coordinated edge sites (4C)	1.591
Three coordinated corner sites (3C)	1.344

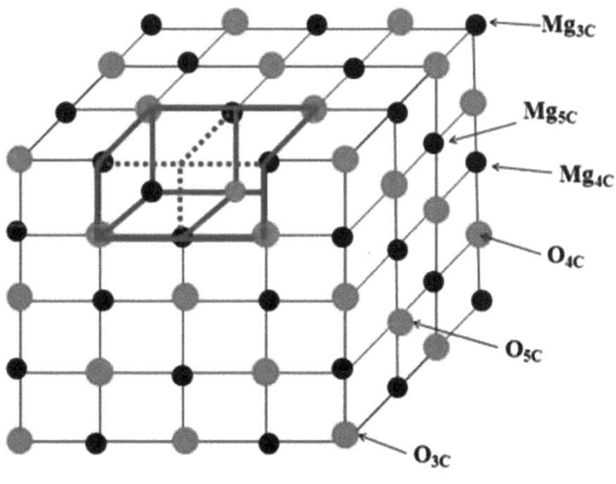

Fig. 1 Low coordinated oxygen and magnesium sites in the crystal structure of magnesium oxide. The dotted red line covered area indicates the oxygen vacancies (F centre) in the MgO structure

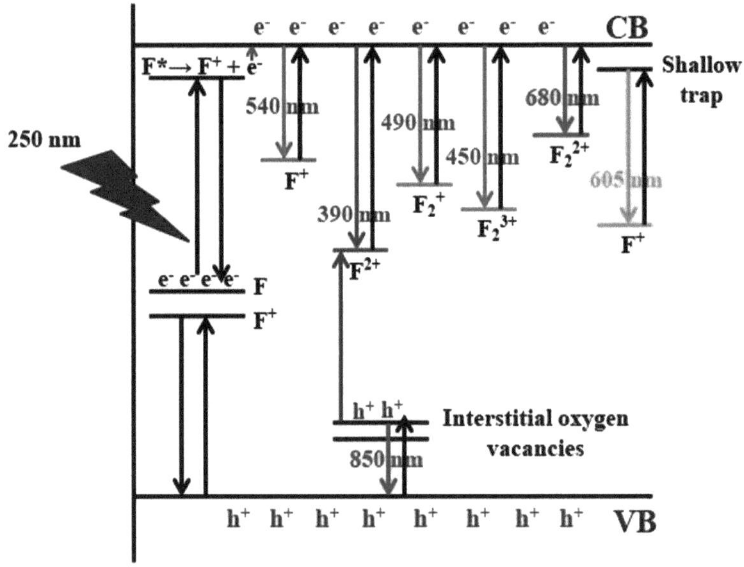

Fig. 2 Light irradiation of 250 nm on MgO (bandgap = 7.8 eV), resulting in the excitation of an electron from the defects. The electron from the conduction band (CB) recombined with the defects (F centres) to emit light with a specific colour for a particular F centre

3 Application of nMgO in Photocatalysis

Reactive oxygen species (ROS) formed during the photocatalytic process are mainly superoxide radical ($O_2^-/HO_2^{\cdot}$) and hydroxyl radical (OH) (Eqs. (1)–(4)) (Armstrong et al., 2015). Redox potential of OH and HO_2^- is 2.8 and 1.65 V versus normal hydrogen electrode (NHE), respectively (Mirzaei et al., 2017).

$$O_{2(aq)} + e^- \leftrightarrow O_2^{-} \quad E_h^o = -0.18\ \text{V} \tag{1}$$

$$O_{2(aq)} + e^- + H^+ \leftrightarrow HO_2^{\cdot} \quad E_h^o = 0.10\ \text{V} \tag{2}$$

$$HO_2^{\cdot}(aq) \leftrightarrow O_2^{-} + H^+ \quad pK_a = 4.8 \tag{3}$$

$$H_2O_2 + e^- + H^+ \rightarrow HO\cdot + H_2O \quad E_h^o = 0.8\ \text{V} \tag{4}$$

Oxidising capacity of the hole depends upon the position of VB of the photocatalyst; a lower VB has a higher oxidising capacity of the hole. H_2O is an electron donor to these holes to form OH (Eqs. (5)–(6)).

$$HO\cdot + e^- + H^+ \rightarrow H_2O \quad E_h^o = 2.73 \tag{5}$$

$$\cdot OH \leftrightarrow O^{\cdot -} + H^+ \quad pK_a = 11.7 \tag{6}$$

3.1 Photocatalytic Activity of nMgO

nMgO has a wide bandgap, which ranges from 3.5 to 6 eV depending upon the defects present in its crystal structure (Jorfi et al., 2016; Mageshwari et al., 2013). The major advantage of a wide bandgap is the availability of high redox potential at VB and low redox potential at CB. This helps to easily perform oxidation and reduction reactions. However, a wide bandgap limits the activity of nMgO in both UV and the visible region (Beranek, 2011; Jorfi et al., 2016; Mageshwari et al., 2013). F centres present in nMgO performed multiple roles in enhancing the photocatalytic activity such as donating electrons present at the F centres to $O_{2(aq)}$ to form O_2^-. F^+ and F^{++} centres trapped excited electrons from VB. nMgO helps to reduce electron–hole recombination and increases the life time of holes in the VB, whereas intermediate F centres transfer electrons to CB of nMgO to perform reduction of O_2 (Fig. 3) (Beranek, 2011; Jorfi et al., 2016; Mageshwari et al., 2013; Naldoni et al., 2018; Xu & Schoonen, 2000).

The position of the OVs/F centres plays a major role in the photocatalytic activity of metal oxides by creating intermediate energy levels which determine the feasibility of the oxidation–reduction reaction (Liu et al., 2019a, 2019b; Naldoni et al., 2018). The position of the OVs as electron donors should be above the single electron reduction of the $O_{2(aq)}$, PS and O_3. F^+ centres and the peroxo group in nMgO can transfer electrons to form O_2^- (Valentin et al., 2006). The surface OH^- group on nMgO reduces the electron–hole recombination, which leads to interaction with the hole in VB to form ·OH radicals (Table 2).

Mageshwari et al. (2013) synthesised template-free MgO nanoflakes at low and high alkaline conditions. At highly alkaline conditions, a large number of defects were observed in the form of F^+ and Mg interstitials. This resulted in higher photocatalytic activity. The bandgap of the MgO nanoflakes is ~5.67 eV, which does not activate nMgO in the UV region. However, intermediate energy levels formed because of defects play a major role in the generation of ROS in the presence of UV light. MgO nanoflakes synthesised at highly alkaline condition degraded methyl orange (MO) and methylene blue (MB) 1.75 and 1.85 times faster than MgO nanoflakes synthesised at low alkaline condition (Mageshwari et al., 2013). One-dimensional MgO nanofibres with size ranges from 40 to 200 nm have a high amount of F^+ defect compared to MgO nanosphere because of incomplete crystallisation during combustion. Thus, the MgO nanofibres degraded 98% reactive yellow (RY) dye ($k = 0.00994\ min^{-1}$) within 80 min, whereas MgO nanospheres degraded only 32% of the RY dye (Mantilaka et al., 2018).

Fujita et al. (2020) synthesised oxygen defective nanocrystalline MgO using high-pressure torsion (HPT) on bulk MgO. They formed positively charged F^+, neutral F

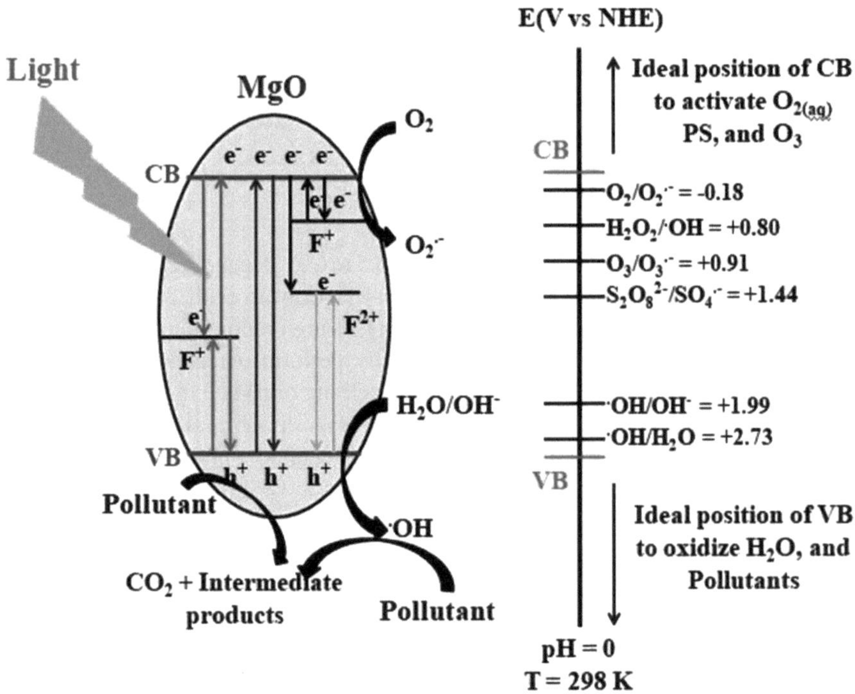

Fig. 3 Photocatalytic activity of nMgO for the degradation of organic pollutants

and F_2^+ (intermediate energy levels) on nMgO and thus reduced the energy required for excitation of electrons from 7.8 to 3.9 eV. Electron paramagnetic resonance (EPR) spectra detect F^+ centre because of its paramagnetic nature; however, dimmer F_2^+ and F centres confirmed with photoluminescence emission (PL) spectra which had peaks at 680 and 540 nm, respectively. HPT applied at room temperature (300 K) showed a high absorbance in the UV region compared to high-temperature HPT at 473 or 673 K (Fujita et al., 2020).

Flower-like MgO (F-MgO) showed highest activity for MB degradation compared to nest-like MgO (N-MgO), spherical MgO (S-MgO), rod-like MgO (R-MgO) and trapezoidal MgO (T-MgO). However, high SSA (142.9 $m^2\ g^{-1}$) and bandgap (5.23 eV) played a major role in the photocatalytic activity of F-MgO. Holes in F-MgO have high oxidation potential of ~4 V versus NHE, which facilitate generation of OH and direct oxidation of MB (Zheng et al., 2019). Electrons in CB of F-MgO have redox potential of ~−1 V versus NHE. This facilitates formation of O_2^-. However, the contribution of the O_2^- is less as high bandgap could not transfer more electrons in the CB of F-MgO. Overall, adsorption played a significant role in the higher photocatalytic activity of F-MgO for MB removal (Zheng et al., 2019).

Photocatalytic activity of nMgO was enhanced by combining it with porous supporting material such as zeolites, carbon-based materials and polymers, which helped to increase the adsorption of pollutants (Hayati et al., 2020; Jorfi et al., 2016).

Table 2 Application of nMgO in photocatalysis for the degradation of organic pollutants

References	Catalyst	Pollutant	Experimental condition	Efficiency
Mageshwari et al. (2013)	MgO	Methyl orange (MO)	Catalyst dose: 0.5 g/L MB = 15 mg/L Reaction time: 210 min UV light (254 nm) = 5 mW/cm^2	99.14% MO removal
Demirci et al. (2015)	MgO–sol gel (MgO–SG) MgO–spray pyrolysis (MgO-FSP)	Methylene blue (MB)	Catalyst dose: 1.66 g/L MB = 6 μM Reaction time: 240 min UV light (254 nm) = 300 W pH = 8	90% of MB degraded by MgO–SG 94% of MB degraded by MgO–FSP
Najafi (2017)	MgO nanoflakes	Methylene blue (MB)	Catalyst dose: 0.4 g/L MB = 10 mg L^{-1} Reaction time: 60 min UV light	90% removal of MB
Mantilaka et al. (2018)	MgO nanofibres	Reactive yellow (RY)	Catalyst dose: 4 g/L RY = 10 mg L^{-1} Reaction time: 100 min UV light = 200 W	98% removal of RY
Zheng et al. (2019)	Microflower-like MgO	Methylene blue (MB) Congo red (CR) Thymol blue (TB)	Catalyst dose: 0.5 g/L Dye = 100 mg L^{-1} Reaction time: 90 min UV light (365 nm) = 400 W	Complete removal of MB, CR and TB

(continued)

Table 2 (continued)

References	Catalyst	Pollutant	Experimental condition	Efficiency
Jorfi et al. (2016)	nMgO	Acid red 73 (AR 73)	Catalyst dose: 0.4 g/L Dye = 200 mg L^{-1} Reaction time: 60 min UV light (350–400 nm) = 20 W	Complete removal of AR73
Fujita et al. (2020)	nMgO	Methylene blue (MB)	Catalyst dose: 1 g/L MB = 15 mg L^{-1} Reaction time: 300 min UV light (289–486 nm)	70% removal of MB

nMgO combined with black carbon (nMgO-C) formed a bimodal mesoporous structure with pore size of 3 and 25 nm and increased SSA from 61 to 216 $m^2\ g^{-1}$. Therefore, nMgO-C showed three times higher activity in degrading the RY dye compared to nMgO (Jorfi et al., 2018). Jorfi et al. (2018) synthesised uniformly distributed nMgO over porous zeolite (nMgO-Z) to control agglomeration of nMgO to improve photocatalytic efficiency. nMgO-Z has a pHpzc of ~8.3; therefore, an increase in pH increased the repulsion between negatively charged nMgO-Z and Azo dye (AR73). This reduced the photocatalytic degradation of AR73. At a pH below 5, the performance of nMgO-Z improved because of the adsorption of the negatively charged AR73 on positively charged nMgO-Z, which facilitated the interaction between the ROS formed at the surface of the catalyst and AR73 (Jorfi et al., 2018). Maximum degradation of AR73 was achieved at pH 3 with ~70% degradation occurring in the presence of 0.4 g L^{-1} of MgO-Z under UVA. An increase in the dose of MgO-Z from 0.4 to 1 g L^{-1} achieved complete removal of AR73 within 60 min (Jorfi et al., 2018).

Carbon nanotubes (CNTs) incorporated nMgO (CNT-MgO) was synthesised and applied for the sonophotocatalytic removal of the antibiotic sulphadiazine (SDZ) (Hayati et al., 2020). Incorporation of CNT (7.5–12.5 wt%) increased the SSA from 66.1 to 149.95 $m^2\ g^{-1}$ and reduced interfacial charge transfer resistance. Photocatalytic activity of the CNT-MgO (0.9 g L^{-1}) achieved ~65% removal of SDZ (75 mg L^{-1}) within 80 min at neutral pH, whereas addition of ultrasound achieved complete removal of SDZ. Interaction between SDZ and CNT-MgO played a major role in the sonophotocatalytic removal of SDZ. SDZ has a positive charge below pH 2.1 ($pK_{a1} = 2.1$, $pK_{a2} = 6.28$), neutral charge between pH 2.11 and 6.28 and negative charge above pH 6.28. Therefore, an increase in pH above 6.28 increased the removal of SDZ because of interaction between positively charged CNT-MgO ($pH_{pzc} = 11.7$) and negatively charged SDZ. OH and holes are major reactive species and O_2^- has a minor role in the degradation of the SDZ (Hayati et al., 2020).

3.2 *Photocatalytic Activity of Doped nMgO*

Doping of nMgO enhanced the defects in the structure, resulting in the red-shift in the bandgap of nMgO (Table 3). It increased the absorbance of light in the UV-visible region and reduced the electron–hole recombination because of intermediate energy levels (Anilkumar et al., 2015; Kumar et al., 2019; Vaizogullar, 2018a, 2018b). Fe^{3+}-doped MgO (Fe-MgO) can be formed by replacing Mg^{2+} from the crystal lattice of nMgO by Fe^{3+} ion because of their similar ionic radius. Eighty-eight per cent of MB dye was removed by Fe^{3+} (4 mol%)-doped nMgO within 120 min under UV light. Fe^{3+} increased the segregation of nanoparticles which enhanced the defects in the crystal structure of nMgO (Anilkumar et al., 2015). Fe^{3+}/Fe^{2+} energy level formed below CB of the Fe-MgO donates electrons to $O_{2(aq)}$ to form O_2^-, whereas Fe^{3+}/Fe^{4+} energy level formed near VB formed OH. Excess Fe^{3+} (5 mol%) in nMgO adversely affects the activation by trapping charge carriers and slow down the mobility of electron–hole pairs (Anilkumar et al., 2015).

Sulphur-substituted oxygen atom from the nMgO lattice formed OVs in sulphur-doped nMgO (S-MgO); S-MgO has reduced band energy of 3.38 eV. This showed higher activity in UV light (Fanaei et al., 2019; Moussavi et al., 2018). Tetracycline (TET) degradation studies using S-MgO reported the generation of hydrocarbon radicals (R·) on S-MgO's surface in the absence of light. Mg^{2+} interstitials and OVs present in S-MgO trapped the hydrocarbon ring present in the TET molecule which formed R, which was responsible for the degradation of the TET (Moussavi et al., 2018).

Rare earth metals have incomplete/partially filled 4f orbitals. This supports f–f electronic transitions within partially filled 4f orbitals. These partially filled orbitals has the ability to trap photogenerated electrons and reduce electron–hole recombination (Kumar et al., 2019). These orbitals act as intermediate energy levels and reduce the band energy required for the generation of charge carriers (Colpani et al., 2018). Substitutional doping of La^{3+} in needle-like nMgO (La-MgO) was achieved because of strong interaction of the La^{3+} with O^{2-} (Fig. 4). The La_2O_3 (Bandgap = 4.24 eV; CB ~−0.8 V and VB ~3.4 V vs. NHE) phase formed in La-MgO is responsible for the red shift in light adsorption because of the reduction in the band energy from 3.88 to 3.62 eV (Vaizogullar, 2018a, 2018b). Seventy-five per cent of FQ, the model pollutant, was removed within 120 min in the presence of La-MgO (2 g L^{-1}).

Table 3 Application of doped nMgO in photocatalysis for the degradation of organic pollutants

References	Catalyst	Pollutant	Experimental condition	Efficiency
Anilkumar et al. (2015)	Fe^{3+}-doped MgO	Methylene blue (MB)	Catalyst dose: 0.24 g/L MB = 10 mg/L Reaction time: 120 min UV light (254 nm) = 30 W	88% MB removal
Vaizogullar (2018a, 2018b)	La^{3+}-doped MgO	Flumequine (FQ)	Catalyst dose: 2 g/L FQ = 10 mg L^{-1} Reaction time: 120 min UV light	75% removal of the FQ
Moradi et al. (2020)	Ag/S-MgO	Diclofenac (DCF)	Catalyst dose: 1 g/L DCF = 50 mg L^{-1} Reaction time: 10 min UV light (365 nm) = 228 $\mu W/cm^2$	86% DCF removal

3.3 Photocatalytic Activity of nMgO Photocomposites

In photocatalysis, generation of charge carriers occurs in femtoseconds, whereas recombination of electron–hole takes place in 10–100 ns (Park et al., 2013). Interfacial charge transfer is slow, with the rate limiting step in the process taking more than 100 ns (Park et al., 2013). Efficiency of photocatalysis can be increased by enhancing the electron–hole recombination time, which would allow sufficient time for the interfacial charge transfer. Doping of nMgO showed promising results to reduce the electron–hole recombination but along with that it reduced the redox potential because of the formation of intermediate energy levels.

nMgO in photo composites improved the photocatalytic activity by introducing additional oxygen defects and acted as an insulating layer to avoid back-transfer of electrons (Table 4) (Bandara et al., 2004; Chen et al., 2010; Li et al., 2011; Vaizogullar, 2018a, 2018b). Commercial Degussa P-25 TiO_2 was combined with a different amount of nMgO and doubled removal rate of the 2,4-dichlorophenol (2,4-DCP) (Bandara et al., 2004). An increase in the content of nMgO beyond the optimum level (3%, w/w) enhanced the insulating effect of nMgO and inhibited the diffusion of charge carriers which inhibit the removal of the 2,4-DCP (Bandara et al.,

Fig. 4 La^{3+}-doped nMgO has energy levels near CB and VB

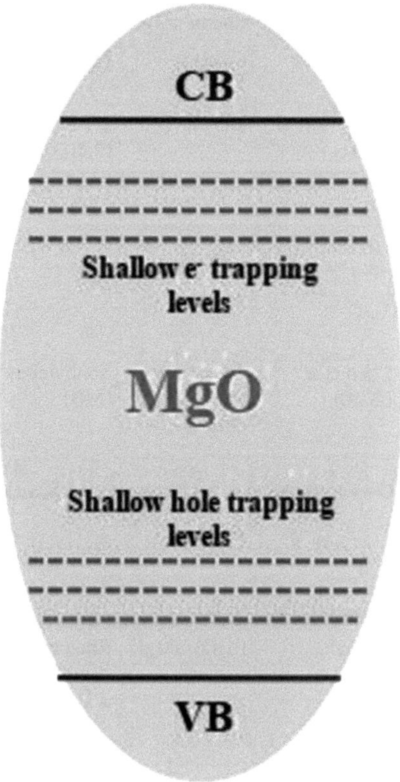

2004). Yield of O_2^- increased from 3.05 to 6.73 μM because of F centres in nMgO, as they played a major role in trapping photogenerated electrons from TiO_2 (Bandara et al., 2004).

nMgO/TiO_2 composites combined with multi-walled carbon nanotubes (MWCNTs) led to additional O_2^- as a result of the transfer of electrons from MWCNTs to the defects in nMgO (Chen et al., 2010). Overall, the nMgO/TiO_2 composite combined with MWCNT increased the flow of electrons from conductive MWCNT to unsaturated defective sites in nMgO to activate O_2 (Chen et al., 2010).

nMgO combined with ZnO resulted in blue shift in the absorbance of the ZnO and upward shift in both CB and VB, which increased reduction and decreased the oxidation capacity of the composite (Abed et al., 2019; Raj et al., 2019). Interstitial defects and OVs in nMgO/ZnO composite are responsible for photocatalytic removal of the 2,6-dichlorophenol (2,6-DCP) (Raj et al., 2019). nMgO and ZnO composites with different mass ratios (2:5, 5:5, 8:5) were synthesised and tested for the degradation of 2,6-DCP. It was found that Mg_5Zn_5 had more defective sites and achieved a maximum removal efficiency of 79.5%.

nMgO/ZnO/In_2O_3 was synthesised in different molar ratio of Mg^{2+}:Zn^{2+}:In^{3+} (1:1:1, 2:1:1, 4:1:1) at 500 and 800 °C (Xiang et al., 2013). nMgO/ZnO/In_2O_3 (2:1:1,

Table 4 Application of nMgO photocomposite for the degradation of organic pollutants

References	Catalyst	Pollutant	Experimental condition	Role of nMgO	Efficiency
Bandara et al. (2004)	TiO_2/nMgO	2,4-Dichlorophenol (2,4-DCP)	Catalyst dose: 1 g/L Pollutant concentration = 1.4 mM UV light = 125 W pH = 5.5	Oxygen defects in MgO trap the electron from conduction band of the TiO_2	Removal of 2,4-DCP using TiO_2/3% MgO with kinetic rate 1.17×10^{-3} M^{-1} h^{-1}
Chen et al. (2010)	TiO_2/MWCNT/nMgO	Methylene blue (MB)	Catalyst dose: 1 g/L MB = 10^{-5} M Reaction time: 120 min UV light (365 nm) = 20 W	Electron transfer from MWCNT and TiO_2 to defective sites at MgO	80% removal of MB
Li et al. (2011)	Bi_2O_3/nMgO	Rhodamine B (RhB)	Catalyst dose: 0.5 g/L MB = 10 mg L^{-1} Reaction time: 60 min Visible light = 250 W	Reduction in bandgap	Bi salt and Mg salt combined in the ratio 2:1 83.17% removal of RhB
Xiang et al. (2013)	nMgO/ZnO/In_2O_3	Methylene blue (MB)	Catalyst dose: 0.5 g/L MB = 10 mg L^{-1} Reaction time: 300 min Visible light = 300 W	Avoid back transfer of electrons	95–98% removal of MB
Vaizogullar (2018a)	nMgO/ZnO	2,6-Dichlorophenol (2,6-DCP)	Catalyst dose: 2 g/L 2,6-DCP = 10 mg L^{-1} Reaction time: 90 min	Generation of the defects in composite structure	79.5% removal of the 2,6-DCP

(continued)

Table 4 (continued)

References	Catalyst	Pollutant	Experimental condition	Role of nMgO	Efficiency
Ramacharyulu et al. (2018)	nMgO/ g-C_3N_4	4-Nitrophenol (4-NP)	Catalyst dose: 1 g/L 4-NP = 20 mg L^{-1} Reaction time: 180 min pH = 3–4 Natural sunlight	Create carbon defects in g-C_3N_4	MgO/g-C_3N_4 in ratio 1:5 removed 99% 4-NP
Wang et al. (2020)	Ag_2S-nMgO/ graphene oxide	Rhodamine B (RhB)	Catalyst dose: 0.1 g/L RhB = 30 mg L^{-1} Reaction time: min pH = 10 UV light (365 nm) = 25 W Visible light = 400 W	Reduction in bandgap	RhB removal: 98.8% and 40% in the presence of UV and visible light, respectively
Moradi et al. (2020)	nMgO/ZnO/ Graphene	Sulphamethoxazole (SMX)	Catalyst dose: 0.8 g/L SUZ = 55 mg L^{-1} Reaction time: 120 min pH = 9 UVA light = 90 W Ultrasound power = 250 W m^{-2}	Reduction in bandgap of the composite	Complete removal of the SMX

MZI2) prepared at 800 °C performed more effectively because of higher SSA and optimum amount of nMgO in inhibiting recombination of photogenerated charge carriers. MB as a model pollutant has LUMO level at −0.75 V versus NHE, which can facilitate transfer of excited electrons from LUMO of MB to CB of the In_2O_3 through nMgO insulating layer (Xiang et al., 2013). Insulating properties of the nMgO was used in nMgO/ZnO/In_2O_3 to avoid back-transfer of the electron through interface from In_2O_3 to MB, which enhanced efficiency. However, high nMgO content (4:1:1, MZI4) inhibited the forward transfer of electron from MB to In_2O_3 and reduced the overall photocatalytic efficiency (Fig. 5) (Xiang et al., 2013). The mechanism of MB

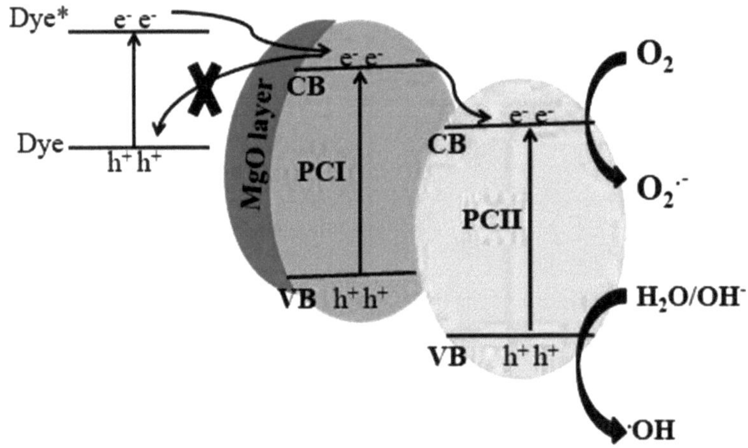

Fig. 5 nMgO acts as an insulating layer in the photocomposite to avoid back-transfer of electrons to the photoexcited dye from CB of the PCI

degradation using MZI2 was mainly governed by the O_2^{-}, whereas OH played a minor role.

Moradi et al. (2020) combined graphene (10 wt%) with the nMgO (90 wt%)/ZnO (10 wt%) composite (MZG) to improve the SSA from 37.83 to 115.6 $m^2\ g^{-1}$. However, higher graphene content (>10 wt%) reduced sonophotocatalytic removal of sulphamethoxazole (SMX) because of the aggregation of ZnO and nMgO. The higher graphene content decreased light penetration to the active sites of the composite. At pH > 10.8, negative charge on the surface of the MZG ($pH_{pzc} = 10.8$) and SMX ($pK_{a1} = 1.7$, $pK_{a2} = 5.6$) reduced the interaction between the two. However, at high pH, generation of OH increased because of the oxidation of OH^{-}. However, still lower removal of SMX confirmed that interaction is the key step in the degradation of SMX (Moradi et al., 2020).

A low activity of g-C_3N_4 for oxygen activation has been observed as a result of lack of defects in pristine g-C_3N_4 (Wu & Liu et al., 2019a, 2019b). nMgO combined with g-C_3N_4 in the ratio of 1:5 increased the contribution of the C_{2P} states of g-C_3N_4 in the LUMO (Ramacharyulu et al., 2018). These C_{2P} states can trap the electrons due to the presence of unpaired electrons and enhanced charge separation. Electrons trapped in C_{2P} are responsible for the photocatalytic degradation of 4-NP, confirmed by the suppression in the degradation of 4-NP from 99 to 11.2% in the presence of electron scavenger $AgNO_3$ (Ramacharyulu et al., 2018).

nMgO prepared using the ultrasound method by the precursor magnesium ethylate in diethylene glycol has a bandgap of 4 eV. It showed moderate activity under visible light and under UVA (Wang et al., 2020). Ag_2S and graphene oxide combined with nMgO reduced the excitation energy required for the generation of the electron–hole from 4 to 2.82 eV. Ag_2S-MgO/GO composite has pH_{pzc} of 6.3; high degradation of rhodamine B dye (RhB) at alkaline pH condition occurred as a result of increased interaction between negatively charged Ag_2S-MgO/GO and positively charged RhB (Wang et al., 2020). Overall, photocatalytic activity of the nMgO depends on oxygen defects in crystal structure, band energy and interaction between the organic pollutant and nMgO.

4 nMgO in Fenton-Like Process

Magnesium peroxide (MgO_2) was reported as an antimicrobial agent in the early 90s (Wu et al., 2019a, 2019b). H_2O_2 is used as an oxidant in photocatalysis and Fenton-like processes to form ROS. Major drawbacks of the application of H_2O_2 are its suitability in acidic pH range, biological toxicity and short lifespan (Vaddadi et al., 2020; Wu et al., 2019a, 2019b). MgO_2 overcomes the drawback of H_2O_2 and has the highest percentage of active oxygen than other peroxides such as CaO_2 and ZnO_2 (Tables 5 and 6) (Gholami et al., 2019; Mosmeri et al., 2018). Therefore, MgO_2 is a promising substitute for the H_2O_2 in the Fenton-like processes (Fig. 6).

$$MgO_2 + H_2O \rightarrow MgO + H_2O_2 \tag{7}$$

MgO_2 was synthesised by reaction between MgO and H_2O_2. Slow release of H_2O_2 is possible under the hydrolysis of MgO_2 (Eq. (7)). Wu et al. (2019a, 2019b) prepared MgO_2 from nMgO synthesised with active oxygen content of 26.9 wt% and purity of 94%. Fe^{3+} to MgO_2 ratio was increased from 0.5:1 to 0.8:1, and it found an increase in the rate of removal of the MB dye from 0.08 to 0.173 min^{-1}. MgO_2 as an oxidant in the Fenton process performed effectively at neutral and slightly alkaline pH compared to H_2O_2; still a decrease in pH increased the efficiency of the process because of the high availability of H_2O_2, high redox potential of ·OH and high solubility of Fe^{3+} (Vaddadi et al., 2020; Wu et al., 2019a, 2019b).

Three-dimensional F-MgO_2 has an SSA of 385 $m^2\ g^{-1}$ and active oxygen content of 24.10%. It degraded 90% TET in Fenton process within 60 min (Liu et al., 2020a, 2020b). However, high dosage of MgO_2 (150 mg L^{-1}) reduced the removal of TET because of co-precipitation of Fe^{3+} on $Mg(OH)_2$ (Liu et al., 2020a, 2020b). Vaddadi et al. (2020) synthesised spherical nano MgO_2 with size in the range of 2–4.1 nm. The release of H_2O_2 from spherical nano MgO_2 was 50 mg L^{-1} in 10 h. Photocatalytic activity of spherical nano MgO_2 (bandgap ~4.4 eV) and activation of the released H_2O_2 under UVC resulted in the complete removal of 17α-ethynylestadiol (EE2) within 60 min. Holes and OH are major contributors for the photocatalytic degradation of EE2. In another study, MgO_2 nanorods with average diameter of ~166 nm and bandgap of ~3.7 eV completely degraded the MO dye at optimum pH 9 (Vaddadi et al., 2020). At a pH of 3, the degradation of MO decreased because of the scavenging effect of the H^+ (El-Shamy, 2020).

Zeng et al. (2020) observed in situ generation of the MgO_2 at the edges of the MgO nanosheet. Oxygen defects present in MgO nanosheets with high specific surface area (147 $m^2\ g^{-1}$) resulted in more exposure of defects to oxidant H_2O_2 which formed

Table 5 Application of the nMgO in Fenton-like processes

References	Catalyst	Pollutant	Experimental condition	Results
Vaddadi et al. (2020)	$nMgO_2$	17α-Ethynylestradiol (EE2)	Dose of oxidant = 0.25 g L^{-1} EE2 = 3 mg L^{-1} Reaction time: 60 min UV light = 4 W	99% removal of EE2
Wu et al. (2019a, 2019b)	$nMgO_2$	Methylene blue (MB)	Fe^{3+}:MgO_2 = 0.8:1 MgO_2 = 50 mg L^{-1} MB = 20 mg L^{-1} Reaction time: 60 min	Complete removal of MB
Liu et al. (2020a, 2020b)	$nMgO_2$	TET	Dose of oxidant = 0.1 g L^{-1} TET = 50 mg L^{-1} Reaction time: 60 min	90% removal of TET
El-Shamy (2020)	$nMgO_2$ nanorod	Methyl orange (MO)	Dose of catalyst = 0.4 g L^{-1} MO = 30 mg L^{-1} Reaction time: 50 min UV light = 300 W pH = 9	Complete removal of MO
Zeng et al. (2020)	In situ $nMgO_2$ nanosheet	MB	Dose of catalyst = 1 g L^{-1} MB = 100 mg L^{-1} H_2O_2 = 20 mL L^{-1} Reaction time: 60 min Visible light	98.1% removal of MB
Zeng et al. (2020)	Graphene oxide/nMgO	MB	Catalyst dose: 1 g/L MB = 100 mg L^{-1} H_2O_2 = 20 mL L^{-1} Reaction time: 45 min	99.4% removal of MB

Table 6 Oxygen content in solid peroxides and solubility product of the metal hydroxide

Peroxides	Purity (wt%)	Active oxygen content (wt%)	Solubility product (K_{sp}) of metal hydroxide (25 °C)
Magnesium peroxide	94.26	26.9	1.8×10^{-11}
Calcium peroxide	70	15.6	5.5×10^{-6}
Zinc peroxide	73–80	12–13	1.2×10^{-17}
Potassium persulphate	99.5	5.9	364
Sodium percarbonate	98	15.1	110

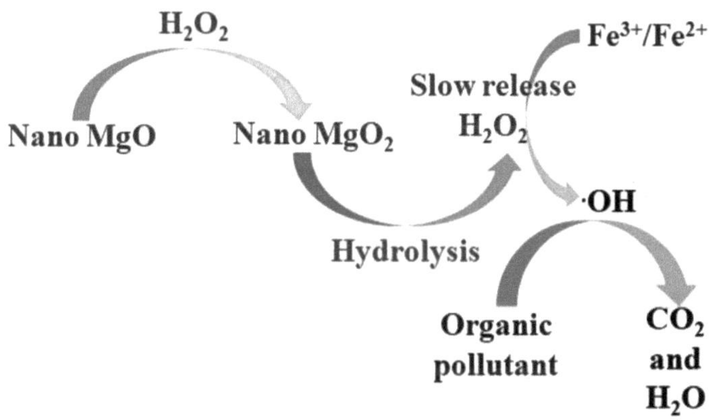

Fig. 6 Role of $nMgO_2$ in Fenton-like process for the degradation of organic pollutants

MgO_2 nanosheets. High amount of MgO_2 was formed at the defective corners of MgO nanosheets in the presence of visible light (34.7%) compared to dark (9.7%) because of the oxidation of the MgO edges in the presence of light. Low coordinated sites at the edges of nMgO performed reduction of the H_2O_2 to form OH. Removal of MB was performed using defect-rich MgO nanosheets in the presence of high dose of H_2O_2 (20 mL L^{-1}). MgO nanosheet achieved ~30% degradation of the MB dye without H_2O_2 in the presence of visible light. Electron–hole was formed because of oxygen defects after visible light irradiation performed single-electron reduction of the H_2O_2 (E_h^o (H_2O_2/OH) = 0.8 V vs. NHE) and O_2 to form OH and O_2^-, respectively (Zeng et al., 2020). Zeng et al. (2020) modified MgO nanosheets by incorporating graphene oxide to reduce electron–hole recombination and bandgap from 5.34 to 4.94 eV. Kinetics of MB degradation was increased from 0.0961 to 0.1761 min^{-1} with the incorporation of graphene oxide in MgO nanosheets (Zeng et al., 2021). OH formed because of the reduction of the H_2O_2 played a major role, whereas O_2^- and holes played a minor role in the photocatalytic removal of MB dye under visible light in both MgO and graphene/MgO nanosheets.

5 Catalytic Activity of nMgO in the Absence of Light

5.1 Activity of nMgO for the Generation of O_2^- in Dark

Activation of the O_2 to form O_2^- was observed on an ultrathin film of nMgO (001) because of the transfer of an electron from defects to adsorbed O_2 on the low energy surface of nMgO (001) in the absence of light (Fig. 7) (Gonchar et al., 2011). E_{OV} is high in non-reducible oxides compared to reducible oxides (Huang et al., 2005). E_{OV} varies with a facet of the structure, nMgO (100) has E_{OV} of 6.18 eV and MgO (111) has 4.89 eV (Hao et al., 2017). OVs formed on nMgO consist of F^+ with a strong ability to donate an electron to reduce O_2 to O_2^- (Hao et al., 2017). O_2 was strongly chemisorbed on defective nMgO (111) compared to defective nMgO (100). This strongly adsorbed O_2 on nMgO (111) formed O_2^- and $^1O_2$ because of electron rearrangement and molecular collision (Hao et al., 2017). The main co-ordination number of the surface OH^- is five, whereas co-ordination number of surface OH^- on (111) facet of nMgO was three. Therefore, nMgO (111) has more tendency to donate electron for the generation of O_2^-. nMgO (111) has shown excellent antibacterial activity and degradation of methyl violet (MV) because of the generation of O_2^- and $^1O_2$ in the absence of light. $Mg(OH)_2$ and commercial nMgO have shown lower activity compared to nMgO (111) and confirmed that hydration and surface basicity have no role in the activation of O_2 in the absence of light (Hao et al., 2017).

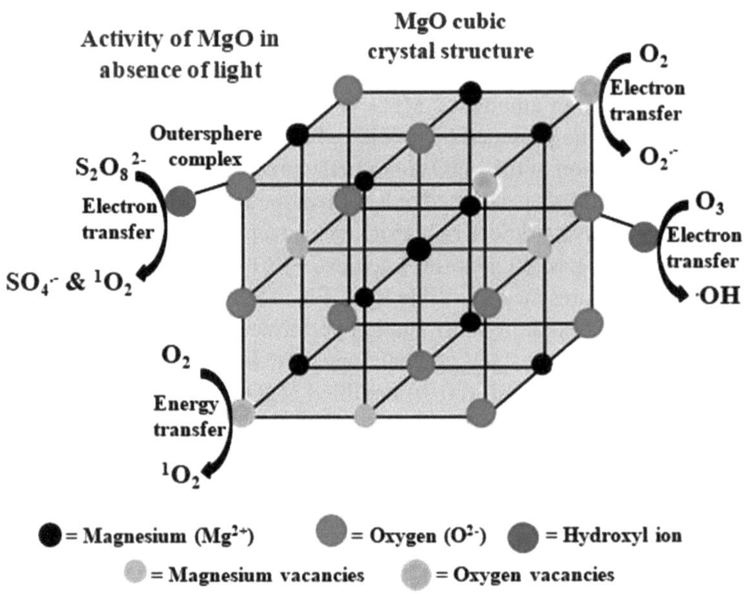

Fig. 7 nMgO mechanism in the activation of the O_2, PS and ozone in the absence of light

It is widely reported that O_2^- is a key in the antibacterial activity of nMgO by damaging the outer protein structure to inhibit the growth of bacteria (Huang et al., 2005). In acidic conditions, O_2^- is unstable and reacts with hydrogen to form H_2O_2 (Eq. (8)) (Sawai et al., 2000). The presence of OH^- ions on the surface of nMgO creates a basic environment, which in turn enhances the stability of O_2^- and its bactericidal activity (Table 7) (Sawai et al., 2000).

$$O_2^- + O_2^- + 2H^+ \leftrightarrow H_2O_2 + O_2 \tag{8}$$

Optimum temperature for the synthesis of nMgO in terms of maximum antibacterial activity performed by El-shear et al. (2020) observed nMgO synthesised at 500 °C has maximum OVs, achieved maximum antibacterial activity for gram positive *Bacillus cereus* compared to nMgO synthesised at temperatures 300, 400, 600 and 700 °C (El-Shaer et al., 2020). MgO nanowires have a high specific surface area of 461 $m^2\ g^{-1}$ and showed antibacterial activity as more edges and corners of nMgO are exposed to form O_2^- (Al-Hazmi et al., 2012). ROS generated as a result of interaction between F centres and O_2 were responsible for the antibacterial activity. Along with that, Mg^{2+} ion interacts with the thiol (^-SH) group of the protein in peptide linkage of bacteria to reduce cell permeability. Mg^{2+} will penetrate more in negatively charged peptidoglycan in G –ve bacteria compared to G +ve bacteria. Therefore, G –ve bacteria is more susceptible to nMgO than G +ve because of the thinner cell wall.

Substitutional doping of lithium (Li^+) in nMgO improved the OVs as Mg^{2+} (0.72 Å) in the crystal is replaced by Li^+ (0.76 Å). Li^+ doping in nMgO increased the exposure of the (111) facet, which contributed to the generation of more ROS (Rao et al., 2013; Tian et al., 2020). Ti and Zn doped nMgO have comparatively low activity because of fewer OVs and the absence of exposed (111) facet (Tian et al., 2020). Substitutional doping of Ag^+ in nMgO (Ag-nMgO) formed oxygen defects and restricted the growth of the nMgO during synthesis. These small-size particles ($d_{avg} = 30$ nm) improved the transfer of electrons from oxygen defects to O_2 significantly because of the reduction in the interfacial charge transfer (Fig. 8) (Cai et al., 2017; Zhu et al., 2016).

Low release of Ag^+ (~0.03 μg mL^{-1}) was observed compared to the minimum inhibitory concentration of 0.5 μg mL^{-1} (Cai et al., 2017). Therefore, a release of Ag^+ from composite is not responsible for the antibacterial activity (Cai et al., 2017; Zhu et al., 2016). Similar to Ag^+, incorporation of Cu^+ and Cu^{2+} in nMgO improved OVs and resulted in the formation of O_2^- for antibacterial activity (Zhang et al., 2018).

Controversial opinions were reported in terms of the mechanism of nMgO as an antibacterial agent. Leung et al. (2014) found non-radical antibacterial activity of nMgO on *E. coli* (Leung et al., 2014). No evidence of oxidative stress or lipid peroxidation was found in bacteria. However, PO_4^{3-} present on the membrane of *E. coli* was attached to nMgO which is responsible for the antibacterial activity. Overall, Leung et al. (2014) concluded that generation of ROS was not necessary for the antibacterial activity of nMgO. Further study on the mechanism of the antibacterial

Table 7 nMgO as catalyst in the activation of oxygen in the absence of light

References	Catalyst	Pollutant	Experimental condition	Activation mechanism	Efficiency
Huang et al. (2005)	nMgO	*Bacillus subtilis* var. niger bacteria Staphylococcus aureus bacteria	Catalyst dose: 10 g/L Conc = 105 cfu mL^{-1} Reaction time: pH = 6.8 No light	Generation of O_2^-	97.5% antibacterial efficiency
Rao et al. (2013)	Li-doped nMgO	*Escherichia coli* (*E. coli*)	Catalyst dose: 0.5 µg/mL Conc = 108 cfu mL^{-1} Reaction time	Generation of O_2^-	100% antibacterial efficiency
Cai et al. (2017)	Ag-doped nMgO	*E. coli*	Catalyst Dose: 50 µg/mL Conc = 107 cfu mL^{-1} Reaction time: 20 min	Generation of O_2^-	Inactivate 7-log bacterial cells
Zhu et al. (2016)	Ag/nMgO	*E. coli*	Catalyst dose: 10 mg/mL Conc = 107 cfu mL^{-1} Reaction time: 60 min	Generation of O_2^-	Inactivate 7-log bacterial cells
Hao et al. (2017)	nMgO (111)	*E. coli* Methyl violet (MV)	Catalyst dose: 1 g/L MV = 20 mg/L Reaction time: 240 min	O_2^- and $^1O_2$	90% removal of MV
Zhang et al. (2018)	Cu-doped nMgO/ Fe_3O_4	*E. coli* *S. aureus*	Catalyst dose: 312 mg/L Conc = 105 cfu mL^{-1} Reaction time: 8 h	O_2^-	99% antibacterial activity
El-Shaer et al. (2020)	nMgO	*Bacillus cereus* Gram positive	Catalyst dose: 0.4 g/L Reaction time: 24 h	ROS	Complete inhibition of bacteria

activity of the nMgO was done by Aničić et al. (2018) who reported that interaction between bacteria and nMgO defects produced ROS as a side product. In this study, it was found that antibacterial activity of nMgO is inversely proportional to the rate of hydrolysis; OVs are superior to low coordinated oxygen (O_{3C} and O_{4C}) present in nMgO for antibacterial activity because of resistance to hydrolysis. Interaction

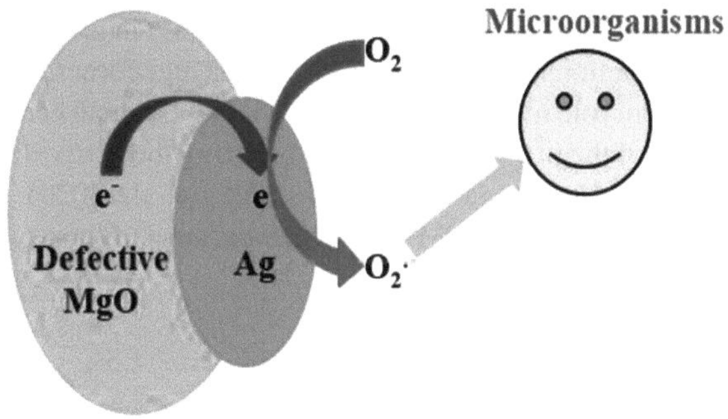

Fig. 8 Mechanism of ROS generation in the Ag/MgO composite for antibacterial activity

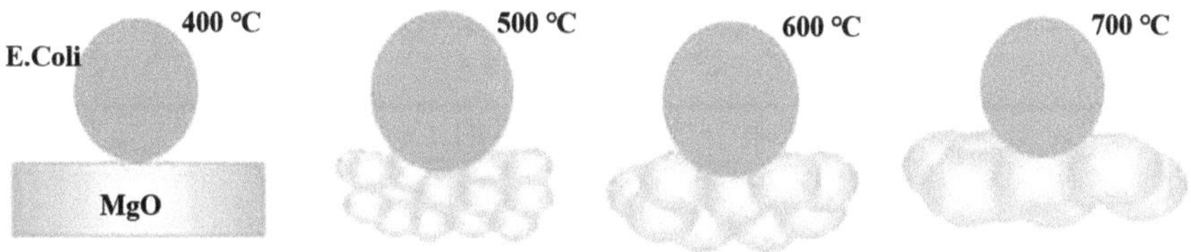

Fig. 9 Morphological changes in the MgO microrod with calcination temperature and effect on the contact between surface of MgO and bacterial

between nMgO and bacteria surface played a major role in bacterial inhibition, whereas ROS had no role in the antibacterial activity. However, F_2 dimmer and F^+ defects directly interact with bacteria, thereby contributing to the antibacterial activity of nMgO.

Contact between nMgO and microorganisms played a major role in disinfection. MgO microspheres synthesised at 500 °C (MgO-500) outperformed in antibacterial activity because of the large mesoporous size and volume, which enhance the contact between *E. coli* and nMgO (Fig. 9) (Purwajanti et al., 2015). In another study by Aničić et al. (2016), effect of the MgO microrod's synthesis temperature on the antibacterial activity found the optimum temperature of 900 °C. During the study, it was found that SSA was high at 700 °C, but surface features are low to obtain optimum contact between bacteria and nMgO, whereas at a high temperature of 1000 °C, surface features are too large, which reduced contact between bacteria and surface (Aničić et al., 2016).

5.2 *nMgO for Persulphate Activation in the Absence of Light*

Recently, carbocatalyst, heteroatom-doped carbocatalyst and heteroatom-doped g-C_3N_4 have been reported to activate PS (peroxymonosulphate (PMS)/ peroxydisulphate (PDS)) in the absence of light (Duan et al., 2018; Gao et al., 2018;

Peng et al., 2020). Region of high and low electron density in graphitic structure and oxygen-containing functional groups present at edges and corners of the carbocatalyst played a significant role in the activation of PMS/PDS (Duan et al., 2018). nMgO activated PS because of the presence of OVs by breaking the peroxide bond (–O–O–) to form SO_4^- and $^1O_2$ (Eqs. (9–(14)) (Table 8) (Peng et al., 2020). OVs present in S-MgO more effectively activated PMS than PDS because of the asymmetric nature of the peroxy bond.

$$HS_2O_8^- \rightarrow S_2O_{8(aq)}^{2-} + H^+ \quad pK_a = -1.3 \tag{9}$$

$$S_2O_{8(aq)}^{2-} + 2H^+ + 2e^- \rightarrow 2HSO_4^- \quad E_h^o = 2.01\ V \tag{10}$$

$$HSO_5^- \rightarrow SO_5^{2-} + H^+ \quad pK_a = 9.2 \tag{11}$$

$$HSO_5^- + 2H^+ + 2e^- \rightarrow HSO_4^- + H_2O \quad E_h^o = 1.8\ V \tag{12}$$

$$S_2O_{8(aq)}^{2-} + e^- \rightarrow SO_4^{\cdot -} + SO_{4(aq)}^{2-} \quad E_h^o = 1.44\ V \tag{13}$$

$$HSO_5^- + h^+ \rightarrow SO_5^- + H^+ \tag{14}$$

Peng et al. (2020) activated the PMS using nMgO for the degradation of the MB dye and achieved complete removal in 20 min at neutral pH. However, nMgO calcinated at 600 °C under argon atmosphere removed surface OH^- and found significant reduction in the activation of PMS. Therefore, it was observed that surface OH^- played a major role in the activation of PMS. PMS interaction with nMgO formed $^1O_2$ and O_2^- for the degradation of the MB dye (Peng et al., 2020).

Fanaei et al. (2019) prepared sulphur-doped nMgO (S-nMgO) for the activation of PMS to generate SO_4^- and OH. Activation of PMS by S-nMgO achieved complete removal and 62.4% mineralisation of acetaminophen (ACT). SO_4^- has made a major contribution in the removal of ACT, whereas OH played a minor role. OVs generated as a result of S-doping are responsible for the activation of the PMS. Due to its protonated surface under acidic and mildly alkaline conditions ($pH_{pzc} = 10.9$), S-nMgO demonstrated consistent removal efficiency across the pH range of 5–10, which was attributed to the electrostatic attraction between PMS and S-nMgO.

At higher pH g > 12, PMS underwent self-decomposition to generate SO_4^- in the presence of OH^-, whereas the activation of PMS decreased by repulsion between negatively charged oxygen in PMS and the nMgO surface (Eqs. (15)–(17)) (Lee et al., 2018). At lower pH < 4, H-bonding between H^+ and peroxy (–O–O–) bond in HSO_5^- inhibits the interaction between HSO_5^- and surface of nMgO (Peng et al., 2020). At lower pH < 1, almost complete dissolution of nMgO reduced activation of PMS (Peng et al., 2020). When the PMS-to-catalyst ratio was increased from 0.025 to 0.07 mM L g^{-1}, the ACT degradation rate rose from 3.1 to 15.7 mg L^{-1} min^{-1}.

Table 8 Application of nMgO in the activation of persulphate for the degradation of organic pollutant

References	Catalyst	Pollutant	Experimental condition	Activation mechanism	Efficiency
Peng et al. (2020)	nMgO	Methylene blue (MB)	Catalyst dose: 0.4 g/L MB = 10 mg/L PMS = 0.8 g/L Reaction time: 20 min pH = 7	Singlet oxygen major reactive species	Complete removal of dye within 20 min
Fanaei et al. (2019)	S-MgO	Acetaminophen (ACT)	Catalyst dose: 1 g/L ACT = 50 mg/L PDS/PMS = 0.5 g/L Reaction time: 30 min pH = 7	SO_4^- and ·OH radical	Complete removal of ACT 62.4% Mineralisation
Jawad et al. (2020)	CuOMgO/Fe_3O_4	4-Chlorophenol	Catalyst dose: 0.2 g/L 4-CP = 40 mg/L PMS = 2 mM Reaction time: 40 min	Singlet oxygen	Complete removal of 4-CP

However, further increase in the PMS to catalyst ratio reduced degradation of ACT because of the recombination of SO_4^- and OH to form SO_4^{2-} and H_2O_2, respectively.

$$(H_3O_{18}S_4)^{5-} \rightarrow HSO_5^- + HSO_4^- + SO_4^{2-} \quad (15)$$

$$HSO_5^- \rightarrow SO_4^- + \cdot OH \quad (16)$$

$$HSO_4^- + \cdot OH \rightarrow SO_4^- + H_2O \quad (17)$$

Cupric oxide (CuO) activates PMS through the inner-sphere complex and forms SO_4^- and OH (Jawad et al., 2020). Incorporation of nMgO in CuO increased OH^- group on the surface of the catalyst, and these densely covered surfaces of OH^- are responsible for strong H-bonding between PMS and nMgOCuO/Fe_2O_3. Therefore, addition of nMgO formed the outer-sphere complex with PMS, which is responsible for the non-radical activation of the PMS (Jawad et al., 2020; Sun et al., 2020). Activation of PMS was inhibited in the presence of PO_4^{3-} because of the strong interaction between PO_4^{3-} and surface OH^-. This confirmed the role of surface OH^- to form the outer-sphere complex with PMS. Along with that, incorporation of nMgO in CuO changes the redox state of Cu^{2+} to Cu^{3+} which offered the highest redox potential (E_h^o (Cu^{3+}/Cu^{2+}) = 2.3 V vs. NHE) for the activation of PMS. Therefore, combined effect of these achieved complete removal of 4-chlorophenol (4-CP) within 40 min. However, removal of 4-CP decreased to 80% at pH 12 because of the repulsion between CuOMgO (pH_{pzc} = 9.3) and PMS. The removal of 4-CP was reduced to 10% in the presence of sodium azide ($^1O_2$ scavenger) and decreased the degradation rate from 0.16 to 0.006 min^{-1}. Therefore, $^1O_2$ played a major role in the removal of 4-CP, whereas OH and SO_4^- made a minor contribution in the removal of 4-CP. $^1O_2$ formed through oxidation of the O_2^- (E_h^o ($O_2^-/^1O_2$) = −0.34 V) formed during interaction between PMS and $CuOMgO/Fe_3O_4^-$—(Eq. (18)) (Jawad et al., 2020).

$$2O_2^{-} + 2H_2O \rightarrow 2^1O_2 + H_2O_2 + 2OH^- \tag{18}$$

Overall, OVs and surface OH^- played a major role in the activation of PS. Still, more in-depth investigations are required on the role of pH, as nMgO increased pH above 10, where PMS undergoes self-activation.

6 Ozone Activation by nMgO Catalyst

Oxidation potential of ozone (O_3) is 2.07 V versus NHE, which is higher than chlorine but lower than OH and SO_4^- (Marcelino et al., 2017). O_3 has sufficient redox potential to oxidise OH^- (E_h^o (OH^-/OH) = 1.9 V vs. NHE) to form OH (Marcelino et al., 2017). Different heterogeneous catalysts such as CNTs, graphene oxide, Fe and Mn oxides and ferrite were used to activate the O_3 to form OH and O_2^- (Yu et al., 2020). Different photocatalysts such as TiO_2, g-C_3N_4 were effectively reducing O_3 to form O_3^- as CB position of both the catalysts can reduce the O_3 (Eq. (19)) (Piera et al., 2000; Xiao et al., 2017). O_3^- converts into highly reactive OH through a series of reactions (Eqs. (20)–(26)) (Xiao et al., 2017). In addition, UV light decomposition of O_3 produces OH and effectively degrades the organic pollutants (Ratpukdi et al., 2010).

$$O_{3(aq)} + e^- \rightleftharpoons O_3^{\cdot-} \quad E_h^o = +1.03\ V \tag{19}$$

nMgO has a polar surface that effectively activated polar O_3 to form OH for the degradation of organic pollutants (Table 9) (Moussavi & Mahmoudi, 2009; Moussavi & Alizadeh, 2010). nMgO is found to be effective in O_3 activation compared to ZnO, Mn_3O_4 and Fe_2O_3, and showed excellent recyclability (Wang et al., 2017a, 2017b). The alkaline nature of nMgO increased the pH of the solution, which is responsible for the decomposition of O_3 to form OH (Moussavi & Mahmoudi, 2009). Metal oxides can activate O_3 by two mechanisms: (i) Hydrogen/electrostatic bonding between negatively charged oxygen in O_3 and protonated surface OH^- and (ii) s^- Lewis acid sites: the central metal directly engages with O_3, while O_3 also interacts with surface hydroxyl groups or adsorbed water molecules coordinated to the central metal (He et al., 2008).

pH plays an important role in the decomposition of O_3; O_3 is decomposed in the presence of OH^- to form different radicals such as OH, O_3^- and $HO_2^\cdot$ (Eqs. (20)–(26)). nMgO formed from the calcination of natural brucite ($Mg(OH)_2 = 94.71\%$, $SiO_2 = 2.85\%$ and remaining other compounds) at 450 °C maintain the pH of aqueous solution around 10.8, favourable for the activation of O_3 through decomposition (He et al., 2008). It was observed that apart from an increase in the solution pH, surface OH^- on nMgO played a role in the activation of O_3. Single ozonation achieved 40% removal of COD, whereas in the presence of the brucite and nMgO, the removal increased to 60 and 90%, respectively. Phenol degradation kinetics was improved for nMgO compared to brucite in the presence of O_3. Phenol ($pK_a = 9.95$) is deprotonated and becomes negatively charged at high pH greater than 10, directly oxidised by electrophilic O_3. Phenol is a weak acid that contributes more OH^- ion and increased OH^- density on nMgO, which helped to enhance O_3 activation (Wang et al., 2017a, 2017b).

$$O_3 + OH^- \rightarrow HO_2^- + O_2 \quad (20)$$

$$HO_2^- + O_3 \rightarrow HO_2^\cdot + O_3^- \quad (21)$$

$$HO_2^\cdot \leftrightarrow H^+ + O_2^- \quad (22)$$

$$O_2^- + O_3 \rightarrow O_2 + O_3^- \quad (23)$$

$$O_3^- + H^+ \rightarrow HO_3^{\cdots} \quad (24)$$

$$HO_3^\cdot \rightarrow \cdot OH + O_2 \quad (25)$$

$$OH + O_3 \rightarrow HO_2^\cdot + O_2 \quad (26)$$

Apart from maintaining the solution's pH in alkaline condition to decompose O_3, the protonated surface OH^- of nMgO activates O_3 to form OH (Zhao et al., 2009). Protonated nMgO ($MgO–OH_2^+$) enhanced electrostatic attraction and hydrogen bonding with O_3 (Eqs. (27)–(28) and (29)–(34)). Electrostatic attraction between positively charged hydrogen and negatively charged oxygen atom in O_3 formed HO_3, unstable centred oxygen radical and H^+ on nMgO surface. This unstable nMgO surface oxidised H_2O to form OH. At low pH (~2), nMgO effectively activates O_3 due to the abundance of protonated surface hydroxyl groups. At pH greater than 6, OH^- played a dominant role in the activation of O_3, whereas efficiency of O_3 activation was reduced between pH 2 and 6 (reduction in positive charge surface

Table 9 Application of nMgO in the activation of O_3 for the degradation of pollutants

References	Catalyst	Pollutant	Experimental condition	Activation mechanism	Efficiency
He et al. (2008)	Natural brucite ($MgOH_2$)/ magnesia (MgO)	Phenol	Catalyst dose: 5 g/L PHE = 100 mg/L Reaction time: 30 min pH = 10.8 Semi-continuous flow	High pH decomposition of O_3 Direct oxidation	Complete removal of phenol
Moussavi and Mahmoudi (2009)	nMgO/O_3 SSA = 118 m^2/g Porous volume = 0.156 cm^3/g	RR198 (Azo dye)	Catalyst dose: 5 g/L RR198 = 200 mg/L Ozonation rate = 200 mg/h Reaction time = 10 min pH = 8	Lewis acid Surface OH^-	Almost complete removal of COD 80% removal of colour
Moussavi et al. (2009)	nMgO/O_3/ H_2O_2 SSA = 118 m^2/g Porous volume = 0.156 cm^3/g	Formaldehyde (FA)	Catalyst dose: 5 g/L FA = 7,000 mg/L H_2O_2 = 0.09 M Reaction time: 120 min pH = 8	Lewis acid Surface OH^-	79% removal of FA 65.5% removal of COD Pretreatment before SBR
Moussavi and Alizadeh (2010)	nMgO nanocrytal/ O_3 SSA = 152 m^2/g Porous volume = 0.182 cm^3/g	Phenol (PHE) in saline wastewater	Catalyst dose: 4 g/L PHE = 1100 mg/L COD = 2500 mg/L Ozonation rate = 250 mg/h Reaction time: 80 min pH = 7	Lewis acid	96% removal of PHE 70% removal of COD Used as prebiological treatment
Moussavi et al. (2014)	nMgO/ GAC/O_3 SSA = 1036 m^2/g	Catechol (CAT)	Catalyst dose: 10 g/L CAT = 1,000 mg/L Ozonation rate = 126 mg/h Reaction time: 30 min pH = 8	Lewis acid Surface OH^-	54% removal of ACT 45.5% removal of TOC

(continued)

Table 9 (continued)

References	Catalyst	Pollutant	Experimental condition	Activation mechanism	Efficiency
Chen et al. (2015)	$nMgO/O_3$	4-Chlorophenol	Catalyst dose: 1 g/L 4-CP = 100 mg/L Ozonation rate = 150 mg/h Reaction time: 30 min pH = 6.2	Surface OH^-	Complete removal of 4-CP by MgO with (111) exposed facet
Mashayekh-Salehi et al. (2017)	Mesoporous $nMgO/O_3$ SSA = 257.3 m^2/g	Acetaminophen (ACT)	Catalyst dose: 2 g/L ACT = 50 mg/L Ozonation rate = 108 mg/h Reaction time: 30 min pH = 5.4	Surface OH^-	71.5% removal of ACT 62.8% removal of TOC
Wang et al. (2017a, 2017b)	$nMgO/O_3$	Phenol (PHE)	Catalyst dose: 40 mg/L PHE = 100 mg/L Ozonation rate = 216 mg/h Reaction time: 20 min	Surface OH^-	Complete removal of phenol
Shen et al. (2017)	nMgO/ ceramic honeycomb	Acetic acid (AC)	Catalyst dose: 20 g/L AC = 100 mg/L Ozonation rate = 2,730 mg/h Reaction time: 30 min pH = 7.5	Surface OH^-	90% removal of acetic acid

OH^- and comparatively low concentration of OH^- ion). Lewis acid sites on nMgO (adsorbed OH^- and H_2O) are responsible for O_3 activation (Moussavi et al., 2009; Zhao et al., 2009). Activation of O_3 with both mechanisms is similar, as the central metal (Lewis acid) atom is not involved in the activation process (Moussavi et al., 2009; Zhao et al., 2009).

Electrostatic attraction

$$MgOH - H^+ + {}^-O - O^+ = O \rightarrow HO_3^{\cdot} + Mg - O - H^+ \quad (27)$$

$$Mg - O - H^+ + H_2O \rightarrow OH + MgOH - H^+ \quad (28)$$

Hydrogen bonding

$$\begin{aligned} &\mathrm{MgOH-H^{+}-^{-}O-O^{+}=O-HH^{+}OMg+H_2O\rightarrow} \\ &\quad \mathrm{OH+HOO^{\cdot}+MgOH-H^{+}} \end{aligned} \quad (29)$$

$$\mathrm{HO_2^{\cdot}\leftrightarrow O_2^{-}+H^{+}\quad pK_a=4.8} \quad (30)$$

$$\mathrm{O_3+OH^{-}\rightarrow O_2+HO_2^{-}} \quad (31)$$

$$\mathrm{HO_2^{-}+O_3\rightarrow HO_2^{\cdot}+O_3^{-}} \quad (32)$$

$$\mathrm{O_3^{-}+H^{+}\rightarrow HO_3^{\cdot}} \quad (33)$$

$$\mathrm{HO_3^{\cdot}\rightarrow OH+O_2} \quad (34)$$

Tert-butanol as OH scavenger has no effect on the degradation of PHE in the MgO/O_3 system (Moussavi & Alizadeh, 2010). It was observed that reactive atomic oxygen formed through hydrogen bonding between oxygen molecule of O_3 and the surface OH^- of nMgO is responsible for the removal of PHE (Moussavi & Alizadeh, 2010). Mashayekh-Salehi et al. (2017) performed oxidation of ACT in the MgO/O_3 system and achieved 62.5% mineralisation. The rate of degradation of ACT in the presence of nMgO was 27.20 mg L^{-1} min^{-1}, which is 18 times higher than in the presence of only O_3. OH generated because of interaction between nMgO and O_3 was released in bulk solution for the degradation of the ACT (Eqs. (35)–(39)) (Chen et al., 2015; Mashayekh-Salehi et al., 2017).

$$\mathrm{MgO-S+O_3\rightarrow MgO-S^{O=O-O}\rightarrow MgO-S^{O\cdot}+O_2} \quad (35)$$

$$\mathrm{MgO-S^{O\cdot}+2H_2O+O_3\rightarrow MgO-S_2^{(OH)}+3\cdot OH+2O_2} \quad (36)$$

$$\mathrm{MgO-S_2^{(OH)}+ACT\rightarrow CO_2+H_2O+intermediates} \quad (37)$$

$$\mathrm{MgO-S^{ACT}+OH\rightarrow CO_2+H_2O+intermediates} \quad (38)$$

$$\mathrm{ACT+OH\rightarrow CO_2+H_2O+intermediates} \quad (39)$$

Chen et al. (2015) studied different facets of nMgO such as (111), (110), (200) and (100) for the activation of O_3. The MgO (111) facet showed the highest catalytic activity in the removal of 4-CP through activation of the O_3. Temperature programmed desorption of CO_2 (CO_2-TPD) profile used to understand surface

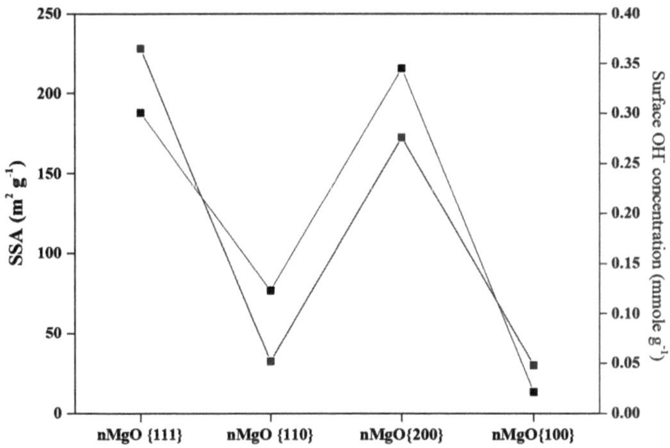

Fig. 10 SSA and surface OH^- concentration of nMgO with different facets

basicity of nMgO with different exposed facets showed peak at a low temperature between 20 and 200 °C and corresponds to low basic sites (OH^- on the surface), peak at a temperature between 200 and 400 °C corresponds to moderate basic sites (oxygen in O^{2-} and Mg^{2+}) and peak at temperature greater than 400 °C corresponds to high basic sites (isolated O^{2-}). CO_2-TDP profile of the MgO (111) facet has high basic site O^{2-} because of the abundance of low coordinated sites in the (111) facet, which promoted activation of O_3. Also, higher SSA of MgO (111) increased exposure of active sites such as corners, edges and kinks. Along with high SSA, nMgO (111) has high surface OH^- concentration compared to (110), (100) and (200) facets of nMgO. Activation of O_3 was observed in the order of nMgO (111) > nMgO (110) > nMgO (200). Therefore, activation of the O_3 is dependent on the exposed facet than SSA and concentration of surface OH^- (Fig. 10).

MgO nanocrystal on granular activated carbon (GAC) in fluidised bed reactor showed 2.5 times higher catalytic activity for the degradation of the catechol than GAC alone. nMgO/GAC is microporous material with specific surface area of ~1036 $m^2\ g^{-1}$. nMgO/GAC dose of 10 g L^{-1} with ozonation (126 mg h^{-1}) achieved 54% degradation and 45.5% mineralisation of the catechol ($C_0 = 1{,}000$ mg L^{-1}) within 30 min (Moussavi et al., 2014). In another study, nMgO was immobilised on a ceramic honeycomb (CH) structure, which performed degradation of acetic acid and achieved 81% removal in 30 min (Shen et al., 2017). nMgO improved the generation of the OH from 0.67 to 4.36 mg L^{-1}, which is responsible for the degradation of acetic acid. CH consists of low basic sites, whereas nMgO/CH has high basicity compared to both nMgO and CH. The high surface basicity of the nMgO improved catalyst composites' protonation to activate O_3 (Shen et al., 2017). Degradation pattern of acetic acid with respect to time found that single ozonation degradation process inhibits after 5 min as a result of fall in pH, and O_3 concentration in solution remained constant after 5 min. However, the MgO/CH catalyst effectively activates O_3 because of high pH (Shen et al., 2017).

$$R_{ct} = \frac{\int [HO\cdot]dt}{\int [O_3]dt} \tag{40}$$

R_{ct} is ratio of OH to O_3 concentration with respect to time (Eq. (40)). R_{ct} parameter is used to understand the effect of OH in the degradation of target pollutants in ozonation. R_{ct} value of MgO/CH/O_3 and single ozonation is 2.63×10^{-7} and 8.1×10^{-8}, respectively. High value of R_{ct} in MgO/CH suggests that OH plays a major role in the degradation of AcOH compared to single ozonation. R_{ct} value of MgO (111) is higher than MgO (100), MgO (200) and MgO (110). MgO (111) has more ability to form OH compared to other exposed facets. Overall, ozonation with nMgO is advantageous in terms of generation of ROS and reduction in the cost of the ozonation, as the required time and ozone dose is reduced.

7 Summary and Knowledge Gaps

Based on our research, the following is a summary of our findings and knowledge gaps.

i. Recent interest in the use of nMgO as a catalyst in AOPs for the degradation of organic pollutants has been encouraged by the investigation of different morphologies and OVs of the nMgO. nMgO is a cost-effective material that can facilitate synthesis in massive quantity due to abundantly available low-cost precursors.
ii. nMgO showed promising results in photocatalysis, PS activation, Fenton-like processes and ozonation. nMgO as a photocatalyst performed moderately in the presence of UV light because of high bandgap. There is a need to explore more on the doping by different metals/non-metals to reduce the band energy for the generation of the charge carriers and reduce charge recombination to improve the performance of nMgO.
iii. Explore different reduction methods to the generation of OVs (F centres) in nMgO and understand the energy level of the F centre in the forbidden energy gap and its thermodynamic feasibility to activate O_2 both in the absence and presence of light. Along with that, there is a need to explore the feasibility of F centres in nMgO to photocatalytically activate the different electron acceptors such as peracetic acid, periodide and H_2O_2 for the degradation of the pollutants.
iv. Different strategies to recover nMgO nanoparticles from an aqueous solution need to be explored. Evaluate leaching effect of magnesium during the treatment process on hardness and pH of the water and its implications.
v. nMgO (111) has high basicity, OVs and low coordinated sites responsible for the high catalytic activity to activate O_2 and O_3 in the absence of light. However, the stability of OVs on the (111) facet of MgO needs to be evaluated to understand the efficiency of the nMgO (111) in the degradation of the organic pollutants in multiple cycles. nMgO (111) can be used to develop a composite

in combination with different visible light active photocatalysts which can be performed both in the presence and absence of light.

vi. nMgO performed exceptionally well in the activation of oxygen and PS in the absence of light. nMgO is used as a disinfectant because of the generation of ROS in the absence of light. Metal oxide semiconductors such as TiO_2, ZnO, WO_3, ZrO_2 and Bi_2O_3 have OVs that cannot generate ROS in the absence of light. It is necessary to evaluate this aspect of nMgO compared to other metal oxide photocatalysts.

vii. nMgO generates O_2^- because of electron transfer to $O_{2(aq)}$ from the OVs in the absence of light. The generation of ROS by nMgO in the absence of light is not highlighted and analysed in the photocatalytic activity of nMgO. On the contrary, nMgO has very little activity in the absence of light during the photocatalytic process, which needs to be evaluated.

viii. PS activation by nMgO through outer-sphere complex generates SO_4, OH and $^1O_2$. Non-radical activation of PS using nMgO through the outer-sphere complex generates $^1O_2$. It selectively degrades the target pollutants because of the selective nature of the electrophilic $^1O_2$ and minimises the effect of the wastewater matrix on the degradation of the pollutant.

ix. nMgO increased pH which accelerates the activation of O_3 to form OH. The surface basicity of nMgO played a major role in the activation of O_3. Surface OH^- group on nMgO performed as a catalytically active site for the activation of the O_3. But more clarity is necessary on the activation mechanism of the O_3, including the role of Mg^{2+}.

x. Role of F centre and oxygen defects of nMgO in the activation of O_3 was not explored yet and need to be understood in terms of an increase in the efficiency of the nMgO/O_3 system.

xi. From an application point of view, most of the studies are carried out with dye and PPCPs at high concentrations in distilled water. Therefore, it is necessary to check the efficiency of the MgO-based AOPs in the degradation of PPCPs present in low concentration ranges from $\mu g\ L^{-1}$ to $ng\ L^{-1}$ in wastewater.

References

Abed, C., Ali, M. B., Addad, A., & Elhouichessssst, H. (2019). Growth, structural and optical properties of ZnO-ZnMgO-MgO nanocomposites and their photocatalytic activity under sunlight irradiation. *Materials Research Bulletin, 110*, 230–238. https://doi.org/10.1016/j.materresbull.2018.10.041

Akintunde, O. O., Linlong, Y., Jinguang, H., Kibria, Md. G., & Gopal, A. (2022). Visible-light driven photocatalytic degradation of 4-chlorophenol using graphitic carbon nitride-based nanocomposites. *Catalysts, 12*(3), 281. https://doi.org/10.3390/catal12030281

Al-Hazmi, F., Alnowaiser, F., Al-Ghamdi, A. A., Al-Ghamdi, A. A., Aly, M. M., Al-Tuwirqi, R. M., & El-Tantawy, F. (2012). A new large-scale synthesis of magnesium oxide nanowires: Structural and antibacterial properties. *Superlattices and Microstructures, 52*(2), 200–209. https://doi.org/10.1016/j.spmi.2012.04.013

Aničić, N., Vukomanović, M., Koklič, T., & Suvorov, D. (2018). Fewer defects in the surface slows the hydrolysis rate, decreases the ROS generation potential, and improves the non-ROS antimicrobial activity of MgO. *Small, 14*(26), 1800205. https://doi.org/10.1002/smll.201870123

Aničić, N., Vukomanović, M., & Suvorov, D. (2016). The nano-texturing of MgO microrods for antibacterial applications. *RSC Advances, 6*(104), 102657–102664. https://doi.org/10.1039/C6RA23058J

Anilkumar, M. R., Nagaswarupa, H. P., Nagabhushana, H., Sharma, S. C., Vidya, Y. S., Anantharaju, K. S., & Gurushantha, K. (2015). Bio-inspired route for the synthesis of spherical shaped MgO: Fe^{3+} nanoparticles: Structural, photoluminescence and photocatalytic investigation. *Spectrochimica Acta Part a: Molecular and Biomolecular Spectroscopy, 149*, 703–713. https://doi.org/10.1016/j.saa.2015.05.003

Armstrong, D. A., Huie, R. E., Koppenol, W. H., Lymar, S. V., Merényi, G., Neta, P., Ruscic, B., Stanbury, D. M., Steenken, S., & Wardman, P. (2015). Standard electrode potentials involving radicals in aqueous solution: Inorganic radicals (IUPAC technical report). *Pure and Applied Chemistry, 87*(11–12), 1139–1150. https://doi.org/10.1515/pac-2014-0502

Bandara, J., Hadapangoda, C. C., & Jayasekera, W. G. (2004). TiO_2/MgO composite photocatalyst: The role of MgO in photoinduced charge carrier separation. *Applied Catalysis b: Environmental, 50*(2), 83–88. https://doi.org/10.1016/j.apcatb.2003.12.021

Beranek, R. (2011). (Photo) electrochemical methods for the determination of the band edge positions of TiO_2-based nanomaterials. *Advances in Physical Chemistry, 1*, 786759. https://doi.org/10.1155/2011/786759

Cai, Y., Wu, D., Zhu, X., Wang, W., Tan, F., Chen, J., & Qiu, X. (2017). Sol-gel preparation of Ag-doped MgO nanoparticles with high efficiency for bacterial inactivation. *Ceramics International, 43*(1), 1066–1072. https://doi.org/10.1016/j.ceramint.2016.10.041

Chen, J., Tian, S., Lu, J., & Xiong, Y. (2015). Catalytic performance of MgO with different exposed crystal facets towards the ozonation of 4-chlorophenol. *Applied Catalysis a: General, 506*, 118–125. https://doi.org/10.1016/j.apcata.2015.09.001

Chen, M. L., Zhang, F. J., & Oh, W. C. (2010). Promoting effect of MgO in the photodegradation of methylene blue over MgO/MWCNT/TiO_2 photocatalyst. *Korean Journal of Materials Research, 20*(7), 345–350. https://doi.org/10.3740/MRSK.2010.20.7.345

Choudhury, B., Basyach, P., & Choudhury, A. (2014). Monitoring F, F^+ and F^{22+} related intense defect emissions from nanocrystalline MgO. *Journal of Luminescence, 149*, 280–286. https://doi.org/10.1016/j.jlumin.2014.01.041

Colpani, G. L., Zanetti, M., Zeferino, R. C. F., Silva, L. L., de Mello, J. M. M., Riella, H. G., & Soares, C. (2018). Lanthanides effects on TiO_2 photocatalysts. In *Photocatalysts-applications and attributes*. IntechOpen

Demirci, S., Öztürk, B., Yildirim, S., Bakal, F., Erol, M., Sancakoğlu, O., Yigit, R. E., Celik, E., & Batar, T. (2015). Synthesis and comparison of the photocatalytic activities of flame spray pyrolysis and sol-gel derived magnesium oxide nano-scale particles. *Materials Science in Semiconductor Processing, 34*, 154–161. https://doi.org/10.1016/j.mssp.2015.02.0299

Devaraja, P. B., Avadhani, D. N., Prashantha, S. C., Nagabhushana, H., Sharma, S. C., Nagabhushana, B. M., & Nagaswarupa, H. P. (2014). Synthesis, structural and luminescence studies of magnesium oxide nanopowder. *Spectrochimica Acta Part a: Molecular and Biomolecular Spectroscopy, 118*, 847–851. https://doi.org/10.1016/j.saa.2013.08.050

Di Valentin, C., Ferullo, R., Binda, R., & Pacchioni, G. (2006). Oxygen vacancies and peroxo groups on regular and low-coordinated sites of MgO, CaO, SrO, and BaO surfaces. *Surface Science, 600*(5), 1147–1154. https://doi.org/10.1016/j.susc.2006.01.009

Duan, X., Sun, H., Shao, Z., & Wang, S. (2018). Nonradical reactions in environmental remediation processes: Uncertainty and challenges. *Applied Catalysis b: Environmental, 224*, 973–982. https://doi.org/10.1016/j.apcatb.2017.11.051

El-Shaer, A., Abdelfatah, M., Mahmoud, K. R., Momay, S., & Eraky, M. R. (2020). Correlation between photoluminescence and positron annihilation lifetime spectroscopy to characterize defects in calcined MgO nanoparticles as a first step to explain antibacterial activity. *Journal of Alloys and Compounds, 817*, 152799. https://doi.org/10.1016/j.jallcom.2019.152799

El-Shamy, A. (2020). Synthesis of new magnesium peroxide (MgO_2) nano-rods for pollutant dye removal and antibacterial applications. *Materials Chemistry and Physics, 243*, 122640. https://doi.org/10.1016/j.matchemphys.2020.122640

Fanaei, F., Moussavi, G., Srivastava, V., & Sillanpää, M. (2019). The enhanced catalytic potential of sulfur doped MgO (S-MgO) nanoparticles in activation of peroxysulfates for advanced oxidation of acetaminophen. *Chemical Engineering Journal, 371*, 404–413. https://doi.org/10.1016/j.cej.2019.04.007

Fujita, I., Edalati, K., Wang, Q., Arita, M., Watanabe, M., Munetoh, S., & Horita, Z. (2020). High-pressure torsion to induce oxygen vacancies in nanocrystals of magnesium oxide for enhanced light absorbance and photocatalysis. *Materialia*, 100670. https://doi.org/10.1016/j.mtla.2020.100670

Gao, Y., Zhu, Y., Lyu, L., Zeng, Q., Xing, X., & Hu, C. (2018). Electronic structure modulation of graphitic carbon nitride by oxygen doping for enhanced catalytic degradation of organic pollutants through peroxymonosulfate activation. *Environmental Science & Technology, 52*(24), 14371–14380. https://doi.org/10.1021/acs.est.8b05246

Gholami, F., Mosmeri, H., Shavandi, M., Dastgheib, S. M. M., & Amoozegar, M. A. (2019). Application of encapsulated magnesium peroxide (MgO_2) nanoparticles in permeable reactive barrier (PRB) for naphthalene and toluene bioremediation from groundwater. *Science of the Total Environment, 655*, 633–640. https://doi.org/10.1016/j.scitotenv.2018.11.253

Gonchar, A., Risse, T., Freund, H. J., Giordano, L., Di Valentin, C., & Pacchioni, G. (2011). Activation of oxygen on MgO: $O_2^{\cdot -}$ radical ion formation on thin, metal-supported MgO (001) films. *Angewandte Chemie International Edition, 50*(11), 2635–2638. https://doi.org/10.1002/anie.201005729

Hao, Y. J., Liu, B., Tian, L. G., Li, F. T., Ren, J., Liu, S. J., & Wang, X. J. (2017). Synthesis of (111) facet-exposed MgO with surface oxygen vacancies for reactive oxygen species generation in the dark. *ACS Applied Materials & Interfaces, 9*(14), 12687–12693. https://doi.org/10.1021/acsami.6b16856

Hao, Y. J., Tian, L. G., Duan, E., Liu, J., Qi, T. Y., Kong, W. Q., & Li, F. T. (2020). Low-temperature methane oxidation triggered by peroxide radicals over noble-metal-free MgO catalyst. *ACS Applied Materials & Interfaces, 12*(19), 21761–21771. https://doi.org/10.1021/acsami.0c04083

Hayati, F., Isari, A. A., Anvaripour, B., Fattahi, M., & Kakavandi, B. (2020). Ultrasound-assisted photocatalytic degradation of sulfadiazine using MgO@CNT heterojunction composite: Effective factors, pathway and biodegradability studies. *Chemical Engineering Journal, 381*, 122636. https://doi.org/10.1016/j.cej.2019.122636

He, K., Dong, Y. M., Li, Z., Yin, L., Zhang, A. M., & Zheng, Y. C. (2008). Catalytic ozonation of phenol in water with natural brucite and magnesia. *Journal of Hazardous Materials, 159*(2–3), 587–592. https://doi.org/10.1016/j.jhazmat.2008.02.061

Hinuma, Y., Toyao, T., Kamachi, T., Maeno, Z., Takakusagi, S., Furukawa, S., & Shimizu, K. I. (2018). Density functional theory calculations of oxygen vacancy formation and subsequent molecular adsorption on oxide surfaces. *The Journal of Physical Chemistry C, 122*(51), 29435–29444. https://doi.org/10.1021/acs.jpcc.8b11279

Hollman, J., Dominic, J. A., & Achari, G. (2020). Degradation of pharmaceuticals in aqueous solutions using UV/peracetic acid process: Kinetics, degradation pathways and comparison with UV/H_2O_2. *Chemosphere, 248*, 125911. https://doi.org/10.1016/j.chemosphere.2020.125911

Huang, L., Li, D. Q., Lin, Y. J., Wei, M., Evans, D. G., & Duan, X. (2005). Controllable preparation of nano-MgO and investigation of its bactericidal properties. *Journal of Inorganic Biochemistry, 99*(5), 986–993. https://doi.org/10.1016/j.jinorgbio.2004.12.022

Izadifard, M., Achari, G., & Langford, C. H. (2008). The pathway of dechlorination of PCB congener by a photochemical chain process in 2-propanol: The role of medium and quenching. *Chemosphere, 73*(8), 1328–1334. https://doi.org/10.1016/j.chemosphere.2008.07.003

Izadifard, M., Achari, G., & Langford, C. H. (2013). Application of photocatalysts and LED light sources in drinking water treatment. *Catalysts, 3*(3), 726–743. https://doi.org/10.3390/catal3030726

Izadifard, M., Achari, G., & Langford, C. H. (2018). Mineralization of sulfolane in aqueous solutions by ozone/CaO_2 and ozone/CaO with potential for field application. *Chemosphere, 197*, 535–540. https://doi.org/10.1016/j.chemosphere.2018.01.072

Jawad, A., Zhan, K., Wang, H., Shahzad, A., Zeng, Z., Wang, J., & Chen, Z. (2020). Tuning of persulfate activation from a free radical to a nonradical pathway through the incorporation of non-redox magnesium oxide. *Environmental Science & Technology, 54*(4), 2476–2488. https://doi.org/10.1021/acs.est.9b04696

Jorfi, S., Barzegar, G., Ahmadi, M., Soltani, R. D. C., Takdastan, A., Saeedi, R., & Abtahi, M. (2016). Enhanced coagulation-photocatalytic treatment of Acid red 73 dye and real textile wastewater using UVA/synthesized MgO nanoparticles. *Journal of Environmental Management, 177*, 111–118. https://doi.org/10.1016/j.jenvman.2016.04.005

Jorfi, S., Pourfadakari, S., & Kakavandi, B. (2018). A new approach in sono-photocatalytic degradation of recalcitrant textile wastewater using MgO@zeolite nanostructure under UVA irradiation. *Chemical Engineering Journal, 343*, 95–107. https://doi.org/10.1016/j.cej.2018.02.067

Jourshabani, M., Dominic, J. A., Achari, G., & Shariatinia Z. (2019). Synergetic photocatalytic ozonation using modified graphitic carbon nitride for treatment of emerging contaminants under UVC, UVA and visible irradiation. *Chemical Engineering Science*, 209. https://doi.org/10.1016/j.ces.2019.115181

Kumar, A., Thota, S., Varma, S., & Kumar, J. (2011). Sol–gel synthesis of highly luminescent magnesium oxide nanocrystallites. *Journal of Luminescence, 131*(4), 640–648. https://doi.org/10.1016/j.jlumin.2010.11.008

Kumar, M. A., Nagaswarupa, H. P., Ravikumar, C. R., Prashantha, S. C., Nagabhushana, H., & Bhatt, A. S. (2019). Green engineered nano MgO and ZnO doped with Sm^{3+}: Synthesis and a comparison study on their characterization, PC activity and electrochemical properties. *Journal of Physics and Chemistry of Solids, 127*, 127–139. https://doi.org/10.1016/j.jpcs.2018.12.012

Langford, C. H., Izadifard, M., Radwan, E., & Achari, G. (2014). Some observations on the development of superior photocatalytic systems for application to water purification by the "adsorb and shuttle" or the interphase charge transfer mechanisms. *Molecules, 19*(12), 19557–19572. https://doi.org/10.3390/molecules191219557

Lee, C., Kim, H. H., & Park, N. B. (2018). Chemistry of persulfates for the oxidation of organic contaminants in water. *Membrane Water Treatment*, *9*(6), 405–419.

Leung, Y. H., Ng, A. M., Xu, X., Shen, Z., Gethings, L. A., Wong, M. T., & Leung, F. C. (2014). Mechanisms of antibacterial activity of MgO: Non-ROS mediated toxicity of MgO nanoparticles towards Escherichia coli. *Small (Weinheim an der Bergstrasse, Germany), 10*(6), 1171–1183. https://doi.org/10.1002/smll.201302434

Li, E. J., Xia, K., Yin, S. F., Dai, W. L., Luo, S. L., & Au, C. T. (2011). Preparation, characterization and photocatalytic activity of Bi_2O_3–MgO composites. *Materials Chemistry and Physics, 125*(1–2), 236–241. https://doi.org/10.1016/j.matchemphys.2010.09.013

Li, F. T., Ran, J., Jaroniec, M., & Qiao, S. Z. (2015). Solution combustion synthesis of metal oxide nanomaterials for energy storage and conversion. *Nanoscale, 7*(42), 17590–17610. https://doi.org/10.1039/C5NR05299H

Li, W., Chen, D., Xia, F., Tan, J. Z., Huang, P. P., Song, W. G., Nursam, N. M., & Caruso, R. A. (2016). Extremely high arsenic removal capacity for mesoporous aluminium magnesium oxide composites. *Environmental Science: Nano, 3*(1), 94–106. https://doi.org/10.1039/C5EN00171D

Ling, L. L., Liu, W. J., Zhang, S., & Jiang, H. (2017). Magnesium oxide embedded nitrogen self-doped biochar composites: Fast and high-efficiency adsorption of heavy metals in an aqueous solution. *Environmental Science & Technology, 51*(17), 10081–10089. https://doi.org/10.1021/acs.est.7b02382

Liu, C., Zhang, A. Y., Si, Y., Pei, D. N., & Yu, H. Q. (2019a). Photochemical protection of reactive sites on defective TiO_{2-x} surface for electrochemical water treatment. *Environmental Science & Technology, 53*(13), 7641–7652. https://doi.org/10.1021/acs.est.9b01307

Liu, J., Yue, Y., Wang, W., Tan, F., Xia, H., Wang, X., & Wong, P. K. (2020). Facile one-step synthesis of 3D hierarchical flower-like magnesium peroxide for efficient and fast removal of tetracycline from aqueous solution. *Journal of Hazardous Materials*, 122877. https://doi.org/10.1016/j.jhazmat.2020.122877

Liu, S., Li, Z., Wang, C., Tao, W., Huang, M., Zuo, M., & Xu, P. (2020b). Turning main-group element magnesium into a highly active electrocatalyst for oxygen reduction reaction. *Nature Communications, 11*(1), 1–11. https://doi.org/10.1038/s41467-020-14565-w

Liu, S., Wang, Z., Han, T., Fei, T., & Zhang, T. (2019b). Mesoporous magnesium oxide nanosheet electrocatalysts for the detection of lead (II). *ACS Applied Nano Materials, 2*(5), 2606–2611. https://doi.org/10.1021/acsanm.9b00600

Liu, W. J., Jiang, H., Tian, K., Ding, Y. W., & Yu, H. Q. (2013). Mesoporous carbon stabilized MgO nanoparticles synthesized by pyrolysis of $MgCl_2$ preloaded waste biomass for highly efficient CO_2 capture. *Environmental Science & Technology, 47*(16), 9397–9403. https://doi.org/10.1021/es401286p

Mageshwari, K., Mali, S. S., Sathyamoorthy, R., & Patil, P. S. (2013). Template-free synthesis of MgO nanoparticles for effective photocatalytic applications. *Powder Technology, 249*, 456–462. https://doi.org/10.1016/j.powtec.2013.09.016

Malkhasian, A., Izadifard, M., Achari, G., & Langford, C. H. (2014). Photocatalytic degradation of agricultural antibiotics using a UV-LED light source. *Journal of Environmental. Science and Health, Part B: Pesticides, Food Contaminants, and Agricultural Wastes, 49*(1), 35–40. https://doi.org/10.1080/03601234.2013.836871

Mantilaka, M. P. G., De Silva, R. T., Ratnayake, S. P., Amaratunga, G., & de Silva, K. N. (2018). Photocatalytic activity of electrospun MgO nanofibres: Synthesis, characterization and applications. *Materials Research Bulletin, 99*, 204–210. https://doi.org/10.1016/j.materresbull.2017.10.047

Mao, N., & Jiang, J. X. (2019). MgO/g-C_3N_4 nanocomposites as efficient water splitting photocatalysts under visible light irradiation. *Applied Surface Science, 476*, 144–150. https://doi.org/10.1016/j.apsusc.2019.01.049

Marcelino, R. B., Leão, M. M., Lago, R. M., & Amorim, C. C. (2017). Multistage ozone and biological treatment system for real wastewater containing antibiotics. *Journal of Environmental Management, 195*, 110–116. https://doi.org/10.1016/j.jenvman.2016.04.041

Mashayekh-Salehi, A., Moussavi, G., & Yaghmaeian, K. (2017). Preparation, characterization and catalytic activity of a novel mesoporous nanocrystalline MgO nanoparticle for ozonation of acetaminophen as an emerging water contaminant. *Chemical Engineering Journal, 310*, 157–169. https://doi.org/10.1016/j.cej.2016.10.096

Medhi, R., Marquez, M. D., & Lee, T. R. (2020). Visible-light-active doped metal oxide nanoparticles: Review of their synthesis, properties, and applications. *ACS Applied Nano Materials, 3*(7), 6156–6185. https://doi.org/10.1021/acsanm.0c01035

Mehrabani-Zeinabad, M., Yu, L., Achari, G., & Langford, C. H. (2016). Mineralization of sulfolane by $UV/O_3/H_2O_2$ in a tubular reactor. *Journal of Environmental Engineering and Science, 11*(2), 44–51. https://doi.org/10.1680/jenes.16.00014

Mirzaei, A., Chen, Z., Haghighat, F., & Yerushalmi, L. (2017). Removal of pharmaceuticals from water by homo/heterogonous Fenton-type processes—A review. *Chemosphere, 174*, 665–688. https://doi.org/10.1016/j.chemosphere.2017.02.019

Moradi, S., Sobhgol, S. A., Hayati, F., Isari, A. A., Kakavandi, B., Bashardoust, P., & Anvaripour, B. (2020). Performance and reaction mechanism of MgO/ZnO/Graphene ternary nanocomposite in coupling with LED and ultrasound waves for the degradation of sulfamethoxazole and pharmaceutical wastewater. *Separation and Purification Technology*, 117373. https://doi.org/10.1016/j.seppur.2020.117373

Mosmeri, H., Gholami, F., Shavandi, M., Alaie, E., & Dastgheib, S. M. (2018). Application of magnesium peroxide (MgO_2) nanoparticles for toluene remediation from groundwater: Batch and column studies. *Environmental Science and Pollution Research, 25*(31), 31051–31061. https://doi.org/10.1007/s11356-018-2920-3

Moussavi, G., & Alizadeh, R. (2010). The integration of ozonation catalyzed with MgO nanocrystals and the biodegradation for the removal of phenol from saline wastewater. *Applied Catalysis b: Environmental, 97*(1–2), 160–167. https://doi.org/10.1016/j.apcatb.2010.03.036

Moussavi, G., & Mahmoudi, M. (2009). Degradation and biodegradability improvement of the reactive red 198 azo dye using catalytic ozonation with MgO nanocrystals. *Chemical Engineering Journal, 152*(1), 1–7. https://doi.org/10.1016/j.cej.2009.03.014

Moussavi, G., Aghapour, A. A., & Yaghmaeian, K. (2014). The degradation and mineralization of catechol using ozonation catalyzed with MgO/GAC composite in a fluidized bed reactor. *Chemical Engineering Journal, 249*, 302–310. https://doi.org/10.1016/j.cej.2014.03.059

Moussavi, G., Mashayekh-Salehi, A., Yaghmaeian, K., & Mohseni-Bandpei, A. (2018). The catalytic destruction of antibiotic tetracycline by sulfur-doped manganese oxide (S–MgO) nanoparticles. *Journal of Environmental Management, 210*, 131–138. https://doi.org/10.1016/j.jenvman.2018.01.004

Moussavi, G., Yazdanbakhsh, A., & Heidarizad, M. (2009). The removal of formaldehyde from concentrated synthetic wastewater using O_3/MgO/H_2O_2 process integrated with the biological treatment. *Journal of Hazardous Materials, 171*(1–3), 907–913. https://doi.org/10.1016/j.jhazmat.2009.06.090

Najafi, A. (2017). A novel synthesis method of hierarchical mesoporous MgO nanoflakes employing carbon nanoparticles as the hard templates for photocatalytic degradation. *Ceramics International, 43*(7), 5813–5818. https://doi.org/10.1016/j.ceramint.2017.01.135

Naldoni, A., Altomare, M., Zoppellaro, G., Liu, N., Kment, S., Zbořil, R., & Schmuki, P. (2018). Photocatalysis with reduced TiO_2: From black TiO_2 to cocatalyst-free hydrogen production. *ACS Catalysis, 9*(1), 345–364. https://doi.org/10.1021/acscatal.8b04068

Olanders, B., & Stroemberg, D. (1995). Reduction of nitric oxide over magnesium oxide and dolomite at fluidized bed conditions. *Energy & Fuels, 9*(4), 680–684. https://doi.org/10.1021/ef00052a016

Pacchioni, G., & Freund, H. (2013). Electron transfer at oxide surfaces. The MgO paradigm: From defects to ultrathin films. *Chemical Reviews, 113*(6), 4035–4072. https://doi.org/10.1021/cr3002017

Park, H., Park, Y., Kim, W., & Choi, W. (2013). Surface modification of TiO_2 photocatalyst for environmental applications. *Journal of Photochemistry and Photobiology c: Photochemistry Reviews, 15*, 1–20. https://doi.org/10.1016/j.jphotochemrev.2012.10.001

Pathak, N., Ghosh, P. S., Gupta, S. K., Kadam, R. M., & Arya, A. (2016). Defects induced changes in the electronic structures of MgO and their correlation with the optical properties: A special case of electron–hole recombination from the conduction band. *RSC Advances, 6*(98), 96398–96415. https://doi.org/10.1039/C6RA21065A

Peng, Q., Tang, X., Liu, K., Luo, X., He, D., Dai, Y., & Huang, G. (2020). High-efficiency catalysis of peroxymonosulfate by MgO for the degradation of organic pollutants. *Minerals, 10*(1), 2. https://doi.org/10.3390/min10010002

Piera, E., Calpe, J. C., Brillas, E., Domènech, X., & Peral, J. (2000). 2,4-Dichlorophenoxyacetic acid degradation by catalyzed ozonation: TiO_2/UVA/O_3 and Fe(II)/UVA/O_3 systems. *Applied Catalysis b: Environmental, 27*(3), 169–177. https://doi.org/10.1016/S0926-3373(00)00149-1

Pilarska, A. A., Klapiszewski, Ł, & Jesionowski, T. (2017). Recent development in the synthesis, modification and application of $Mg(OH)_2$ and MgO: A review. *Powder Technology, 319*, 373–407. https://doi.org/10.1016/j.powtec.2017.07.009

Possato, L. G., Pereira, E., Gonçalves, R. G., Pulcinelli, S. H., Martins, L., & Santilli, C. V. (2020). Controlling the porosity and crystallinity of MgO catalysts by addition of surfactant in the sol-gel synthesis. *Catalysis Today, 344*, 52–58. https://doi.org/10.1016/j.cattod.2018.10.027

Purwajanti, S., Zhou, L., Ahmad Nor, Y., Zhang, J., Zhang, H., Huang, X., & Yu, C. (2015). Synthesis of magnesium oxide hierarchical microspheres: A dual-functional material for water remediation. *ACS Applied Materials & Interfaces, 7*(38), 21278–21286. https://doi.org/10.1021/acsami.5b05553

Radwan, E., Yu, L., Achari, G., & Langford, C. H. (2016). Photocatalytic ozonation of pesticides in a fixed bed flow through UVA-LED photoreactor. *Environmental Science and Pollution Research, 23*(21), 21313–21318. https://doi.org/10.1007/s11356-016-7346-1

Raj, B. R., Umadevi, M., & Parimaladevi, R. (2019). Enhanced photocatalytic degradation of textile dyeing wastewater under UV and visible light using ZnO/MgO nanocomposites as a novel photocatalyst. *Particulate Science and Technology, 38*(7), 812–820.

Ramacharyulu, P. V. R. K., Abbas, S. J., Sahoo, S. R., & Ke, S. C. (2018). Mechanistic insights into 4-nitrophenol degradation and benzyl alcohol oxidation pathways over MgO/gC_3N_4 model catalyst systems. *Catalysis Science & Technology, 8*(11), 2825–2834. https://doi.org/10.1039/C8CY00431E

Rao, Y., Wang, W., Tan, F., Cai, Y., Lu, J., & Qiao, X. (2013). Influence of different ions doping on the antibacterial properties of MgO nanopowders. *Applied Surface Science, 284*, 726–731. https://doi.org/10.1016/j.apsusc.2013.08.001

Ratpukdi, T., Siripattanakul, S., & Khan, E. (2010). Mineralization and biodegradability enhancement of natural organic matter by ozone–VUV in comparison with ozone, VUV, ozone–UV, and UV: Effects of pH and ozone dose. *Water Research, 44*(11), 3531–3543. https://doi.org/10.1016/j.watres.2010.03.034

Ruiz Puigdollers, A., Schlexer, P., Tosoni, S., & Pacchioni, G. (2017). Increasing oxide reducibility: The role of metal/oxide interfaces in the formation of oxygen vacancies. *ACS Catalysis, 7*(10), 6493–6513. https://doi.org/10.1021/acscatal.7b01913

Sawai, J., Kojima, H., Igarashi, H., Hashimoto, A., Shoji, S., Sawaki, T., & Shimizu, M. (2000). Antibacterial characteristics of magnesium oxide powder. *World Journal of Microbiology and Biotechnology, 16*(2), 187–194. https://doi.org/10.1023/A:1008916209784

Shen, T., Wang, Q., & Tong, S. (2017). Solid base MgO/ceramic honeycomb catalytic ozonation of acetic acid in water. *Industrial & Engineering Chemistry Research, 56*(39), 10965–10971. https://doi.org/10.1021/acs.iecr.7b02469

Sterrer, M., Berger, T., Diwald, O., & Knözinger, E. (2003). Energy transfer on the MgO surface, monitored by UV-induced H_2 chemisorption. *Journal of the American Chemical Society, 125*(1), 195–199. https://doi.org/10.1021/ja028059o

Sun, M., Lei, Y., Cheng, H., Ma, J., Qin, Y., Kong, Y., & Komarneni, S. (2020). Mg doped CuO–Fe_2O_3 composites activated by persulfate as highly active heterogeneous catalysts for the degradation of organic pollutants. *Journal of Alloys and Compounds, 825*, 154036. https://doi.org/10.1016/j.jallcom.2020.154036

Tian, Q., Ye, J., Yuan, W., Zhang, S., Shi, L., Zhong, J., & Ning, G. (2020). Highly effective antibacterial activity of lithium-doped magnesium oxide particles synthesized by the microwave-assisted hydrothermal route. *Powder Technology, 371*, 130–141. https://doi.org/10.1016/j.powtec.2020.05.031

Vaddadi, L. P., Avisar, D., Vadivel, V. K., Menashe, O., Kurzbaum, E., Cohen-Yaniv, V., & Mamane, H. (2020). LP-UV-Nano MgO_2 pretreated catalysis followed by small bioreactor platform capsules treatment for superior kinetic degradation performance of 17α-ethynylestradiol. *Materials, 13*(1), 83. https://doi.org/10.3390/ma13010083

Vaizogullar, A. I. (2018a). Heterostructured MgO/ZnO photocatalysts: Synthesis, characterization and UV light-induced photocatalytic activity using model pollutant 2,6-dichlorophenol. *Kinetics and Catalysis, 59*(4), 418–427. https://doi.org/10.1134/S0023158418040146

Vaizogullar, A. İ. (2018b). Needle-like La-doped MgO photocatalyst: Synthesis, characterization and photodegradation of flumequine antibiotic under UV irradiation. *Journal of Electronic Materials, 47*(11), 6751–6758. https://doi.org/10.1007/s11664-018-6591-0

Wan, S., Lin, J., Tao, W., Yang, Y., Li, Y., & He, F. (2019). Enhanced fluoride removal from water by nanoporous biochar-supported magnesium oxide. *Industrial & Engineering Chemistry Research, 58*(23), 9988–9996. https://doi.org/10.1021/acs.iecr.9b01368

Wang, B., Xiong, X., Ren, H., & Huang, Z. (2017a). Preparation of MgO nanocrystals and catalytic mechanism on phenol ozonation. *RSC Advances, 7*(69), 43464–43473. https://doi.org/10.1039/C7RA07553G

Wang, H., Li, G., & Fakhri, A. (2020). Fabrication and structural of the Ag_2S-MgO/graphene oxide nanocomposites with high photocatalysis and antimicrobial activities. *Journal of Photochemistry and Photobiology B: Biology*, 111882. https://doi.org/10.1016/j.jphotobiol.2020.111882

Wang, H., Zheng, W., Li, W., Tian, F., Kuang, S., Bu, Y., & Ao, J. P. (2017b). Control the energy band potential of ZnMgO solid solution with enhanced photocatalytic hydrogen evolution capacity. *Applied Catalysis b: Environmental, 217*, 523–529. https://doi.org/10.1016/j.apcatb.2017.04.079

Wu, D., Bai, Y., Wang, W., Xia, H., Tan, F., Zhang, S., & Wong, P. K. (2019a). Highly pure MgO_2 nanoparticles as robust solid oxidant for enhanced Fenton-like degradation of organic contaminants. *Journal of Hazardous Materials, 374*, 319–328. https://doi.org/10.1016/j.jhazmat.2019.04.058

Wu, Y., Liu, L. M., An, X., & Wei, T. (2019b). New insights into interfacial photocharge transfer in TiO_2/C_3N_4 heterostructures: Effects of facets and defects. *New Journal of Chemistry, 43*(11), 4511–4517. https://doi.org/10.1039/C9NJ00027E

Xiang, X., Xie, L., Li, Z., & Li, F. (2013). Ternary MgO/ZnO/In_2O_3 heterostructured photocatalysts derived from a layered precursor and visible-light-induced photocatalytic activity. *Chemical Engineering Journal, 221*, 222–229. https://doi.org/10.1016/j.cej.2013.02.030

Xiao, J., Rabeah, J., Yang, J., Xie, Y., Cao, H., & Brückner, A. (2017). Fast electron transfer and ·OH formation: Key features for high activity in visible-light-driven ozonation with C_3N_4 catalysts. *ACS Catalysis, 7*(9), 6198–6206. https://doi.org/10.1021/acscatal.7b02180

Xu, Y., & Schoonen, M. A. (2000). The absolute energy positions of conduction and valence bands of selected semiconducting minerals. *American Mineralogist, 85*(3–4), 543–556. https://doi.org/10.2138/am-2000-0416

Yao, Y., Gao, B., Inyang, M., Zimmerman, A. R., Cao, X., Pullammanappallil, P., & Yang, L. (2011). Removal of phosphate from aqueous solution by biochar derived from anaerobically digested sugar beet tailings. *Journal of Hazardous Materials, 190*(1–3), 501–507. https://doi.org/10.1016/j.jhazmat.2011.03.083

Yu, G., Wang, Y., Cao, H., Zhao, H., & Xie, Y. (2020). Reactive oxygen species and catalytic active sites in heterogeneous catalytic ozonation for water purification. *Environmental Science & Technology, 54*(10), 5931–5946. https://doi.org/10.1021/acs.est.0c00575

Zeng, W., Gao, M., Liu, K., Li, C., Cao, N., Zhao, X., & Wei, T. (2021). Boosting charge separation and surface defects for superb photocatalytic activity of magnesium oxide/graphene nanosheets. *Applied Surface Science, 535*, 147658. https://doi.org/10.1016/j.apsusc.2020.147658

Zeng, W., Yin, Z., Gao, M., Wang, X., Feng, J., Ren, Y., Wei, T., & Fan, Z. (2020). In-situ growth of magnesium peroxide on the edge of magnesium oxide nanosheets: Ultrahigh photocatalytic efficiency based on synergistic catalysis. *Journal of Colloid and Interface Science, 561*, 257–264. https://doi.org/10.1016/j.jcis.2019.11.122

Zhang, X., Wang, W., Zhang, Y., Zeng, T., Jia, C., & Chang, L. (2018). Loading Cu-doped magnesium oxide onto surface of magnetic nanoparticles to prepare magnetic disinfectant with enhanced antibacterial activity. *Colloids and Surfaces b: Biointerfaces, 161*, 433–441. https://doi.org/10.1016/j.colsurfb.2017.11.010

Zhao, L., Sun, Z., & Ma, J. (2009). Novel relationship between hydroxyl radical initiation and surface group of ceramic honeycomb supported metals for the catalytic ozonation of nitrobenzene in aqueous solution. *Environmental Science & Technology, 43*(11), 4157–4416. https://doi.org/10.1021/es900084w

Zheng, Y., Cao, L., Xing, G., Bai, Z., Huang, J., & Zhang, Z. (2019). Microscale flower-like magnesium oxide for highly efficient photocatalytic degradation of organic dyes in aqueous solution. *RSC Advances, 9*(13), 7338–7348. https://doi.org/10.1039/C8RA10385B

Zhu, X., Wu, D., Wang, W., Tan, F., Wong, P. K., Wang, X., & Qiao, X. (2016). Highly effective antibacterial activity and synergistic effect of Ag-MgO nanocomposite against Escherichia coli. *Journal of Alloys and Compounds, 684*, 282–290. https://doi.org/10.1016/j.jallcom.2016.05.179

Nanobiochar and Biochar Nanocomposites in Wastewater Treatment

Sohom Chatterjee and Remya Neelancherry

1 Introduction

Biochar is a carbon-rich material produced from biomass such as wood, agricultural residues, organic waste, solid waste or animal litters through pyrolysis (heating in the absence of oxygen) (Chausali et al., 2021) Being a stable form of carbon, biochar persists for hundreds to thousands of years in soil and aids in sequestration of carbon. Biochar is highly porous, with a large surface area because of the internal pore structure formed during pyrolysis. This porosity enables biochar to absorb and retain water, nutrients and various organic and inorganic compounds. Biochar is relatively stable and resistant to decomposition compared to its raw biomass precursor. Its stability allows for long-term sequestration of carbon when incorporated into soils, contributing to mitigation efforts with regard to climate change. The carbon content of biochar typically ranges from 50 to 90% by weight, depending on the feedstock and pyrolysis conditions. This high carbon content helps to sequester carbon in soils. Due to its porous structure, biochar has high affinity for retaining nutrients such as nitrogen, phosphorus, potassium and various micronutrients. This property of biochar can enhance soil fertility and nutrient supply for plant growth. Biochar can act as a buffer, helping to stabilise soil pH levels over time. This can be particularly beneficial in acidic soils by raising pH towards neutral levels. Biochar provides a habitat for soil microorganisms because of its porous structure, which can enhance soil microbial activity and diversity. This microbial activity can contribute to improved soil health and nutrient cycling. The porosity of biochar allows it to effectively retain water within soil pores, increasing the water-holding capacity of soil and reducing water loss through leaching and evaporation. This property can be especially valuable in drought-prone regions or sandy soils with poor water retention. The production

S. Chatterjee · R. Neelancherry (✉)
School of Infrastructure, Indian Institute of Technology Bhubaneswar, Bhubaneswar, Odisha, India
e-mail: remya@iitbbs.ac.in

© Capital Publishing Company, New Dehli, India. 2025

R. Neelancherry and S. M. Kumar, *Nanotechnology in Wastewater Treatment*,
https://doi.org/10.1007/978-3-031-94246-4_10

of biochar through pyrolysis can result in reduced emissions of greenhouse gases compared to the direct combustion of biomass. By converting biomass into a stable form of carbon, production of biochar can mitigate emissions of carbon dioxide and other greenhouse gases. Biochar can improve soil structure and tilth, leading to enhanced soil aeration, drainage and root penetration. Its incorporation into soil can also help remediate degraded soils, reduce soil erosion and promote overall soil health and productivity. Biochar is applied in agriculture (as a soil amendment), environmental remediation (such as in wastewater treatment and contaminated soil remediation) and carbon sequestration initiatives (as a means to mitigate climate change).

Different variants of biochar production systems include top-lit updraft (TLUD) gasification, torrefaction, slow pyrolysis, fast pyrolysis, microwave pyrolysis and hydrothermal carbonisation.

(i) Top-lit updraft (TLUD) gasifiers are simple, low-cost devices that can be used to produce biochar at small to medium scales. Biomass is loaded into a vertical chamber, and combustion is initiated at the top. As the fire burns downward, pyrolysis of the biomass occurs below the flame front, producing biochar.
(ii) Torrefaction is carried out at a low temperature of ~200–300 °C at a low heating rate of ≥60 °C /min for hours to days in the absence of oxygen (Chen et al., 2014). The volatiles and moisture content feedstock are released at a slow rate, resulting in the rigid structure of carbonaceous biochar.
(iii) Slow pyrolysis reactors operate at lower temperatures (400–700 °C) and longer residence times compared to gasifiers. These reactors can be batch-wise, semi-continuous or continuous in operation and are designed to optimise the yield and quality of biochar.
(iv) Fast pyrolysis reactors operate at higher temperatures (500–800 °C) and shorter residence times, resulting in the production of bio-oil as the primary product, along with biochar and syngas (a mixture of hydrogen and carbon monoxide). Biochar produced by fast pyrolysis has different properties from that produced by slow pyrolysis.
(v) Microwave pyrolysis is an emerging technology that uses microwave radiation to rapidly and uniformly heat biomass, leading to efficient pyrolysis reactions. This method consumes less energy and has faster processing time.
(vi) Hydrothermal carbonisation (HTC) involves the conversion of biomass into biochar under high temperature and pressure in the presence of water. This process can be particularly useful for wet or low-energy-density feedstocks, such as sludge or algae.

Among the various methods of biochar production, pyrolysis is commonly used. Particle size of biochar varies from micrometres to centimetres based on the type of feedstock, pyrolysis and operating conditions (Narde & Remya, 2022). Biochar with particle size ranging from 0.04 to 20 mm is generally utilised for agricultural purposes and soil conditioning. Decrease in the particle size of biochar to 10–600 μm improved surface-active sites for adsorption and subsequent increase in adsorption capacity (Naghdi et al., 2019). Further reduction in the particle size of biochar to the nanoscale (100 nm or smaller) resulted in greater porosity, higher surface to volume ratio, improved specific surface area, increased surface energy, improved adsorption and biological effectiveness (Ramanayaka et al., 2020a). Nanobiochar is gaining interest these days, as it amalgamates the advantages of biochar with nanomaterial.

2 Methods of Nanobiochar Production

Nanobiochar is produced from bulk biochar through physical (Fig. 1) or chemical modification (Fig. 2). Physical modification includes milling, grinding, sonication and flash heating (Lyu et al., 2018) while chemical modification comprises steam activation, CO_2 activation, chemical oxidation and acid/base treatment (Neha et al., 2022).

The ball milling technique reduces bulk biochar to nanobiochar by causing it to collide with metallic balls. Ball to biochar ratio, milling speed and milling time are regulated to attain the desired particle size of nanobiochar (Yuan et al., 2020). The other factors that help in achieving the desired particle size are controlling aggregation, adjusting the ball-to-sample ratio, varying the milling time and choosing between dry and wet ball milling techniques. Wet ball milling, a less labour-intensive and eco-friendly approach, results in nanobiochar with superior dispersivity and higher surface functionality. Ceramic mortar is used for grinding bulk biochar for the production of nanobiochar (Elbehiry et al., 2022). Sonication employs high-energy ultrasound waves to disintegrate a bulk biochar dispersed in a solvent. The solvent media acts as the carrier for ultrasound, which impinges on particle surfaces by microcavitation which produces short-lived, very high temperature shockwaves (Chaubey et al., 2023). Opening of clogged pores and exfoliation of carbon structure due to shock waves results in uniform biochar with improved porosity (Liu et al., 2018; Yuan et al., 2020). Following the sonication method, larger particles are removed by filtering, gravity settling or centrifugation to obtain a supernatant that contains biochar nanoparticles. Conversely, direct nanobiochar manufacturing by flash heating results in graphitic nanosheets (Ramanayaka et al., 2020b). In steam activation, partial gasification of trapped volatiles from bulk biochar (Eq. (1)), partial devolatilisation, formation of crystalline carbon and increased porosity are attained by steam exposure. Conversely, CO_2 purging results in reaction of CO_2 with amorphous carbon in bulk biochar in the limited oxygen atmosphere to form CO (Eq. (2)). Evolution of CO augments the surface area of biochar because of corrosion and increases pore volume and microporous structure (Bushra & Remya, 2020).

$$C + H_2O \rightarrow CO + H_2 \tag{1}$$

$$C + CO_2 \rightarrow 2CO \tag{2}$$

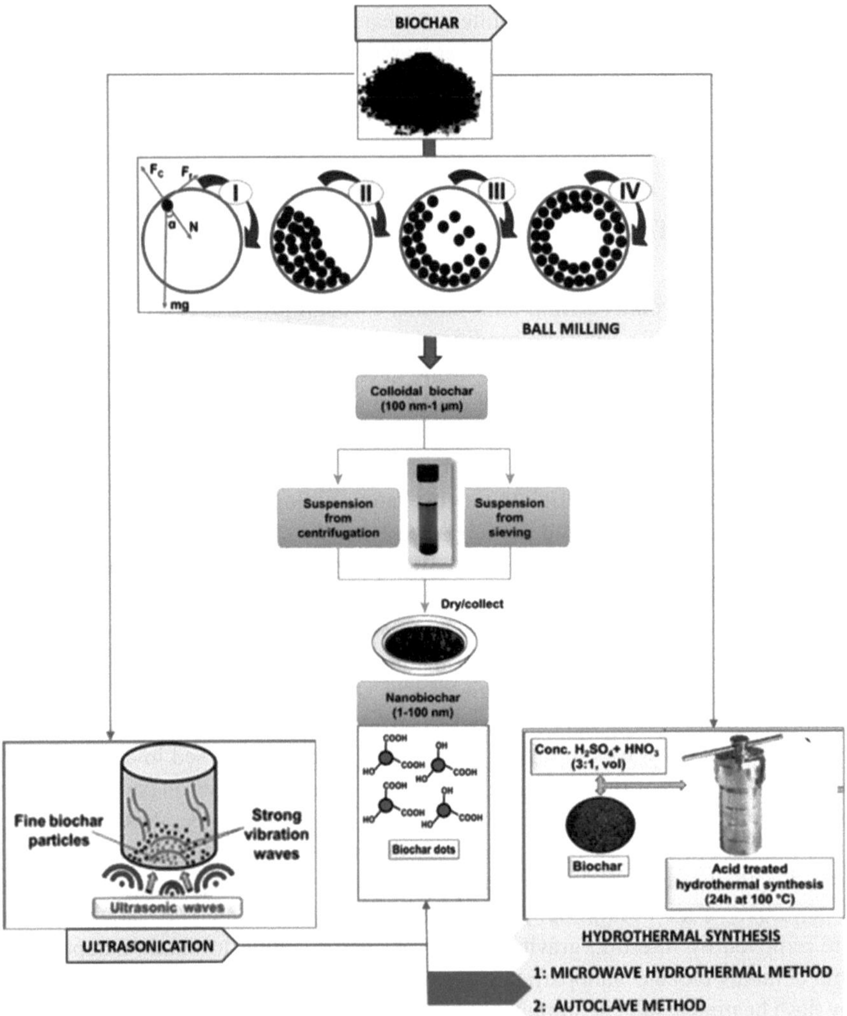

Fig. 1 Production of nanobiochar from bulk biochar through physical modification

Biochar nanodots of $\leq$5 nm were obtained by chemical treatment of bulk biochar using sulphuric acid and nitric acid in a hydrothermal reactor under high pressure followed by filtration with 0.22 μm membrane and drying at 65 °C (Liu et al., 2018). Alkali modification of biochar is performed using NaOH, KOH and NH_4OH for chemical reduction of unpyrolysed organic matter and ash content from the pores of bulk biochar resulting in high fixed carbon content of the biochar. Alkali modification enhances the hydroxyl functional group (OH^-) on biochar surface which facilitates improved hydrophobicity and adsorption of non-polar pollutants (Vijayaraghavan, 2019). Acid modification of biochar employs HCl, HNO_3, H_2O_2 and H_3PO_4 for

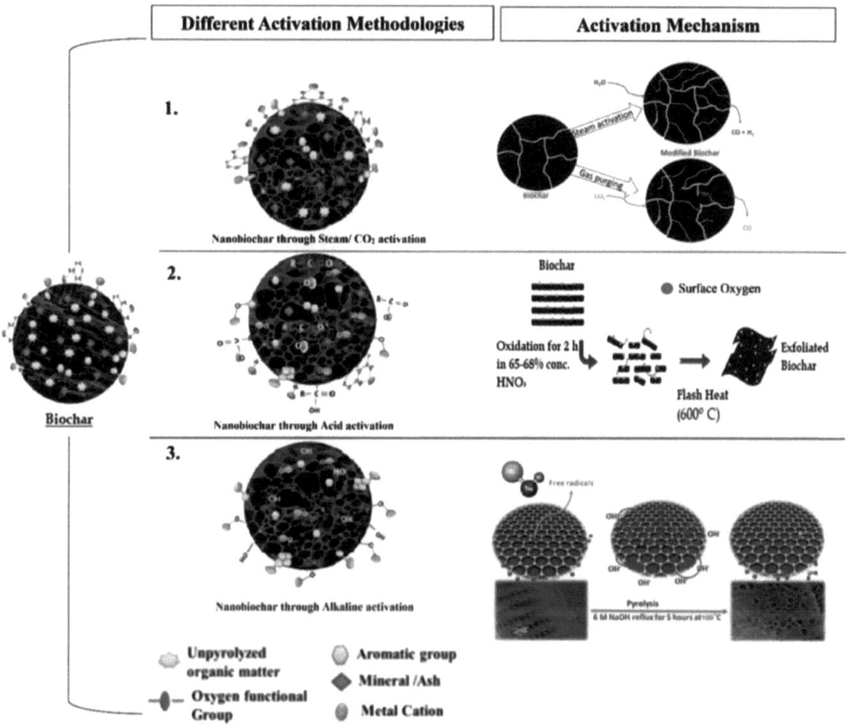

Fig. 2 Production of nanobiochar from bulk biochar through chemical modification

chemical oxidation of volatiles in biochar, thus improving oxygenated functional groups like carboxyl group (−COOH) on biochar surface (Yang et al., 2019). HNO_3 modification resulted in higher functional group on biochar surface compared to other acids (Bushra & Remya, 2020). Acid-modified biochar exhibited improved hydrophilicity and adsorption potential of polar compounds. Acid pretreatment using 1 M HNO_3 at 75 °C for 4 h, pyrolysis and post-treatment by flash heating at 950 °C for 45 s resulted in the production of biochar nanosheet (Genovese et al., 2015).

3 Production of Biochar Nanocomposites

Biochar nanocomposites are synthesised by combining biochar with natural or engineered nanomaterials to create a hybrid material with improved properties (Fig. 3). Incorporation of nanomaterial in bulk biochar and modification of nanobiochar with nanomaterials are performed to produce biochar-supported nanocomposites and biochar nanocomposites, respectively (Neha et al., 2022). In the former, particle size of biochar is in the range of microns, whereas in the latter, particle size of biochar is <100 nm. Biochar nanocomposites leverage the unique properties of both nanomaterials and biochar, such as enhanced reactivity, increased surface area, mechanical strength and conductivity tailored to specific applications. Nanomaterials are incorporated in nanobiochar to obtain oxide/hydroxide biochar nanocomposites, functional biochar nanocomposites, metal biochar nanocomposites and magnetic biochar nanocomposites.

Incorporation of nanomaterial on nanobiochar is performed through pretreatment of feedstock before biochar production or post-treatment after nanobiochar preparation. Pretreatment involves the fortification of target element through bioaccumulation or immersing the feedstock in a solution of target element/compound. Conversely, post-treating involves target nanomaterial impregnation after the preparation of nanobiochar (Chaubey et al., 2023). Biochar nanocomposites infused with

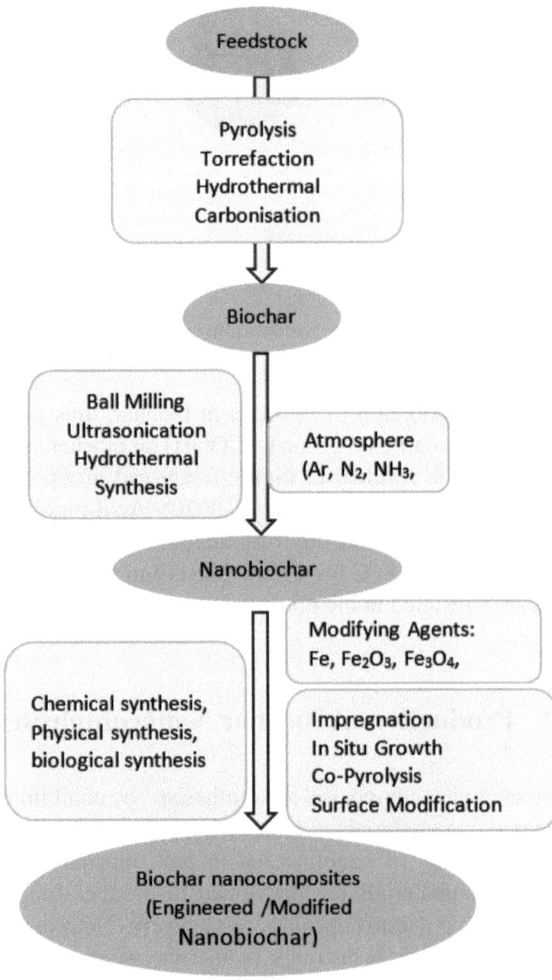

Fig. 3 Production of biochar nanocomposite through pre-/post-treatment of feedstock/biochar

metals, metal oxides and metal hydroxides were created using these methods to improve the removal of contaminant and for various other applications.

Impregnation of nanometal oxide/hydroxide is done to obtain oxide/hydroxide biochar nanocomposites. Nanomaterials like graphene, graphene oxide, carbon nanotubes, chitosan, graphitic carbon nitride, etc. are supported on biochar to obtain functional nanocomposites. Metal-impregnated biochar is produced by coprecipitating metals either on the surface of biomass (pretreatment) or biochar (post-treatment). The resulting biochar nanocomposite exhibits a significantly improved adsorption capacity for pollutants compared to standard biochar because of the presence of two solid phases: the nanomaterial and the biochar matrix (Zhang et al., 2012a). Both phases enhance the overall adsorption capacity of the composites through various mechanisms, including precipitation, hydrogen bonding and ligand exchange (Fig. 4). In addition, impregnation by magnetic metals like iron enhances the separability of powdered biochar after the treatment process through magnetic separation.

4 Application of Nanobiochar and Biochar Nanocomposites in the Treatment of Wastewater

Nanobiochar and biochar nanocomposites are used in the treatment of wastewater for the removal of pollutants through adsorption or the degradation of pollutants through chemical oxidation, photocatalysis, etc. Biochar nanocomposites are employed for the removal of heavy metals (Table 1), pharmaceuticals and personal care products (Table 2), dye (Table 3), nutrients like phosphate and nitrate (Table 4) and phenolic compounds (Table 5).

The removal of lead (Pb) from wastewater has been investigated with several biochar nanocomposites like Fe-loaded biochar (Elbehiry et al., 2022), biochar/graphene oxide composite (Cruz et al., 2020), ZnO–biochar (Cruz et al., 2020), biochar/nano-ZnO (Cruz et al., 2020), biochar/TiO_2 (Mandal et al., 2020), CaO–biochar (Wang et al., 2018a, 2018b) and biochar/CuO (Wang et al., 2018a). Their surface areas range from 250 to 350 m^2/g, with functional groups like –OH, –COOH, C=O and Zn – O, among others. Operating conditions include pH levels between 3 and 10, initial concentrations from 50 to 200 mg/L and temperatures around 25 °C. Removal efficiencies range from 76 to 95%, using mechanisms like ion exchange, complexation, π–π interaction, hydrogen bonding, adsorption, photocatalysis, photodegradation and surface precipitation. The removal of chromium (Cr-VI) from wastewater is investigated by using different nanobiochars like $CoFe_2O_4$–nano biochar (Elbehiry et al., 2022), nano-silica/biochar (Elbehiry et al., 2022), TiO_2–Ag nanocomposite (Ghosh et al., 2020), Fe_2O_3/biochar (Mandal et al., 2020), Fe–biochar (Mandal et al., 2020), CuO–biochar (Zhao et al., 2018) and biochar/nano-MgO (Elbehiry et al., 2022). Their surface areas range from 270 to 340 m^2/g, with functional groups such as Co–O, Fe–O, Si–O and Mg–O. The removal efficiencies

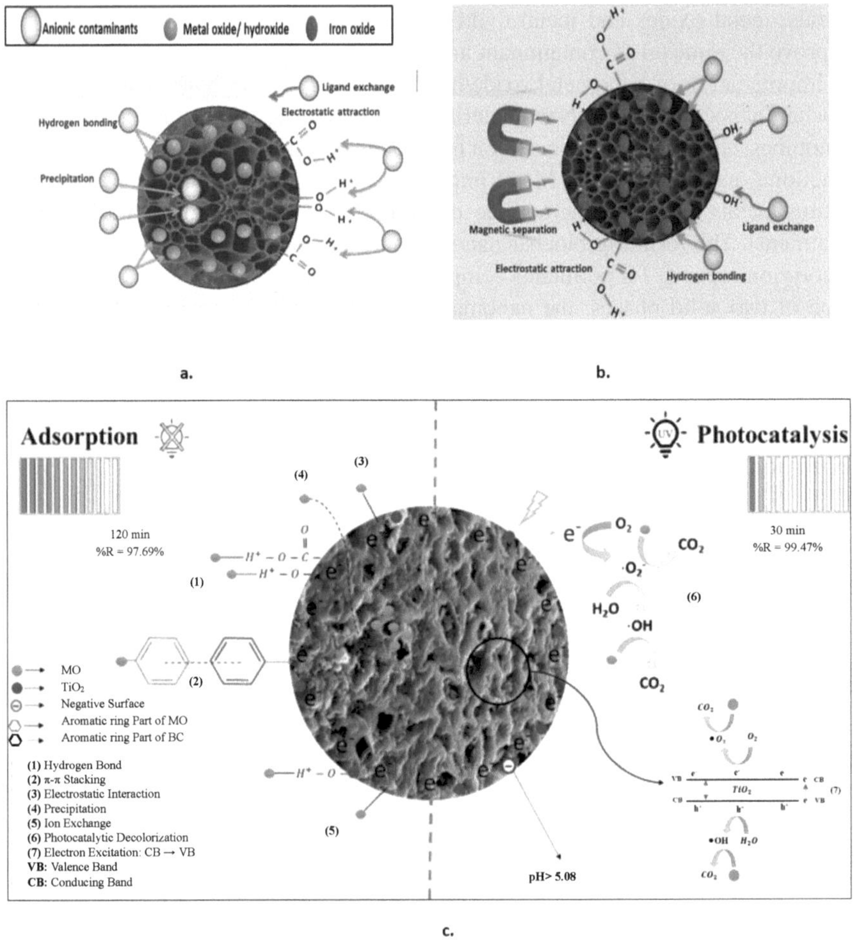

Fig. 4 Mechanism of pollutant removal by nanobiochar composites: **a** oxide/hydroxide biochar nanocomposite, **b** magnetic biochar nanocomposite, **c** functional biochar (TiO_2/ZnO/graphene oxide) nanocomposite

vary between 87 and 95% under different conditions including pH 2–10, 50–100 mg/L initial concentrations and 25 °C temperature, utilising adsorption, reduction, electrostatic attraction and surface complexation. The removal of Cadmium (Cd II) with biochar–ZnS (Elbehiry et al., 2022), $NiFe_2O_4$–biochar (Mandal et al., 2020) and biochar/$MnFe_2O_4$ (Wang et al., 2018b), with surface areas from 270 to 310 m^2/g has been looked into. The functional groups include Zn–S, Ni–O, Fe–O and Mn–O. The removal efficiencies range from 85 to 91% with pH levels of 4 to 9, initial concentrations of 50 mg/L and temperature of 25 °C, using adsorption, ion exchange and magnetic separation. The removal of Arsenic (As V) is studied so far with biochar–CeO_2 (Saffan et al., 2022), biochar/$CuFe_2O_4$ (Saffan et al., 2022), biochar/ZnO (Cruz

Table 1 Heavy metal removal by nanobiochar/biochar nanocomposite

Pollutant	Nanobiochar/ biochar nanocomposite	Surface area (m^2/g)	Functional groups	Biochar dosage (g/L)	Operating conditions (pH, initial concentration, temperature)	Removal efficiency (%)	Removal mechanism	References
Pb(II)	Fe-loaded biochar	250	–OH, –COOH, C=O	0.5	pH 5, 100 mg/L, 25 °C	93	Ion exchange, complexation	Elbehiry et al. (2022)
	Biochar/graphene oxide composite	350	C=O, –OH	0.5	Neutral pH, 200 mg/L	95	7i–7i interaction, hydrogen bonding	Cruz et al. (2020)
	ZnO–biochar	300	Zn–O	0.5	pH 7, 150 mg/L	92	Adsorption, photocatalysis	Cruz et al. (2020)
	Biochar/nano-ZnO	310	Zn–O	0.5	pH 7, 100 mg/L	85	Adsorption, photocatalysis	Cruz et al. (2020)
	Biochar/TiO_2	320	Ti–O	0.4	UV light, pH 7, 50 mg/L	80	Photodegradation	Mandal et al. (2020)
	CaO–biochar	315	Ca–O	0.5	pH 3–10, 50 mg/L, 25 °C	76	Adsorption, surface precipitation	Wang et al. (2018a)
	Biochar/CuO	295	Cu–O	0.3	pH 4–8, 100 mg/L, 25 °C	82	Adsorption, photodegradation	Wang et al. (2018a)
Cr(VI)	$CoFe_2O_4$–nano biochar	280	Co–O, Fe–O	0.3	pH 2–7, 100 mg/L, 25 °C	94	Adsorption, reduction	Elbehiry et al. (2022)
	Nano-silica/ Biochar	340	Si–O	0.3	pH 5–6, 50 mg/L, 25 °C	95	Adsorption, surface complexation	Elbehiry et al. (2022)

(continued)

Table 1 (continued)

Pollutant	Nanobiochar/ biochar nanocomposite	Surface area (m^2/g)	Functional groups	Biochar dosage (g/L)	Operating conditions (pH, initial concentration, temperature)	Removal efficiency (%)	Removal mechanism	References
	NM	300	Zn–O, Mo–O	0.2	pH 2–7, 50 mg/L, 25 °C	90	Electrostatic attraction, reduction	Ghosh et al. (2020)
	Fe_2O_3/biochar	330	Fe–O	0.4	pH 3–10, 50 mg/L, 25 °C	90	Adsorption, reduction	Mandal et al. (2020)
	Fe–biochar	300	Fe–O	0.4	pH 2–6, 50 mg/L, 25 °C	94	Adsorption, reduction	Mandal et al. (2020)
	CuO–biochar	290	Cu–O	0.3	pH 5–7, 50 mg/L, 25 °C	88	Adsorption, degradation	Zhao et al. (2018)
	Biochar/ nano-MgO	320	Mg–O	0.4	pH 3–9, 100 mg/L, 25 °C	87	Adsorption, surface precipitation	Elbehiry et al. (2022)
Cd(II)	Biochar–ZnS	300	Zn–S	0.4	pH 5–8, 50 mg/L, 25 °C	85	Adsorption, ion exchange	Elbehiry et al. (2022)
	$NiFe_2O_4$–biochar	270	Ni–O, Fe–O	0.4	pH 4–9, 50 mg/L, 25 °C	90	Adsorption, magnetic separation	Mandal et al. (2020)
	Biochar/MnFeO	310	Mn–O, Fe–O	0.4	pH 4–9, 50 mg/L, 25 °C	91	Adsorption, magnetic separation	Wang et al. (2018b)
As(V)	Biochar–CeO	310	Ce–O	0.5	pH 6–8, 50 mg/L, 25 °C	83	Adsorption, ion exchange	Saffan et al. (2022)

(continued)

Table 1 (continued)

Pollutant	Nanobiochar/biochar nanocomposite	Surface area (m^2/g)	Functional groups	Biochar dosage (g/L)	Operating conditions (pH, initial concentration, temperature)	Removal efficiency (%)	Removal mechanism	References
	Biochar/$CuFe_2O_4$	315	Cu–O, Fe–O	0.5	pH 3–7, 50 mg/L, 25 °C	93	Adsorption, magnetic separation	Saffan et al. (2022)
	Biochar/ZnO	310	Zn–O	0.5	pH 7, 100 mg/L	87	Adsorption, photocatalysis	Cruz et al. (2020)
	Fe_3O_4/biochar composite	320	Fe–O	0.6	pH 3–7, 50 mg/L, 25 °C	99	Adsorption, magnetic separation	Zhong et al. (2018)
Hg(II)	Biochar/FeS	325	Fe–S	0.3	pH 3–7, 50 mg/L, 25 °C	97	Adsorption, reduction	Mandal et al. (2020)
	Biochar/Ag nanoparticles	300	Ag–O	0.4	Neutral pH, 100 mg/L	90	Adsorption, reduction	Mandal et al. (2020)
U(VI)	MnO–biochar	330	Mn–O	0.4	pH 3–6, 100 mg/L, 25 °C	90.4	Adsorption, ion exchange	Mandal et al. (2020)

Table 2 Nanobiochar/biochar nanocomposite for degradation of pharmaceutical and personal care product

Pollutant	Biochar/biochar nanocomposite	Biochar dosage (g/L)	Operating conditions (initial concentration, pH, temperature)	Removal/ degradation efficiency (%)	Removal mechanism	References
Sulfamethoxazole (SUL)	Biochar/MgO nanocomposite	2	50 mg/L, pH 5.0, 25 °C,	90	Adsorption, precipitation	Azzam et al. (2023)
Triclosan (TRI)	Biochar/MgO nanocomposite	2	10 mg/L, pH 6.0, 25 °C	85	Adsorption, precipitation	Azzam et al. (2023)
Ibuprofen (IBU)	$MnFe_2O_4$/biochar nanocomposite			91	Adsorption, complexation	Mansouri et al. (2021)
Acetaminophen (ACE)	$MnFe_2O_4$/biochar nanocomposite	0.5	10 mg/L, pH 7.0, 25 °C	88	Adsorption, complexation	Mansouri et al. (2021)
Carbamazepine (CAR)	$MnFe_2O_4$/biochar nanocomposite			85	Adsorption, complexation	Mansouri et al. (2021)
Tetracycline (TET)	Magnetic biochar ($NiFe_2O_4$/biochar)	1	20 mg/L, pH 6.5, 25 °C	89	Adsorption, Magnetic separation	Wang et al. (2021)
	Ag/biochar	–	pH 5, 100 mg/L, temp 25 °C	90	Adsorption, antimicrobial activity	Zhao et al. (2021)
Diclofenac (DIC)	Magnetic biochar ($NiFe_2O_4$/biochar)	1	15 mg/L, pH 7.0, 25 °C	87	Adsorption, magnetic separation	Wang et al. (2021)
Amoxicillin (AMO)	Biochar/graphene oxide nanocomposite	0.75	30 mg/L, pH 5.5, 30 °C	93	Adsorption, electrostatic attraction	Mansouri et al. (2021)

(continued)

Table 2 (continued)

Pollutant	Biochar/biochar nanocomposite	Biochar dosage (g/L)	Operating conditions (initial concentration, pH, temperature)	Removal/ degradation efficiency (%)	Removal mechanism	References
Metformin (MET)	Biochar/graphene oxide nanocomposite	0.75	20 mg/L, pH 6.0, temp 30 °C	93	Adsorption, electrostatic attraction	Mansouri et al. (2021)
Ciprofloxacin (CIP)	Fe_3O_4/biochar nanocomposite	3–4	pH 6.0, 20 mg/L, 25 °C	89	Adsorption, electrostatic attraction	Mansouri et al. (2021)
Norfloxacin (NOR)	Fe_3O_4/biochar nanocomposite	3–4	pH 6.0, 20 mg/L, temp 25 °C	95.2	Adsorption, magnetic separation	Zhang et al. (2018)
Ofloxacin (OFL)	Fe_3O_4/biochar nanocomposite	3–4	pH 6.0, 20 mg/L, temp 25 °C	92.1	Adsorption, magnetic separation	Zhang et al. (2018)

Table 3 Dye removal by nanobiochar/biochar nanocomposite

Pollutant	Nanobiochar/ biochar nanocomposite	Surface area (m^2/g)	Functional group	Biochar dosage (g/L)	Operating conditions (initial concentration, pH, temperature)	Removal efficiency	Removal mechanism	References
Methylene Blue	Ag–nanoparticle biochar (Ag–nBC)	412	OH, COOH, C=O	0.5	100 mg/L, pH 7, 25 °C	96.8	Adsorption through electrostatic attraction and interactions	Shaikh et al. (2022)
	Ball-milled biochar (BM-BC)	350	C=O, COOH	2	200 mg/L, pH 6, 20 °C	85.7	Adsorption through surface complexation and pore-filling mechanism	Goswami et al. (2022)
Congo Red	Magnetic biochar/ Fe_3O_4 (MBC–Fe_3O_4)	312	Fe–O, COOH	1	50 mg/L, pH 3, 30 °C	89	Adsorption via magnetic separation and electrostatic interactions	Goswami et al. (2022)
Rhodamine B	ZnO/biochar nanocomposite (ZnO–BC)	275	Zn–O, OH	0.3	150 mg/L, pH 5, 25 °C	93	Photocatalytic degradation and adsorption	Perveen et al. (2022b)
Tartrazine	Zirconium ferrite biochar (BC–$ZrFe_2O_s$)	430	Fe–O, Zr–O, OH	0.5	100 mg/L, pH 2, 30 °C	89.2	Chemisorption involving complexation with hydroxyl, Fe and Zr groups	Perveen et al. (2022a)

Table 4 Phosphate and nitrate removal by nanobiochar/biochar nanocomposite

Pollutant	Biochar/ biochar nanocomposite	Surface area (m^2/g)	Functional groups	Biochar dosage (g/L)	Operating conditions (initial concentration, pH, temperature)	Removal efficiency (%)	Removal mechanism	References
Phosphate	Biochar–MgAl LDH nanocomposite	132.5	–OH COs^{2-}, Al–O, Mg–O	1	10 mg/L [for kinetic study, three different initial concentrations considered: 10, 30 and 50 mg/L], pH 7, 25 °C	90.6	Ion exchange, electrostatic attraction	Alagha et al. (2020)
	MgO–biochar nanocomposite	289.3	–OH, Mg–O, C=O	0.5	20 mg/L, pH 6.5, 25 °C	92.1	Chemisorption, precipitation	Zhang et al. (2012b)
Nitrate	Biochar–MgAl LDH nanocomposite	132.5	–OH, COs^{2-}, Al–O, Mg–O	1	10 mg/L [for kinetic study, three different initial concentrations considered: 10, 30 and 50 mg/L], pH 7, 25 °C	75.3	Ion exchange, electrostatic attraction	Alagha et al. (2020)
	MgO–biochar nanocomposite	289.3	–OH, Mg–O, C=O	0.5	20 mg/L, pH 6.5, 25 °C	85.5	Chemisorption, precipitation	Zhang et al. (2012b)

Table 5 Phenolic compound removal/degradation by nanobiochar/biochar nanocomposite

Pollutant	Biochar/biochar nanocomposite	Surface area (m^2/g)	Functional group	Biochar dosage (g/L)	Operating conditions (pH, initial concentration, temperature)	Removal efficiency (%)	Removal mechanism	References
Phenol	Biochar/TiO_2 nanocomposite	–	Hydroxyl	0.5	pH 7, 10 mg/L, 30 °C	80	Photocatalytic degradation and adsorption	Turco et al. (2018)
	Fe–Mn oxide modified biochar	450	Hydroxyl, carboxyl	0.1	pH 6–7, 100 mg/L, 25 °C	94	Adsorption and catalytic oxidation	Ramírez et al. (2017)
	Biochar-based geopolymer nanocomposite	225	Silanol, aluminol	0.2	pH 6, 50 mg/L, 25 °C	85	Adsorption and pore filling	Turco et al. (2018)
2-Chlorophenol	Activated carbon/biochar nanocomposite	980	Hydroxyl, carboxyl	0.05	pH 5–6, 50 mg/L, 25 °C	96	Adsorption through pore filling and interactions	Ramírez et al. (2017)

et al., 2020) and Fe_3O_4 /biochar (Zhong et al., 2018) composites, having surface areas of 310–320 m^2/g. Their functional groups include Ce–O, Cu–O, Zn–O and Fe–O. Removal efficiencies range from 83 to 99%, with pH levels of 3–8, initial concentrations of 50–100 mg/L and temperatures around 25 °C, employing adsorption, ion exchange, photocatalysis and magnetic separation.

The removal of Mercury (Hg II) by biochar/FeS (Mandal et al., 2020) and biochar/Ag (Mandal et al., 2020) nanoparticles, having surface areas of around 300–325 m^2/g, is being investigated till date. The functional groups are Fe–S and Ag–O. Removal efficiencies are 90 and 97% under conditions including pH 3–7, 50–100 mg/L initial concentrations and 25 °C, using adsorption and reduction mechanisms. The removal of Uranium (U VI) is carried out using MnO_2–biochar (Mandal et al., 2020), which has a surface area of 330 m^2/g with Mn–O functional groups. The removal efficiency is 90.4% at pH 3–6, 100 mg/L initial concentration and 25 °C, employing adsorption and ion exchange mechanisms.

The removal of various pharmaceutical and personal care products from wastewater is investigated with the removal of biochar nanocomposites like sulphamethoxazole (SUL) being achieved using biochar/MgO nanocomposite with 90% efficiency (Azzam et al., 2023). Triclosan (TRI) is removed with 85% efficiency using the same composite (Azzam et al., 2023). Ibuprofen (IBU) and acetaminophen (ACE) are removed using $MnFe_2O_4$/biochar nanocomposite with efficiencies of 91% and 88%, respectively (Mansouri et al., 2021). The removal of Carbamazepine (CAR) achieves 85% efficiency with $MnFe_2O_4$/biochar (Mansouri et al., 2021). Tetracycline (TET) and diclofenac (DIC) are removed with 89% and 87% efficiency, respectively, using magnetic biochar (Ni Fe_2O_4/biochar) (Wang et al., 2021). Amoxicillin (AMO) and metformin (MET) are removed with 93% and 89% efficiency, respectively, using biochar/graphene oxide nanocomposite (Mansouri et al., 2021). Ciprofloxacin (CIP), Norfloxacin (NOR) and Ofloxacin (OFL) achieved removal efficiencies of 95.2, 92.1 and 88.7% using the Fe_3O_4/biochar nanocomposite (Zhang et al., 2018).

Removal of dye from wastewater is achieved for methylene blue with 96.8% efficiency using Ag–nanoparticle biochar (Ag–nBC), which has a surface area of 412 m^2/g and functional groups like –OH, –COOH and C=O (Shaikh et al., 2022). Ball-milled biochar (BM-BC) removes methylene blue with 85.7% efficiency, having a surface area of 350 m^2/g and functional groups like C=O and COOH (Goswami et al., 2022). Congo red is removed with 89% efficiency using magnetic biochar/Fe_3O_4 (MBC–Fe_3O_4), which has a surface area of 312 m^2/g with Fe–O and COOH groups (Goswami et al., 2022). Rhodamine B is removed with 93% efficiency using ZnO/biochar nanocomposite (ZnO–BC) with a surface area of 275 m^2/g and functional groups like Zn–O and OH (Perveen et al., 2022b). Tartrazine is removed with 89.2% efficiency using zirconium ferrite biochar (BC–$ZrFe_2O_5$) with a surface area of 430 m^2/g and functional groups like Fe–O, Zr–O and OH (Perveen et al., 2022a).

Removal of phosphate using biochar–MgAl LDH nanocomposite was done with 90.6% efficiency with a surface area of 132.5 m^2/g and functional groups like OH, CO_3^{2-}, Al–O and Mg–O (Alagha et al., 2020). MgO–biochar nanocomposite removes phosphate with 92.1% efficiency, having a surface area of 289.3 m^2/g and functional groups like OH, Mg–O and C=O (Zhang et al., 2012b). Removal of nitrate using biochar–MgAl LDH nanocomposite achieved 75.3% efficiency (Alagha et al., 2020), while MgO–biochar nanocomposite removes nitrate with 85.5% efficiency under similar conditions (Zhang et al., 2012b).

Removal of phenolic compound was investigated with biochar/TiO_2 nanocomposite, Fe–Mn oxide modified biochar and biochar-based geopolymer nanocomposite, achieving efficiencies of 80%, 94% and 85%, respectively (Ramírez et al., 2017; Turco et al., 2018). Removal of 2-Chlorophenol using activated carbon/biochar nanocomposite achieves 96% efficiency (Ramírez et al., 2017). These materials have various functional groups like hydroxyl, carboxyl, silanol and aluminol, and use mechanisms like adsorption, catalytic oxidation, pore filling and π–π interactions.

5 Mechanism of Pollutant Removal by Nanobiochar and Biochar Nanocomposites

Nanobiochar and biochar nanocomposites are used in the treatment of wastewater for the removal of pollutants. Various mechanisms such as pore filling, precipitation, hydrogen bonding, electrostatic attraction, ligand exchange and catalytic degradation are responsible for pollutant adsorption and degradation in a nanobiochar/biochar nanocomposite system (Fig. 3).

6 Challenges in the Application of Nanobiochar/Biochar Nanocomposites

Nanobiochar and biochar nanocomposites showed promising results in the treatment of wastewater for the removal of pollutants. Even though biochar is produced from agricultural residues or solid waste, the production process of biochar, conversion to nanobiochar and incorporation of nanomaterials into biochar can increase the cost significantly. Certain biochar nanocomposite syntheses employ potentially toxic elements which can cause secondary pollution upon release into the environment. The potential accumulation of pollutant and its by-products in biochar and indirect release of these pollutants from the nanobiochar/biochar nanocomposites have to be properly addressed to prevent future contamination. Noteworthy, low-temperature pyrolysis can produce highly toxic biochar because of inadequate operating conditions, releasing polycyclic aromatic hydrocarbons (PAHs) and persistent free radicals (PFRs). These can generate reactive oxygen species, posing significant agro-environmental and health risks. Biochar may also release heavy metals, which are harmful to living organisms. Carbon nanoparticles in biochar are difficult to degrade and can cause oxidative stress, mechanical damage and enzyme disruption. In addition, while biochar can enhance plant growth, increasing dry biomass and root/shoot lengths, the accumulation of carbon nanoparticles in soil can alter the balance of toxic metals, enhancing their translocation into plants and potentially disrupting the soil ecosystem (Yang et al., 2022).

Biochar-based catalysts have gained significant attention for their outstanding performance in addressing water and soil pollution. However, their potential environmental impacts are still largely unknown and not well comprehended. Application of biochar in the preparation of biochar-based catalysts involves acids, alkalis and oxidants, which can be a source of pollution. When biochar-based catalysts are used for environmental modification, metal components may be released, causing secondary pollution. Semiconductors like TiO_2, CuO and ZnO are generally toxic, while iron oxides (Fe_3O_4 and Fe_2O_3) are less harmful. For instance, TiO_2 biochar nanocomposite negatively impacts vertebrate health, damaging cardiac and reproductive systems and threatening invertebrate recruitment. Several environmental factors can alter the structure of metal oxide nanoparticles with increasing toxicity. For example, ZnO nanoparticles become more toxic to mammalian cells when transformed into larger, porous particles in the presence of phosphate. So, future research should examine the effects of structural changes in nanoparticles. Optimising the fabrication process of biochar-based materials is essential to enhance their stability and minimise the release of toxic contaminants into the environment. A highlighted risk with biochar-based catalysts is that, it can create more toxic intermediates from pollutants. UV-AOPs often only partially oxidise pollutants into transformation products (TPs), complicating water remediation and threatening health at low concentrations. Generally, degradation reduces toxicity, as seen with Cr (III) being less toxic than Cr (VI) and TC photocatalysis intermediates being less toxic than TC. However, the bio-accumulation of TC intermediates remains a concern (Yang et al., 2022).

References

Alagha, O., Manzar, M. S., Zubair, M., Anil, I., Mu'azu, N. D., & Qureshi, A. (2020). Comparative adsorptive removal of phosphate and nitrate from wastewater using biochar-MgAl LDH nanocomposites: Coexisting anions effect and mechanistic studies. *Nanomaterials, 10*(2), 336. https://doi.org/10.3390/NANO10020336

Azzam, A. B., Tokhy, Y. A., Dars, F. M. E., & Younes, A. A. (2023). Heterogeneous porous biochar-supported nano $NiFe_2O_4$ for efficient removal of hazardous antibiotic from pharmaceutical wastewater. *Environmental Science and Pollution Research International, 30*(56), 119473–119490. https://doi.org/10.1007/s11356-023-30587-5

Bushra, B., & Remya, N. (2020). Biochar from pyrolysis of rice husk biomass—Characteristics, modification and environmental application. In *Biomass conversion and biorefinery*. https://doi.org/10.1007/s13399-020-01092-3

Chaubey, A. K., Pratap, T., Preetiva, B., Patel, M., Singsit, J. S., Pittman, C. U., & Mohan, D. (2023). Definitive review of nanobiochar. ACS Omega. https://doi.org/10.1021/acsomega.3c07804

Chausali, N., Saxena, J., & Prasad, R. (2021). Nanobiochar and biochar based nanocomposites: Advances and applications. *Journal of Agriculture and Food Research*. https://doi.org/10.1016/j.jafr.2021.100191

Chen, D., Zhou, J., Zhang, Q., Zhu, X., & Lu, Q. (2014). Upgrading of rice husk by torrefaction and its influence on the fuel properties. *BioResources*. https://doi.org/10.15376/biores.9.4.5893-5905

Cruz, G. J. F., Mondal, D., Rimaycuna, J., Soukup, K., Gómez, M. M., Solis, J. L., & Lang, J. (2020). Agrowaste derived biochars impregnated with ZnO for removal of arsenic and lead in water. *Journal of Environmental Chemical Engineering, 8*(3), 103800. https://doi.org/10.1016/j.jece.2020.103800

Elbehiry, F., Darweesh, M., Al-Anany, F. S., Khalifa, A. M., Almashad, A. A., El-Ramady, H., El-Banna, A., Rajput, V. D., Jatav, H. S., & Elbasiouny, H. (2022). Using biochar and nanobiochar of water hyacinth and black tea waste in metals removal from aqueous solutions. *Sustainability, 14*(16), 10118. https://doi.org/10.3390/SU141610118

Genovese, M., Jiang, J., Lian, K., & Holm, N. (2015). High capacitive performance of exfoliated biochar nanosheets from biomass waste corn cob. *Journal of Materials Chemistry A*. https://doi.org/10.1039/C4TA06110A

Ghosh, M., Mondal, M., Mandal, S., Roy, A., Chakrabarty, S., Chakrabarti, G., & Pradhan, S. K. (2020). Enhanced photocatalytic and antibacterial activities of mechanosynthesized TiO_2–Ag nanocomposite in wastewater treatment. *Journal of Molecular Structure, 1211*, 128076. https://doi.org/10.1016/j.molstruc.2020.128076

Goswami, L., Kushwaha, A., Kafle, S. R., & Kim, B. S. (2022). Surface modification of biochar for dye removal from wastewater. *Catalysts, 12*(8), 817. https://doi.org/10.3390/catal12080817

Liu, G., Zheng, H., Jiang, Z., Zhao, J., Wang, Z., Pan, B., & Xing, B. (2018). Formation and physicochemical characteristics of nano biochar: Insight into chemical and colloidal stability. *Environmental Science and Technology*, 10369–10379. https://doi.org/10.1021/acs.est.8b01481

Lyu, H., Gao, B., He, F., Zimmerman, A. R., Ding, C., Huang, H., & Tang, J. (2018). Effects of ball milling on the physicochemical and sorptive properties of biochar: Experimental observations and governing mechanisms. *Environmental Pollution*, 54–63. https://doi.org/10.1016/j.envpol.2017.10.037

Mandal, S., Pu, S., Shangguan, L., Liu, S., Ma, H., Adhikari, S., & Hou, D. (2020). Synergistic construction of green tea biochar supported nZVI for immobilization of lead in soil: A mechanistic investigation. *Environment International, 135*, 105374. https://doi.org/10.1016/j.envint.2019.105374

Mansouri, F., Chouchene, K., Roche, N., & Ksibi, M. (2021). Removal of pharmaceuticals from water by adsorption and advanced oxidation processes: State of the art and trends. *Applied Sciences, 11*(14), 6659. https://doi.org/10.3390/app11146659

Naghdi, M., Taheran, M., Pulicharla, R., Rouissi, T., Brar, S. K., Verma, M., & Surampalli, R. Y. (2019). Pine-wood derived nanobiochar for removal of carbamazepine from aqueous media: Adsorption behavior and influential parameters. *Arabian Journal of Chemistry*, 5292–5301. https://doi.org/10.1016/j.arabjc.2016.12.025

Narde, S. R., & Remya, N. (2022). Biochar production from agricultural biomass through microwave-assisted pyrolysis: Predictive modelling and experimental validation of biochar yield. *Environment, Development and Sustainability, 24*(9), 11089–11102. https://doi.org/10.1007/s10668-021-01898-9

Neha, S., Rajput, P., & Remya, N. (2022). Biochar from microwave co-pyrolysis of food waste and polyethylene using different microwave susceptors—Production, modification and application for metformin removal. *Environmental Research*, 210. https://doi.org/10.1016/j.envres.2022.112922

Perveen, S., Nadeem, R., Nosheen, F., Asjad, M. I., Awrejcewicz, J., & Anwar, T. (2022). Biochar-mediated zirconium ferrite nanocomposites for tartrazine dye removal from textile wastewater. *Nanomaterials, 12*(16), 2828. https://doi.org/10.3390/nano12162828

Perveen, S., Nadeem, R., Nosheen, F., Tongxiang, L., & Anwar, T. (2022). Synthesis of biochar-supported zinc oxide and graphene oxide/zinc oxide nanocomposites to remediate tartrazine dye from aqueous solution using fixed-bed column reactor. *Applied Nanoscience (Switzerland), 12*(5), 1491–1505. https://doi.org/10.1007/s13204-021-02323-3

Ramanayaka, S., Tsang, D. C. W., Hou, D., Ok, Y. S., & Vithanage, M. (2020a). Green synthesis of graphitic nanobiochar for the removal of emerging contaminants in aqueous media. *Science of the Total Environment*. https://doi.org/10.1016/j.scitotenv.2019.135725

Ramanayaka, S., Vithanage, M., Alessi, D. S., Liu, W. J., Jayasundera, A. C. A., & Ok, Y. S. (2020b). Nanobiochar: Production, properties, and multifunctional applications. *Environmental Science: Nano*. https://doi.org/10.1039/D0EN00486C

Ramírez, E. E. P., Asunción, M. d. l. L., Rivalcoba, V. S., Santos, A. L. M. H., C. V., Ramírez, E. E. P., Asunción, M. d. l. L., Rivalcoba, V. S., & Santos, A. L. M. H., & C. V. (2017). Removal of phenolic compounds from water by adsorption and photocatalysis. In *Phenolic Compounds—Natural Sources, Importance and Applications*. https://doi.org/10.5772/66895

Saffan, M. M., Koriem, M. A., El-Henawy, A., El-Mahdy, S., El-Ramady, H., Elbehiry, F., Omara, A. E. D., Bayoumi, Y., Badgar, K., & Prokisch, J. (2022). Sustainable production of tomato plants (Solanum lycopersicum L.) under low-quality irrigation water as affected by bio-nanofertilizers of selenium and copper. *Sustainability (Switzerland), 14*(6), 3236. https://doi.org/10.3390/su14063236

Shaikh, W. A., Kumar, A., Chakraborty, S., Islam, R. U., Bhattacharya, T., & Biswas, J. K. (2022). Biochar-based nanocomposite from waste tea leaf for toxic dye removal: From facile fabrication to functional fitness. *Chemosphere, 291*, 132788. https://doi.org/10.1016/j.chemosphere.2021.132788

Turco, A., Monteduro, A. G., Mazzotta, E., Maruccio, G., & Malitesta, C. (2018). An innovative porous nanocomposite material for the removal of phenolic compounds from aqueous solutions. *Nanomaterials, 8*(5), 334. https://doi.org/10.3390/nano8050334

Vijayaraghavan, K. (2019). Recent advancements in biochar preparation, feedstocks, modification, characterization and future applications. *Environmental Technology Reviews, 8*(1), 47–64. https://doi.org/10.1080/21622515.2019.1631393

Wang, C., Wang, H., & Cao, Y. (2018). Pb(II) sorption by biochar derived from Cinnamomum camphora and its improvement with ultrasound-assisted alkali activation. *Colloids and Surfaces a: Physicochemical and Engineering Aspects, 556*, 177–184. https://doi.org/10.1016/j.colsurfa.2018.08.036

Wang, Q., Shi, Y., Lv, S., Liang, Y., & Xiao, P. (2021). Peroxymonosulfate activation by tea residue biochar loaded with Fe_3O_4 for the degradation of tetracycline hydrochloride: Performance and reaction mechanism. *RSC Advances, 11*(30), 18525–18538. https://doi.org/10.1039/D1RA01640G

Wang, Y. Y., Ji, H. Y., Lu, H. H., Liu, Y. X., Yang, R. Q., He, L. L., & Yang, S. M. (2018). Simultaneous removal of Sb(III) and Cd(II) in water by adsorption onto a $MnFe_2O_4$–biochar nanocomposite. *RSC Advances, 8*(6), 3264–3273. https://doi.org/10.1039/C7RA13151H

Yang, B., Dai, J., Zhao, Y., Wu, J., Ji, C., & Zhang, Y. (2022). Advances in preparation, application in contaminant removal, and environmental risks of biochar-based catalysts: A review. In *Biochar* (Vol. 4, Issue 1). Springer. https://doi.org/10.1007/s42773-022-00169-8

Yang, X., Zhang, S., Ju, M., & Liu, L. (2019). Preparation and modification of biochar materials and their application in soil remediation. *Applied Sciences (Switzerland)*, 223–241. https://doi.org/10.3390/app9071365

Yuan, Y., Zhang, N., & Hu, X. (2020). Effects of wet and dry ball milling on the physicochemical properties of sawdust derived-biochar. *Instrumentation Science and Technology., 48*(4), 287–300. https://doi.org/10.1080/10739149.2019.1708751

Zhang, L., Guo, J., Huang, X., Wang, W., Sun, P., Li, Y., & Han, J. (2018). Functionalized biochar-supported magnetic $MnFe_2O_4$ nanocomposite for the removal of Pb(II) and Cd(II). *RSC Advances, 9*(1), 365–376. https://doi.org/10.1039/C8RA09061K

Zhang, M., Gao, B., Yao, Y., Xue, Y., & Inyang, M. (2012a). Synthesis of porous MgO-biochar nanocomposites for removal of phosphate and nitrate from aqueous solutions. *Chemical Engineering Journal*, 210. https://doi.org/10.1016/j.cej.2012.08.052

Zhang, M., Gao, B., Yao, Y., Xue, Y., & Inyang, M. (2012b). Synthesis of porous MgO-biochar nanocomposites for removal of phosphate and nitrate from aqueous solutions. *Chemical Engineering Journal, 210*, 26–32. https://doi.org/10.1016/j.cej.2012.08.052

Zhao, M., Wang, W., Huang, C., Dong, W., Wang, Y., Cheng, S., Wang, H., & Qian, H. (2018). Facile synthesis of UCNPs/$Zn_xCd_{1-x}S$ nanocomposites excited by near-infrared light for photochemical reduction and removal of Cr(VI). *Chinese Journal of Catalysis, 39*(7), 1240–1248. https://doi.org/10.1016/S1872-2067(18)63061-X

Zhao, W., Li, Y., Zhao, P., Zhang, L., Dai, B., Xu, J., Huang, H., He, Y., & Leung, D. Y. C. (2021). Novel Z-scheme Ag-C_3N_4/SnS_2 plasmonic heterojunction photocatalyst for degradation of tetracycline and H_2 production. *Chemical Engineering Journal, 405*, 126555. https://doi.org/10.1016/j.cej.2020.126555

Zhong, D., Zhang, Y., Wang, L., Chen, J., Jiang, Y., Tsang, D. C. W., Zhao, Z., Ren, S., Liu, Z., & Crittenden, J. C. (2018). Mechanistic insights into adsorption and reduction of hexavalent chromium from water using magnetic biochar composite: Key roles of Fe_3O_4 and persistent free radicals. *Environmental Pollution, 243*, 1302–1309. https://doi.org/10.1016/j.envpol.2018.08.093

Index

A
Activated Carbon (AC), 54
Adsorbent nanomaterials, 54–56
Adsorbent nanoparticles, 28
Adsorption isotherm analysis, 92
  Langmuir equation, 93
Adsorption kinetic analysis, 91–92
  EBT dye adsorption quantity, 92
  EBT dye, chemisorption characteristic of, 92
Adsorption method, 82
Advanced Oxidation Processes (AOPs), 1, 3
  copper-based nanocatalysts, 175
  coupling biological treatment with, 9
  electrochemistry, 174
  electro-fenton, 185
  fenton reaction, 181–182
  fundamentals of, 172–174
  hybrid systems, 186–188
  nanomaterials, application of, 176–188
  nanomaterials for, 175–176
  nanotechnology, use of, 13–14
  oxygen reactive radicals/reactive oxygen species (ROS), 172–173
  ozonation, 179–181
  photocatalysis, 176–178
  photocatalysis using semiconductors, 174
  photo-fenton processes, 182–185
  sonocatalysis, 178–179
  sulphate radical, mechanism for, 173
Advanced water treatment technologies, 8–9
  AOPs, 8–9
  carbon nanotubes (CNTs), 9
  emerging contaminants (ECs), 9
  graphene oxide (GO), 9
  microfiltration (MF), 8
  nanofiltration (NF), 8
  persistent organic pollutants (POPs), 9
  RO, 8
  UF, 8
AFM-PFM analysis, 77
AgNPs, applications in degradation of pollutants, 137–138
  emerging contaminants (ECs) in aqueous solutions, 138
  for disinfection, 137
AgNPs synthesis
  extraction methods of plants, 129–131
  maceration extraction, 130
  microwave-assisted extraction, 131
  solvent-based extraction, 130
  ultrasound-assisted extraction, 131
Analysis of Variance (ANOVA), 97–99
Aquaporin-based NF membranes, 162
Arsenate, 109
Arsenic
  acute poisoning, 108
Arsenic chemistry in aqueous solutions, 109–110
  arsenate (As V) species, 109
  arsenite (As III) species, 109
  inorganic arsenic, 109
Arsenic pollution
  water and wastewater purification, nano-technology for, 110–111
Arsenic removal technologies, 114–121
  adsorption, 117–118
  advantages and limitations, 121
  chemical precipitation, 119–120
  chemi–sorption, 117
  membrane filtration, 120–121
  nanofiltration (NF), 120

© Capital Publishing Company, New Dehli, India. 2025

R. Neelancherry and S. M. Kumar, *Nanotechnology in Wastewater Treatment*,
https://doi.org/10.1007/978-3-031-94246-4

nano–photocatalysts, 117
oxidation, 114
photocatalysis, 115–117
RO membrane, 120
Artificial nanoparticles, 57
Atomic Force Microscopy (AFM), 4, 5, 76
AuNP–modified electrodes, 113
Azo dye, 81

B
Biochar, 239–241
acid modification, 242
fast pyrolysis reactors, 240
Hydrothermal Carbonisation (HTC), 240
microwave pyrolysis, 240
production through pyrolysis, 240
slow pyrolysis reactors, 240
top-lit updraft (TLUD) gasifiers, 240
torrefaction, 240
Biochar nanocomposites, 243–245
nanomaterial, incorporation of, 244
nanometal oxide/hydroxide, impregnation of, 245
Biochar Nanoparticles (BCNPs), 42
Biofunctionalisation, 51–52
chitosan, 51
Block copolymer NF membranes, 161
Brunauer–Emmett–Teller (BET), 4, 5

C
$CaFe_2O_4$ nanoparticles, 83
synthesis of, 83
$CaFe_2O_4$-PANI
active functional groups, 88
$CaFe_2O_4$-PANI nanocomposite
characterisation studies, 86–89
development of, 84–85
experimental design based on central composite design, 86
performance, evaluation of, 100–101
physiochemical analysis of, 85–86
$CaFe_2O_4$-PANI nanocomposite dose
contact time, 90–91
EBT dye concentration, 90
solution pH, influence of, 89
Carbonaceous materials
porous nature of, 66–67
Carbonaceous Nanocomposites (CNCs), 42
Carbon-based
carbon dots (Cdots), 55
nanoadsorbent, 54
nanomaterials, 28, 40–43
Carbon Nanomaterials (CNMs), 42, 112
–based NF membranes, 159–160
GO nanosheets, 160
Carbon Nanotubes (CNTs), 40–41, 47–48, 55
Carbon Quantum Dots (CQDs), 42
Catalytic NF membranes, 161
Central composite design
experimental design, 86
Ceramic materials, 153
Ceramic membrane, 57, 153
catalyst immobilisation, 22
Chemical functionalisation, 50
CNT-based
catalysts, 41
disinfection methods, 41
membranes, 41
Coexisting ions
influence of, 96–97
Composite nanomaterials, 113
Conventional NF membranes, 152–153
Conventional water treatment techniques, 8
Copper Nanoparticles (CuNPs), 36
Cyclic Voltammetry (CV), 73–74

D
Dielectric materials, 20
Doped nMgO, photocatalytic activity of, 207
Dynamic Light Scattering (DLS), 4, 5

E
EBT dye uptake
regeneration study of, 99–100
Effective water governance frameworks, 2
Electrochemical Impedance Spectroscopy (EIS), 74–76
Mott-Schottky analysis, 75–76
Nyquist plot, 74
Electrochemical sensor technology
nanoparticle-modified electrodes, 113
Electron microscopy, 66–67
Emerging Contaminants (ECs), 9
Emerging Organic Pollutants (EOPs), 65, 171–172
Energy Dispersive X-ray Spectroscopy (EDS/EDX), 4, 66
Environmental engineering
nanotechnology, role of, 5, 6t–7t
Equilibrium isotherm analysis, 92–95
Eriochrome Black-T (EBT) dye

batch adsorption experiment of, 85

F
Factors affecting AgNP synthesis
pH, 136
reaction time, 137
temperature, 136
Fourier Transform Infrared Spectroscopy (FTIR), 4, 69
Fullerenes
based membranes, 43
Functionalisation
classification based on, 49–53, 50t

G
Global water management, 1
Graphene, 55
based nanomaterials, 41
oxide (GO) nanosheets, 41
quantum dots (GQDs), 41
Green synthesis route
AgNPs production from plant extracts, 132–134
irradiation method, 133
microwave-assisted synthesis, 133
polysaccharide method, 132
sodium borohydride ($NaBH_4$), 132
tollens synthesis method, 133

H
Heavy metal, 108
Hierarchical nanostructures, 46
Hybrid metallic nanoparticles (Cu, Ag, Fe), 37–38
Hybrid nanocomposites, 47
Hydrogen peroxide
direct or indirect photolysis, 174

I
Industrial revolution, 1
Inorganic nanofibres, 49
Ionic and molecular self-assembly, 3
Iron-doped zirconium oxide nanoparticles, 15
Iron Nanoparticles (FeNPs), 37
Iron Oxide Nanoparticles (FeONPs), 39

J
JOEL JSM-7600F Field Emission Gun-Scanning Electron Microscope (FEG-SEM) device, 85

M
Manganese Oxide Nanoparticles (MnO_2NPs), 40
Membrane-based nanomaterials, 28, 56–57
Membrane processes, 1
Metal
organic framework (MOF)-based NF membranes, 160
oxide nanoparticles (MONPs), 38, 40
oxide nanotubes, 48
Metal-based nanoparticles, 28, 111–112
Metal ferrites' dye, 82
Metal/Metal oxide–based NF membranes
Al_2O_3 nanoparticles, 158
Fe/Fe-oxide nanoparticles, 158
fouling of, 158
polyamide and cellulose acetate membranes, 158
silica nanoparticles, 158
TiO_2 nanoparticles, 157
Metal nanoparticles, 36–38
Microwave
based tertiary wastewater treatment, 18
discharge electrodeless lamps (MDEL), 22
induced catalytic degradation (MICD), 18
Miniature robots, 59
Mixed-Matrix Membranes (MMMs), 156
Multiwalled CNTs (MWCNTs), 159
MW absorber Metal-Organic Framework (MOF)-derived catalysts, 20
MW-based catalytic process, 22
MW-responsive catalysts, 20
MXene-based NF membrane, 162

N
Nanobiochar, 241–243
ball milling technique, 241
physical modification, 241
wet ball milling, 241
Nanobiochar and biochar nanocomposites
application in wastewater treatment, 245–256
challenges in the application, 256–257
dye removal, 252t, 255
heavy metal removal, 247t–249t

pharmaceutical and personal care product, degradation of, 250t–251t, 255
phenolic compound removal/ degradation, 254t, 256
phosphate and nitrate removal, 253t
Nanocatalysts, 3
Nanocomposite nanomaterials, 28
Nanocomposites, 47, 112
Nano-enabled coatings, 14
Nanofibres, 48–49
Nanofiltration (NF), 147
fundamentals of, 148–156
Nano–infused sensors, 113
Nanomaterial, 3
adsorbent's zeta potential, 78
-based physico-chemical treatment technologies, 65
bottom-up methods, 3
characterisation of, 65
chemical characterisations of, 69–71
classifications, applications, advantages, challenges, 30t–35t
electrical characterisations of, 73–76
-enhanced membranes, 3
environmental impact, 58
-infused microfluidic devices, 59
long-term stability, 58
structural characterisations, 66–69
surface characterisations of, 76–78
synthesis and characterization, 3–5
top-down approach, 3–4
toxicity, 58
Nanomaterial morphologies, 44
Nanoparticle-based adsorbents
decentralised water treatment systems, 13
Nanoparticles, 2–3, 44–45
Nanopolymers, 28
Nanorods, 45
Nanoscale
magnetic metal ferrites, 82
membranes, 46
semiconductors, 53
transistors, 3
Nano sensors, 13
pollutants and microbial pathogens, 14
Nanospheres/spherical nanoparticles, 45
Nanostructured materials, 28, 46
Nanostructures, 3
Nanotechnology, 2, 107
cost and scalability, 5
enhanced filtration, 1
health and safety concerns, 5
regulation and standardisation, 5
Nanotubes, 48
NF membrane
cantilever/tip assembly, 155
characterisation of, 155
electron beam irradiation, 154
interfacial polymerisation, 154
LBL method, 154
membrane modules, 155
methodologies for preparation, 153
phase inversion method, 153
UV/photo-grafting, 154
NF separation mechanisms, 149–152
NMgO, 200
catalytic activity in the absence of light, 216–222
generation of O_2 - in dark, 216–219
in fenton-like process, 213–215
in photocatalysis, application of, 202–213
persulphate activation in absence of light, 219–222
photocatalytic activity of, 203–207
physicochemical properties, 200–202
properties of, 200
three-dimensional F-MgO_2, 213
NMgO catalyst
ozone activation by, 222–228
NMgO photocomposites
photocatalytic activity of, 208–213
Novel nanomaterial-based NF membranes, 156–163
NZVI-based adsorbents, 56

O

Oxide-based nanoadsorbents, 56

P

Pb-based materials
toxicity effects of, 17
Perovskites, 3
Persistent Organic Pollutants (POPs), 9, 172
Photocatalysis, 14
challenges to be addressed, 15
compared with other AOPs, inherent advantages, 14
Photocatalytic
heterojunctions, 71
nanomaterials, 28
NPs, 112

Photocatalytic nanomaterials
titanium dioxide (TiO_2), 53
zinc oxide, 54
Photolithography, 4
Photoluminescence (PL) spectroscopy, 71
Piezocatalysis
water remediation, 17
Piezocatalytic
efficiency, 17
heterojunctions, 71
materials, 17
Piezocatalytic process
redox reactions, 17
Piezoelectric Force Microscopy (PFM), 76
Piezoelectric ZnO nanorods, 77
Polyaniline (PANI), 83
Polyaniline-based calcium ferrite nanocomposites
reagents used, 83
Polymeric
nanocomposite membrane, 56
nanofibres, 48
nanomaterials (PNPs), 43–44
Polymer–nanomaterial composites, 47
Polyvinylidene Fluoride (PVDF), 18

Q
Quantum dots, 3

R
Reactive Oxygen Species (ROS), 13–14
Real water sample, 95–96
effect of, 95–96
Reduced Graphene Oxide (rGO), 41
Responsive nanomaterials, 52–53
smart nanogels, 52
temperature-responsive variants, 52–53
Reverse Osmosis (RO), 1
membranes, 148
RSM-based optimization, 97–99

S
Scanning Electron Microscopy (SEM), 4, 66–67
Secondary electrons, 66
Silver Nanoparticles (AgNPs), 128
characterisation of, 134
FTIR spectroscopy, 134
physicochemical techniques, 128
synthesis of, 128–129
TEM, 135
UV–Vis spectroscopy, 134
XRD, 135
Single-Walled Carbon Nanotubes (SWCNTs), 40–41, 159
Surface modifications, 14
Sustainable water management, 2

T
Tannic acid–based NF membranes, 162
Thin-film
composite membranes, 153
nanocomposite (TFN), 65
Titanium Oxide Nanoparticles (TiO_2NPs), 38
TMDC nanosheets, 56
Transmission Electron Microscopy (TEM), 4, 67

U
Ultrafiltration (UF), 1
Ultrasonication-based piezoexcitation, 18
Ultraviolet
diffused reflectance spectroscopy (UV-DRS), 72
photo-electron spectroscopy (UPS), 71
UV-visible photocatalyst ZnO, 72
UV–vis spectroscopy, 4–5

W
Wastewater treatment, 10–13
advanced technologies, 10–12
biological nutrient removal (BNR) processes, 12
membrane filtration technologies, 12
mitigating nutrient pollution, 12
Water
contaminants, 81
pollution, 81
quality, 2
resources, 2
stewardship, 2
treatment, 8
Water and wastewater treatment
environmental considerations, 23
nanomaterial production, economic feasibility of, 23
nanotechnology, 13–22
nanotechnology-enhanced membranes, fouling of, 23
transparent communication, 23–24
treatment processes, deployment in, 23

X
X-Ray Diffraction (XRD), 4, 67–69
  Williamson–Hall plot, 69
  ZnO, morphologies of, 68
X-ray Photoelectron Spectroscopy (XPS), 69–71
  carbon correction, 69–71

Z
Zeta potential, 77–78
Zinc Oxide Nanoparticles (ZnONPs), 39

MIX
Papier aus verantwortungsvollen Quellen
Paper from responsible sources
FSC® C105338

If you have any concerns about our products,
you can contact us on
ProductSafety@springernature.com

In case Publisher is established outside the EU,
the EU authorized representative is:
Springer Nature Customer Service Center GmbH
Europaplatz 3, 69115 Heidelberg, Germany

Printed by Libri Plureos GmbH
in Hamburg, Germany